Springer Series in Fashion Business

Series editor

Tsan-Ming Choi, The Hong Kong Polytechnic University, Hung Hom, Hong Kong

More information about this series at http://www.springer.com/series/15202

Jochen Strähle
Editor

Green Fashion Retail

Springer

Editor
Jochen Strähle
School of Textiles and Design
Reutlingen University
Reutlingen
Germany

ISSN 2366-8776 ISSN 2366-8784 (electronic)
Springer Series in Fashion Business
ISBN 978-981-10-2439-9 ISBN 978-981-10-2440-5 (eBook)
DOI 10.1007/978-981-10-2440-5

Library of Congress Control Number: 2016949605

Printed on acid-free paper

This Springer imprint is published by Springer Nature
The registered company is Springer Nature Singapore Pte Ltd.
The registered company address is: 152 Beach Road, #22-06/08 Gateway East, Singapore 189721, Singapore

Acknowledgments

I would like to thank all the authors of this book for their contribution. It is always a huge challenge to focus on individual topics and to deliver in-depth results. I am very happy to have worked together with this team and I think everyone can be proud of the final outcome.

I especially would like to thank Prof. Tsan-Ming Choi from the Hong Kong Polytechnic University for giving us the opportunity to publish this book within the book series Fashion Business.

Many thanks also to all my research partners and academics worldwide who have inspired me with their works, research results and have given me constructive feedback throughout the last years.

Last but not least, I would like to thank my Ph.D. Student Mr. Deniz Köksal for organizing the publication procedure, formatting, and aligning the content. Without him this book could not have been published.

Contents

List of Figures

List of Tables

Chapter 1
Green Fashion Retail

Jochen Strähle

Abstract The purpose of this paper is to provide an overview over the fashion retail landscape from a sustainability perspective. It draws the outline of the key influencing developments, like digitaliziation on the one hand and changing consumer behavior on the other hand. It shows that new business models such as insitutionalized second hand formats may be on chance to provide an answer to the changing market rules.

Keywords Sustainability · Fashion retail · Business models · Going green

1.1 The Green Future of Fashion Retail

The discussion about sustainability has increased over the last decades. The triple-bottom line with the focus on the future generations (Brundtland 1987) have become a basic framework for academics and marketers alike to shape the economy towards a more sustainable lifestyle. The main challenge is to balance the different perspectives of people, profit and planet or with other words the social, economical and ecological aspects (Elkingto 2002). Companies are responding by integrating sustainability aspects into their strategies (Maignan et al. 2005) and consider going green as a way of differentiation in their market (Yang et al. 2010).

The fashion industry has gathered widespread attention in public especially because of the Rana Plaza incident in 2013 (Kozlowski et al. 2015). However, even before that it is visible that attention towards the negative impact of the fashion system on sustainability was rising even before the collapse of the factory building (Beard 2008). The production of fashion products involves several steps which usually include different actors in the supply chain (Karthik and Gopalakrishnan 2014). Due to the different types of products coming from various sourcing countries which are then distributed globally, fashion retailers face difficulties to

J. Strähle (✉)
School of Textiles and Design, Reutlingen University, Reutlingen, Germany
e-mail: jochen.straehle@reutlingen-university.de

J. Strähle (ed.), *Green Fashion Retail*, Springer Series in Fashion Business,
DOI 10.1007/978-981-10-2440-5_1

have full transparency over their supply chain (Kogg 2009). However, fashion companies strive for competitive prices on the retail floor, so the reduction of labour costs by producing in developing countries for the production is a main lever to decrease the cost of goods (Nagurney and Yu 2012). But not only labour costs are a trigger, but also the ecological side is more and more criticized because the cultivation requires a high degree of natural resources (Draper et al. 2007), mainly of course the intense use of water, e.g. for cotton production, or the use of chemicals (Chen and Burns 2006; Piegsa 2010).

Consumers buy fashion products mostly because of an emotional rather a rational need (Cao et al. 2014). Socializing and relevancy to peers play a dominant role within the decision process (Jørgensen and Jensen 2012) which leads to an excessive overconsumption which can be seen with fast fashion retailers (Morgan and Birtwistle 2009). Even though sustainability has become also a recognizable part of the decision process, it is still neglected within the buying decision in the mainstream (Hiller Connell and Kozar 2014). However it can be observed that it does become a more influencing factor for fashion consumers (Han 2013).

Digitalization changed the structure of the fashion retail market in recent years (Strähle 2015), especially the way of the distribution (Buvari et al. 2014). The change towards the intense use of smartphones for information and transaction purposes changes the interaction between consumers and fashion companies (Bruce 2012; Daurer et al. 2012). Consumers have changed their traditional channel orientation towards a consumer based omni-channeling (Strang 2013), where the customer considers a brand more like a cloudy construct rather than a fixed point of sale (Strähle 2015). Therefor companies will have to understand the buying process more in detail, which means especially understanding the different stages of information demand during this process (Solomon and Rabolt 2004). The rise of e-commerce technology has enabled companies to interact with their communities through social media (Zolkepli and Kamarulzaman 2015), which is considered as one of the most powerful innovations in the twenty-first century (Diaz 2015). Social media has nowadays proliferated from a consumer-to-consumer communication to a possibility of interaction between consumer-to-company and viceversa (Johnson 2014). This interaction is ideal for building trust through transparency on the business (Hajli 2014). Especially for green-orientated customers, the self-selection of communication channels enables companies to focus on their dedicated target group in an efficient way (Minton et al. 2012).

The new technologies do not only alter the interaction between companies and consumers, but also lead to innovative approaches regarding the business models as such. Consumers have now the possibility to become creators themselves which is known as the prosumer concept as fusion of production and consumption (Ritzer 2015). The outdoor fashion industry already has proven to be very progressive here by integrating professional users in the development process of their products (Strähle and Wagner 2016). As this changes the traditional value chain of fashion houses, adaption of the classical concepts lead to new business approaches based on the new technologies (Ribiere and Tuggle 2010). Companies like Tinkercad or Betabrand for example are based on the co-creation (Ramirez 1999) of design, which means that design comes from consumers themselves and the company only covers the production and distribution part (Betabrand—Crowdfunded Clothing 2016; Tinkercad-Create 3D digital designs with online CAD 2016). Another area is the evolution of the second-hand market. As young fashion consumers show a high frequency of shopping fashion products in the traditional fast fashion environment, they have been targeted from a different angle: Developing from the classical boot-sales or flea-markets (Han 2013), online retailers like Kleiderkreisel (Janauskas 2016) use the Internet to overcome the local focus to deliver a supra-regional infrastructure for the exchange of used products and work more as a mediator or platform than a traditional retailer (Scholl et al. 2013).

As a consequence of the developments mentioned, new fashion retail business model arise. Highly visible in the second-hand market (Guiot and Roux 2010) new start-ups are trying to find their position profiting from this development. Social media play a crucial role in this context and already some start-ups seem to start grasping the opportunities of this new green fashion retail order.

1.2 Leveraging the Opportunities by Going Green

Fashion marketers will have to react to the changing society and customer needs. The following articles provide insights on the most relevant aspects in this area. The aim is to provide theoretical and practical insight on how going green may positively influence a fashion retailer's strategy. Companies will have to act differently in their field if they want to be successful in the future. This may require a total change of their current business model or, at least, some major adaptions.

The book is structured in fourteen articles. Based on distinct research questions, the reader will be able to dig deep into the individual levers for a possible adaption. It thus can provide a solid understanding on how to integrate green aspects into any fashion retailer's business model.

The first articles serve as an introduction to deliver a solid understanding about sustainability in the fashion context. After this introductory article,the second article conducts a literature review on the key aspects of sustainability in the fashion retail. The following paper contributes with a definition about the closed-loop production, where the most relevant theoretical frameworks will be discussed. The next article discusses the different impacts of production and consumption on sustainability. It will be shown that it is not only the companies which should be made responsible towards, e.g. pollution of the environment, but is much more the consumer's role and responsibility to act. This will affect also the long-term strategies of the fashion eco-system, as companies will have to deliver other systems which are accepted by consumers.

The next papers cover the the supply side of the industry. The fifth article focuses on the manufacturing standards and on how these impact retail brands and their communication in particular. The sixth article discusses the prosumer concept. As the integration of consumers in the production process changes not only the value creation process but also has positive sustainable impacts, this is considered as a new way of fashion production and consumption.

The following articles go deep into alternative business models for fashion retailers by extending the loop. The seventh article covers an analysis of the second-hand market for fashion products. It reveals that there do exist already several models in the second-hand industry but still on a very low scale. The eigth article discusses alternatives to the classical fast fashion consumption which may be focusing on using rather than buying. The idea of a shared economy is a central aspect in this topic. As second-hand products are not really professionalized yet, the next article broadens the perspective by looking at an industry which already has seemed to live through the same development: the car industry. It will be shown that second-hand products are now a vital part of the business model and it will be shown how this industry managed to adapt their value chain to the new requirements. Consequently, the next article focuses on the value chain of a second-hand store in the fashion industry and how it differs from a traditional one. Finally this block will be concluded with a case study of Swedish fashion brand Filippa K, which has shown to be a forerunner in the integration of sustainable aspects, shared economy and even running an own branded second-hand store within their business model.

The last articles try to focus the information demand of sustainable consumers and how companies are able to achieve this through different tools and media. The twelvth article discusses the role of social media in this context. It is clear that the prosumer concept requires interaction between a fashion retailer and its consumers and especially social media seems appropriate for this. The thirteenth article highlights the potentials of a dedicated fashion app which may be used for additional information about a product in a local retail environment. This may include manufacturing details, history or even a transparent communication about the whole supply chain. The book closes with a dedicated case study of www.honestby.com. The Belgian fashion retailer already communicates almost everything, starting from listing all the suppliers, the used materials and up to open up even the calculation of each product.

References

Beard ND (2008) The branding of ethical fashion and the consumer: a luxury niche or mass-market reality? Fash Theory J Dress Body Cult 12(4):447–467

Betabrand| Crowdfunded| Clothing A global community of collaborators (2016). Retrieved 8 Jan 2016, from https://www.betabrand.com/

Bruce A (2012) Multi-Channeling der Zukunft – Multi-Channel-Erfolgsfaktoren im wachsenden Markt aus Sicht von Google. In: Heinemann G, Schleusener M, Zahalria S (eds) Modernes Multi-channeling im Fashion-Handel: Konzepte, Erfolgsfaktoren, Praxisbeispiele. Dt. Fachverl, Frankfurt, pp 50–69

Brundtland GH (1987) Report of the world commission on environment and development: our common future. Oslo, United Nations. http://www.un-documents.net/wced-ocf.htm

Buvari R, Dosé T, Vonstad B (2014). A channel approach to fashion
Cao H, Chang R, Kallal J, Manalo G, McCord J, Shaw J, Starner H (2014) Adaptable apparel: a sustainable design solution for excess apparel consumption problem. J Fash Market Manag 18 (1):52–69. doi:10.1108/JFMM-08-2012-0046
Chen H-L, Burns LD (2006) Environmental analysis of textile products. Cloth Text Res J 24 (3):248–261. doi:10.1177/0887302X06293065
Daurer S, Molitor D, Spann M (2012) Digitalisierung und Konvergenz von Online- und Offline-Welt: Einfluss der mobilen Internetsuche auf das Kaufverhalten. Zeitschrift für Betriebswirtschaft 82(S4):3–23. doi:10.1007/s11573-012-0580-1
Diaz A (2015) Defining Social Media. https://www.boundless.com/
Draper S, Murray V, Weissbrod I (2007) Fashioning sustainability—a review of the sustainability impacts of the clothing industry. Forum for the future
Elkington J (2002) Cannibals with forks: the triple bottom line of 21st century business (Reprint). Capstone, Oxford
Guiot D, Roux D (2010) A second-hand shoppers' motivation scale: antecedents, consequences, and implications for retailers. J Retail 86(4):355–371. doi:10.1016/j.jretai.2010.08.002
Hajli MN (2014) A study of the impact of social media on consumers. Int J Market Res 56(3):387–404. doi:10.2501/IJMR-2014-025
Han J (2013) Understanding second-hand retailing: a resource based perspective of best practices leading to business success—viewcontent.cgi. Iowa State University. http://lib.dr.iastate.edu/cgi/viewcontent.cgi?article=4643&context=etd
Hiller Connell K, Kozar J (2014) Environmentally sustainable clothing consumption: knowledge, attitudes, behaviour. Roadmap to sustainable textiles and clothing environmental and social aspects of textiles and clothing supply chain. Singapore, Springer Singapore, pp 41–63
Janauskas J (2016) kleiderkreisel.de. http://www.kleiderkreisel.de/. Retrieved 2 Jan 2016
Johnson LA (2014) Connecting sustainability reporting and social media. Natl Provis 228(1):30–35
Jørgensen MS, Jensen CL (2012) The shaping of environmental impacts from Danish production and consumption of clothing. Ecol Econ 83:164–173. doi:10.1016/j.ecolecon.2012.04.002
Karthik T, Gopalakrishnan D (2014) Environmental analysis of textile value chain: an overview. In: Roadmap to sustainable textiles and clothing: environmental and social aspects of textiles and clothing supply chain. Springer, Singapore, pp 153–158
Kogg B (2009) Responsibility in the supply chain: interorganisational management of environmental and social aspects in the supply chain; case studies from the textile sector. International Institute for Industrial Environmental Economics, Lund
Kozlowski A, Searcy C, Bardecki M (2015) Corporate sustainability reporting in the apparel industry: an analysis of indicators disclosed. Int J Product Perform Manag IJPPM 64(3):377–397
Maignan I, Ferrell OC, Ferrell L (2005) A stakeholder model for implementing social responsibility in marketing. Eur J Market 39(9/10):956–977. doi:10.1108/03090560510610662
Minton E, Lee C, Orth U, Kim C-H, Kahle L (2012) Sustainable marketing and social media. J Advert 41(4):69–84. doi:10.2753/JOA0091-3367410405
Morgan LR, Birtwistle G (2009) An investigation of young fashion consumers' disposal habits. Int J Consum Stud 33(2):190–198. doi:10.1111/j.1470-6431.2009.00756.x
Nagurney A, Yu M (2012) Sustainable fashion supply chain management under oligopolistic competition and brand differentiation. Int J Prod Econ 135:532–540. doi:10.1016/j.ijpe.2011.02.015
Piegsa E (2010) Green Fashion: ökologische Nachhaltigkeit in der Bekleidungsindustrie. Diplomica Verlag, Hamburg, p 2010
Ramirez R (1999) Value co-production: intellectual origins and implications for practice and research. Strateg Manag J 20(1):49–65
Ribiere VM, Tuggle FD (Doug) (2010) Fostering innovation with KM 2.0

Ritzer G (2015) Automating prosumption: the decline of the prosumer and the rise of the prosuming machines. J Consum Cult 15(3):407–424. doi:10.1177/1469540514553717

Scholl G, Gossen M, Grubbe M, Brumbauer T (2013) Vertiefungsanalyse 1: alternative Nutzungskonzepte—Sharing, Leasing und Wiederverwendung. Institute für Ökologische Wirtschaftsforschung. http://www.ioew.de/publikation/alternative_nutzungskonzepte_sharing_leasing_und_wiederverwendung/

Solomon MR, Rabolt NJ (2004) Consumer behavior in fashion. Prentice Hall, Upper Saddle River

Strähle J (2015) Emotionalizing the fashion retail. In: Emotionalizing fashion retail. Norderstedt, BoD, pp 23–27

Strähle J, Wagner Y (2016) Prosumer concepts in the outdoor fashion industry. Int J Econ Commer Manag 4(2):59–83

Strang R (2013) Retail without boundaries. www.scmr.com

Yang C-L, Lin S-P, Chan Y, Sheu C (2010) Mediated effect of environmental management on manufacturing competitiveness: an empirical study. Int J Prod Econ 123(1):210–220. doi:10.1016/j.ijpe.2009.08.017

Zolkepli IA, Kamarulzaman Y (2015) Social media adoption: the role of media needs and innovation characteristics. Comput Hum Behavi 43:189–209. doi:10.1016/j.chb.2014.10.050

Chapter 2
Key Aspects of Sustainability in Fashion Retail

Jochen Strähle and Viola Müller

Abstract The purpose of this paper is to explain the key aspects and growing relevance of sustainability in fashion retail and to evaluate the possibilities of fashion retailers to act sustainable in supply chain management as well as carving out the challenges they have to deal with. The research methodology applied for this purpose is a critical literature review examining books and articles. The findings demonstrate the rising importance of sustainability in fashion retail. In this regard, fashion retailers play a key role and responsibility for sustainability in the fashion supply chain, from the beginning up to the end. This paper mainly analyzes sustainability in the fashion supply chain. It does not analyze topics like second-hand shopping or social media sustainability.

Keywords Sustainability · Fashion retail · Supply chain management · Five-R analysis green washing

2.1 Introduction

2.1.1 Landscape of Textile and Fashion Industry

The current world population has passed over the number of 7.3 billion people. Basic needs of these people are food and clothing. So the textile and fashion industry plays a major role all over the world. The textile and fashion industry has experienced considerable growth and success over the last two decades. This has led to new arising topics in economic, environmental and social areas within the industry. Facing challenges of global textile and fashion supply chains requires global solutions (Shishoo 2012).

J. Strähle (✉) · V. Müller
School of Textiles and Design, Reutlingen University, Reutlingen, Germany
e-mail: jochen.straehle@reutlingen-university.de

V. Müller
e-mail: ViolaMueller@gmx.net

J. Strähle (ed.), *Green Fashion Retail*, Springer Series in Fashion Business,
DOI 10.1007/978-981-10-2440-5_2

As the fashion industry becomes more and more competitive, most of the fashion retailers are trying to differentiate themselves in the market. They hope to design an optimal business model, the most efficient production patterns and strong corporate values. According to the researches carried out by the Sustainable Brand Insights (Sustainable Brands Insights 2012), "understanding the latest sustainability trends, consumer insights and industry-specific best practices" is necessary to maintain the competitiveness in the market.

Sustainability has become an omnipresent word in recent years. Organizations around the world are increasingly incorporating the requirements of sustainability into their day-to-day operations. They do so not only because of the social and environmental risks and the governance challenges the supply chain poses, but also because of the benefits supply chain sustainability can deliver. Sustainable supply chain management can be a strong driver of value and success, for business as much as for society. There are many definitions about sustainability but there is little known about the actual drivers and barriers to adopting environmentally sustainable practices within organizations.

The textile and fashion industry is considered to be one of the most polluting industries in the world. Large fashion retailers can play a key role in promoting sustainability because they are intermediaries between producers and manufacturers on the one hand and customers on the other hand. Durieu (2003) for example, argued that retailers "can greatly influence changes in production processes and consumption patterns and are positioned to exert pressure on producers in favour of more sustainable consumer choices". Sustainability in the textile and fashion industry can be controlled along the supply chain. From fibre production up to the end consumer use, in each stage there are different factors influencing the environment. In Fig. 2.1, the social and environmental impacts along the clothing supply chain are shown.

This work is determined to give an overview about the topic of sustainability in fashion retail. It contains the different social and environmental impacts along the fashion supply chain. Therefore, a literature review was conducted and the findings are presented in the following order. First, a definition of fashion retail and sustainability will be given to set the basis for the following sections. Then a closer view of the reasons to "go green", the relevance of sustainable supply chain

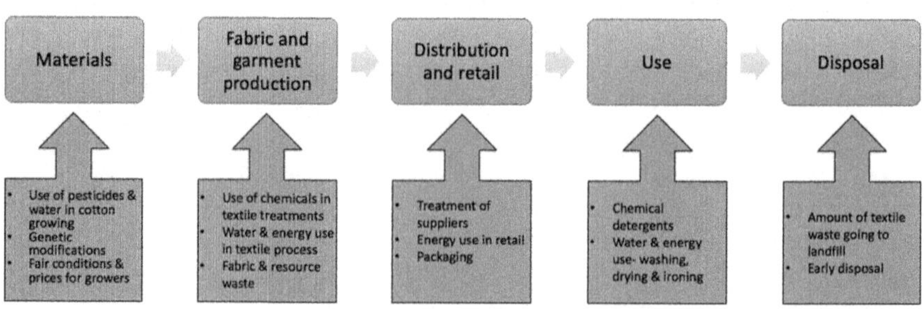

Fig. 2.1 Clothing supply chain. Adapted from Gwilt (2014)

management and in connection with that the challenges of sustainable fashion value chains are given. The third section is dedicated to define and explain the term "green washing" and the "seven sins". The following section shows an exemplary process for a fashion retailer, who wants to implement a sustainability and energy management system in his retail stores and an example of a fashion company that is already using such sustainability and an energy system is given. In the end of this paper, a conclusion will be given and the limitations of this work will be pointed out in order to make suggestions for further research.

2.1.2 Fashion Retail

The textile and clothing industry consists of companies that are involved with fibres, primary textile manufacturers, who produce yarns and fabrics, as well as add coloration, apparel manufacturers and retailing enterprises, where the textile product interfaces with the consumer (University of NorthernIowa 2015). It is an international and highly globalized industry, with clothing often designed in one country, manufactured in another and sold in a third. This industry has been radically evolving in recent years due to retail consolidation, globalization and e-commerce. Christopher et al. (2004) defines fashion markets as typically exhibiting the following characteristics: short life cycles, high volatility, low predictability and high impulse purchasing.

In this paper, the definition of the whole textile and clothing industry is used for the term "Fashion Retail".

2.1.3 Sustainability

> Sustainability is not singularly about minimizing negative impact, but also maximising positive impact, allowing individuals, communities and economic systems to flourish. To work sustainably is to question the status quo, challenge convention and find new ways of working that achieve ecological, social and cultural balance that is in tune with human behaviour (Williams et al. 2009, p. 8).

Sustainability has become an omnipresent word in recent years but there is no standard in terms of what identify a sustainable product or process. Most people associate sustainability with environmental protection, being ecological, green or organic. But they do not think about the economy and society. There are more than 70 definitions of sustainability existing in the literature (Pearce et al. 1989). One of the most often-cited definition is given by the Brundtland report. Sustainability as stated in the Brundtland report means to meet "the needs of the present without compromising the ability of future generations to meet their own needs" (Hariembrundtland 1985).

The interaction of this definition and fashion retailers who implemented a whole culture and ethos of sustainability can be found in the so-called "triple bottom line"

Table 2.1 The triple bottom line

	Economic	Environmental	Social
Typical measures	Sales, profit, ROI	Pollutants emitted	Health and safety record
	Taxes paid	Carbon footprint	Community impacts
	Monetary flows	Recycling and reuse	Human rights; privacy
	Jobs created	Water and energy use	Product responsibility
	Supplier relations	Product impacts	Employee relations

Adapted from Savitz and Weber (2006)

(Wilson 2015). This model consists of three categories: people, profit and planet. In other words, it can be said that businesses have to measure their success not only by financial performance, but also by their impact on the economy, the environment and the society in which they operate (Elkington 2002). The goal of fashion retailers is to find a balance between these three factors. The triple bottom line can be seen as a kind of balanced scorecard that determines the degree to which a business is creating value for the economy, the environment and the society (Savitz and Weber 2006). Typical measures for the economic, environmental and social elements are shown in Table 2.1. A product or fashion retailer can only be called sustainable when his whole "cradle-to-grave" life cycle is sustainable (Blackburn and Textile Institute (Manchester, England) 2009).

2.2 Literature Review

2.2.1 Reasons to "Go Green"

Already in the 1960s people started to think critical of the industrialized society. They worry about consumption and large-scale technologies as well as industrialism; this results in the so-called "eco movement". Also in the 1970s and 1980s people started new movements. Due to the fact that many political issues were discussed, criticized and publicized, people changed their everyday behaviour and a new eco attitude was noticed. The eco movement peaked out at the end of the 1970s. Consumers wanted to wear eco-friendly fashion and sign out their disclaimer against the current fashion. Pieces of clothing were practicable and comfortable and not very fashionable because consumers wanted to draw attention to the bad influences of the fashion industry. This development influenced the attitude of today's society. Sustainability is often associated with the slovenly look of the past. But designers and producers nowadays work hard to change the image of sustainable fashion in the customer's mind (Diekamp and Koch 2010).

Due to the large-scale outsourcing of manufacturing to low-cost countries, long lead times and forecast-driven apparel value chains, there are increased forecasting errors, mark downs and lost sales as well as an excess of inventory, more discounted merchandise and greater consumer dissatisfaction (Mattila et al. 2002). This leads fashion retailer to rethink their actual business and take a look at the topic sustainability. As the European Commission (Europäische Kommission 2009) says "retailers are increasingly recognizing sustainability as a significant opportunity for their businesses to grow, compete and innovate". Therefore, they try to integrate specific, comprehensive sustainability into their business strategies for internal operations and external relationships (Maignan et al. 2005). A lot of companies see sustainability not only as a chance to contribute to social goals, but also as a powerful source of competitive advantage and a matter of corporate survival (Yang et al. 2010). Due to the increasing demand for sustainable products and transparent supply chains the decision of a fashion retailer to "go green" can arise from many reasons but they can be categorized into four distinct groups: social well-being, environmental stewardship, economic prosperity and governance.

Social well-being includes the improvement of labour standards and conditions, enhance communities and the creation and delivery of socially responsible products and services (Mahler 2007). Fashion retailers have to broaden their mind beyond the shop floor and to the consumers, like utilizing point-of-sale literature to educate consumers and to promote more sustainable consumption. Consumers today are not well informed about the topic sustainability. They have to learn how to act in a sustainable way and adopt knowledge.

Fashion retailers must also emphasize and develop their corporate social responsibility (CSR) goals further. Because this leads to completely satisfied communities, customers and stakeholders within and beyond the society in which they operate. Being sustainable in terms of social well-being needs to incorporate socially responsible values and transparency into company's supply chains. The rising transparency leads to rising scrutiny, therefore collaboration across the supply chain is outstanding and organizations are inevitably realizing and gain the benefits of working together (Pui-Yan Ho and Choi 2012). Collaboration also means more than just having a relationship across the supply chain, it means rather working together towards a common goal (Shedroff 2009). Social well-being demands an increase in the quality of life, social justice and social coherence (Lang and Murphy 2014).

Environmental stewardship stands for the environmental responsibility fashion retailers have. It can consist of, for example, conserving energy and resources, consuming more renewable and less-polluting, increasing recycling, minimizing packaging and reducing the retailer's carbon footprint (Mahler 2007). In other words Vezzoli and Manzini (2008, p. 6) state "The term environmental sustainability refers to systemic conditions where neither on a planetary nor on a regional level do human activities disturb the natural cycles more than planetary resilience allows, and at the same time do not impoverish the natural capital that has to be shared with future generations". Companies can also achieve an advantage by renewed target marketing or create eco-market space and product differentiation by experimenting and innovation, such as looking for sustainable resources and new raw materials. Furthermore it is possible or moreover necessary joining with other companies and to form partnerships to exposure and highlight the importance of environmental stewardship (Pui-Yan Ho and Choi 2012).

In 1958, companies in general had an average tenure of 61 years in the Fortune 500. The Fortune 500 is an annual list compiled and published by Fortune magazine that ranks 500 of the largest US corporations by total revenue for their respective fiscal years. By 1980 this had reduced to 25 years and it is today only 18 years. Main reasons to this development are increasing competition and the notion of creative destruction (Foster 2012). To achieve long-term growth companies have to be highly sensitive and responsive to market forces because as one retailer develops a new and successful strategy the others have to react and develop me-too or differentiation strategies (Porter 2004). De Geus (1988) proposed "The ability to learn faster than your competitors may be the only sustainable competitive advantage". In other words, it is important that a learning company is based upon ecological principles. This offers opportunities to reach major advantages. As Mahler (2007) says economic prosperity comprises of promoting profits, creating jobs, attracting customers, reducing costs, anticipating and managing long-term risks and fostering long-term competitiveness. Economic sustainability requires a long-term perspective regarding the handling of financial assets and human resources. Furthermore, it focuses on the carrying capacity and diversity of industry structure and consumption patterns (Lang and Murphy 2014).

The fourth reason to "go green" is governance. This should not be mistaken with the term government. Governance is not as formal and is more an arrangement for member groups who have accepted each other as stakeholders: public and private, profit and non-profit, national and trans-national, expert and amateur, producer and consumer, large and small. All these members have an enhanced awareness of interdependence. These arrangements can set up standards and monitoring systems for a sustainable business programme (Lang and Murphy 2014). These member groups are often more accepted from consumers. Fashion retailers who hold on these standards become more competitive and increase their image from a consumer's point of view. Therefore, it is necessary to create a strategic business plan with measurable success factors.

2.2.2 Sustainable Supply Chain Management

From the traditional point of view, it was supposed that the most important players in the development process are the company and the customer. But the most effective organizations have learned to consider input, needs and cooperation with suppliers, distributors, retailers and other business partners throughout the whole supply chain. The more systems-oriented an organization is and the more it considers the full spectrum of sustainable issues (managing and using human, natural and financial capital), the wider is the circle of concerns and actors to involve (Shedroff 2009).

Cooper et al. (1997, p. 1) defines the supply chain management as "The process of planning, implementing, and controlling the efficient, cost-effective flow and storage of raw materials, in-process inventory, finished goods, and related information flow from point-of-origin to point-of-consumption for purpose of conforming to customer requirements."

Supply chain management concepts have been implemented to improve operating performance, provide new sources of competitive advantage while offering a better value to consumers, and ultimately develop excellent managed organizations and inter-organizational relationships (Trent 2004).

Characteristics of supply chain management especially in fashion retail are retail-led, usually stochastic systems with all kinds of inherent uncertainties (on demand service and value), relatively short product life cycles and decisions in fashion supply chains are highly consumer demand driven (Choi 2014). In addition to the definition of general supply chain management the following definition is related to the fashion retail supply chain management:

> Fashion Retail Supply Chain Management encompasses the planning and management of all activities involved in sourcing and procurement, conversion, and all logistic management activities in the fashion retail supply chain. It includes coordination and collaboration with supply chain partners. In essence, Fashion Retail Supply Chain Management integrates supply and demand management within and across the fashion retail supply chain with a goal of satisfying the customer requirements under the leadership of the retailer (Choi 2014, p. 2).

The trend driven by the consumers' increased level of awareness and concerns on environmental protection as well as the fact that the traditional fashion apparel supply chain use toxic chemicals, consumes high numbers of water and electricity, produces a lot of waste and air emissions, long transportation ways and excessive packaging, highly affects the sustainable supply chain management practices in fashion apparel. It leads to a new sustainability mind-set (Five-R model) and the use of environmental management systems (Choi 2013; Muthu 2014). Even though the motives of fashion retailers vary, the key target is generally to include an interest in achieving sustainable streams of products, services, information and funds to provide maximum value to all involved stakeholders (Salzmann et al. 2005). Sustainable supply chain management can be defined as the traditional supply chain management practices which integrate "environmental, social and economic impacts throughout the lifecycles of goods and services" (Global Compact Office 2011).

To arrange environmental operations and work flows, the ISO14000 standard can help fashion retailers to identify processes, develop a performance measurement system, measure the supply chain system, prioritize processes, develop alternatives to processes, select approaches and establish auditing and improvement procedures (Pui-Yan Ho and Choi 2012).

There are many models for fashion retailers to adapt in their quest for sustainability. This paper discusses the so-called Five-R model more specific. It is formed by the five-R's: recycle, reuse, reduce, re-design and re-imagine (Fig. 2.2).

Recycling is the process by which materials are collected, processed into reusable forms and subsequently used as raw materials for new products. In general, there are two types of waste: post-industrial waste and post-consumer waste. At the post-industrial waste, the content and composition of the product and material is

Fig. 2.2 5R's of sustainability. Own figure based on Esty and Winston (2006)

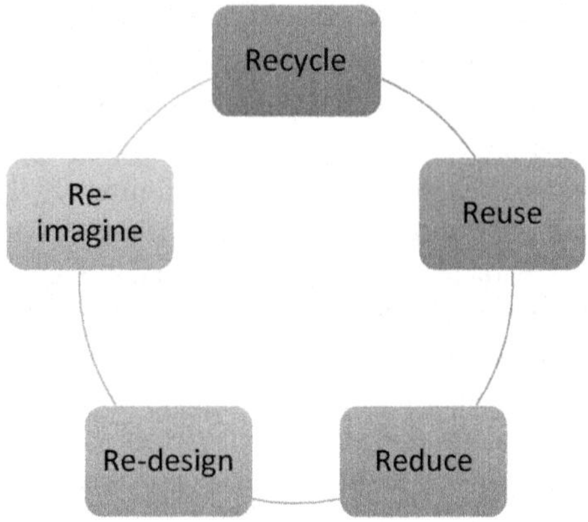

known (Pui-Yan Ho and Choi 2012). From Shedroff (2009) point of view recycling is an important principle of sustainability, but to be really effective, products need to be easily disassembled into component parts and separated by material. Therefore, the post-consumer waste it is a more complicated separation process, because the product has undergone many changes since the initial industrial process. This leads to a more resource and capital intensive recycling process, despite the higher value in doing so (Pui-Yan Ho and Choi 2012).

Reuse is another way to close the loop. The term reuse refers to the fact to use an item or product in its original form for several times instead of discarding it. Through this behaviour the usage life of a product can be extended. In the fashion branch this could be cloak pins, garment packaging or sewing needles (Pui-Yan Ho and Choi 2012). Instead of the extending the use of garments and make them last longer, Shedroff (2009) recommended to deign the company's operation and components to be easily exchanged so that the majority of components stay in use. Additionally to this there has been a growth in online sales or exchange of garments through retailers such as eBay which has helped to increase the flow and accessibility of second-hand clothing (Muthu 2014).

The third R is reducing. This implies a source reduction and a waste prevention. Waste prevention includes strict avoidance of waste generation, qualitative and quantitative reduction at source, and reuse of products (pre-waste 2015). In fashion, this could for example mean to look more carefully on the fabric sourcing. Obtaining fabrics locally instead of a global procurement leads to shorter transport distances and thereby to lower environmental pollution. Each transportation sends different levels of pollution into the environment and affects different populations and ecosystems around the world (Muthu 2014). If it is necessary to source global, fashion retailers could try to minimize their safety stock ordered to prevent fabric procurement. During "the garment manufacturing process, sophisticated lay planning software can help map out the ideal layout for pattern pieces which would result in less fabric wastage following the fabric cutting process" (Pui-Yan Ho and Choi 2012, p. 168). The idea of eco-effectiveness defines not only to reduce the waste, but also to eliminate the concept of waste (Shedroff 2009).

A re-designed product means that the item was developed in a way that reduces environmental impacts for someone somewhere in its life-cycle journey from supplier inputs to product to end-of-life disposal (Esty and Winston 2006). The re-design process includes different perspectives. The first one is on product design. Product design involves topics like the use of materials. The second perspective refers to the process re-design. One example for the process of re-design is the so-called business process re-engineering (BPR) scheme. Companies can try to streamline their existing business processes to enhance their efficiency. In context with supply chain management and BPR commonly, the change or simplification of business processes to reduce logistic-related wastes as well as enhancing workflows with reduction of wastes is planned (Pui-Yan Ho and Choi 2012).

The fifth and last R is re-imagine. Before the process of re-designing starts, fashion retailers should also keep in mind the re-imagination of their products and processes. Re-designing and re-imagining have the aim of thinking direct environmentally creative, heartening to seek new opportunities to add value to what companies do and realize that re-designing and re-imagining precludes the former Three-R's in generating greater profit and long-lasting company value and vision (Pui-Yan Ho and Choi 2012). As Powell (2010) explains the so-called concept "Design for Environment", designers create products by taking the environment into account. It is less expensive to design a low-impact product than to manage or to rebuild a high-impact product.

2.2.3 *Challenges of Sustainable Fashion Value Chains*

The barriers of sustainable fashion value chains can be summarized in four challenges: fashion logistics challenges, challenges of overproduction, challenges of irresponsible consumption and challenges of fulfilling social responsibility. Each challenge will be explained in the following.

2.2.3.1 Fashion Logistics

The biggest challenge of fashion retailers to an efficient logistic in a globalized network are long lead times from the order up to the delivery. This is also known as "the lead time gap" (Fernie and Sparks 2004). Muthu (2014) says that lead times in a traditional fashion business (up-front buying based on seasonal forecasting and planning) can be up to 36 weeks. The result of a research od key performance measures of major European fashion retailers (traditional product design and slow response), made by Cachon and Swinney (2011) is that traditional fashion companies such as Mango (Spanish branded retailer), Lindex (Swedish branded retailer) or John Smedley (UK branded retailer) have production lead times up to 26 weeks. The majority of this time is non-value adding and generates costs. Long lead times are also caused by the the geographical extension and attended with that the necessity of long-range forecasts ahead of sales seasons (Christopher et al. 2004). This fact influences the financial performance in a negative way, caused by imprecise accuracy leading to loss of revenues and profits, surplus of inventory and hence a large number of products which must be sold on discount prices as well as the risk that customers not finding what they want in the shop (Mattila et al. 2002). Today a 48 weeks lead time is common and assumed that it causes sales forecast errors of about 40 %. Furthermore, the radical relocation of the production sites towards Far East lead to higher transportation costs, loss of employment in the manufacturing sector and higher carbon footprints (Allwood 2006). There is no concrete definition of the term carbon footprint, but there are some key characteristics that are accepted. This results in an open definition from Peters et al. (2010, p. 23): "The carbon footprint of a functional unit is the climate impact under a specified metric that considers all relevant emission sources, sinks and storage in both consumption and production within the specified spatial and temporal system boundary."

Another important and challenging element of fashion logistic in the broadest sense is the packaging of a product. Caused by short life cycles, the amount of packaging on the markets is almost equal to the amount of packaging waste, whereof most of the packaging is composite packaging. This contains materials like laminated foils, aluminium foil or polyethylene. These materials are difficult to recycle and causes high costs and low resale value. To palliate the public concerns, politicians have supported recycling and reuse strategies, to reduce the demand for raw materials and decrease the quantity of waste going into landfills (Xie et al. 2013).

2.2.3.2 Overproduction Caused by Forecasting Error

Forecasting error has serious consequences on sustainable fashion value chains. Mattila et al. (2002) says that fashion retailers on average only sell two-third of their seasonal fashion products at the full price. The rest of the products are sold on a discount price. There are different business strategies to improve retail performance measures such as the optimization of the flow of information and merchandise between value chain members to maximize consumer satisfaction (Ko and Kincade 1997). By realizing different technologies like sharing POS information, EDI, electronic transmission of orders and invoices, computer-aided design (CAD), the use of computer technology and manufacturing and electronic point of sale that means collecting sales information at the cash register from barcodes. This leads to efficiently reduce safety stocks, avoid overproduction and minimize unsold merchandise. The reduction of overproduction has a big impact on the environment, keeping in mind the fact that textile and apparel raw materials production are henpecked by energy-intensive processes, the use of toxic chemicals and the high water consumption (Christopher et al. 2004). "At the operational level it is crucial to take timely decisions on what to buy, what to move, and what to make, vital supply chain planning to counter demand uncertainty" (Muthu 2014, p. 233).

2.2.3.3 Irresponsible Consumption and Uninformed Consumer

The high precariousness in the demand pattern of fashion products that leads to an increase in unsustainability is not only caused by the production system. The manufacturing process generates pollutants as well as the consumption of goods. Consumers are today as well irresponsible, considering the fact that German women have on average 118 pieces of clothing in their closet and men 73 pieces. Every fifth piece of clothing is almost unworn. Clothing consumption contains the acquisition, storage, usage, maintaining and throwaway (Grooten et al. 2012). Most consumers do not keep in mind the environmental impacts of their clothing purchases. During the different stages of a product life cycle, everything from the manufacturing of fibres to the disposal of garments results can worsening the ecosystem health. "As long as the unsustainable consumption of clothing products persists, environmental degradation will continue as well" (Muthu 2014, p. 43). This implicates that an overall sustainability can only be achieved if fashion retailers produce more environmentally friendly products but at the same time the clothing consumption behaviour of individuals change and become more environmentally responsible and good-natured. The change in consumer behaviour assumes that the knowledge of sustainability is present. To expand the understanding of environmentally significance a focus on increasing fundamental knowledge and how factors like information, incentives and constraints combine and interact with personal values, attitudes and beliefs, to inform and shape the consumer decision-process is necessary (Daneshvary et al. 1998).

2.2.3.4 Fulfilling Social Responsibility

Fashion retailers have to be aware of their social responsibility. They have to determine their corporate social responsibility (CSR) along their fashion value chain. This can be spilt up into three main sectors: wages, working hours and working conditions (Perry and Towers 2013). The whole fashion branch is imputed of paying low wages. As Carson (2013) says, there are traditional monitoring methods such as codes of conduct and inspections in place but fashion retailers have failed to pay a living wage to their workers, they accepted child labour, have misuse human rights and assert minimum labour standards in the workplace. Diverse scandals due to the infringement of ethical values in the fashion value chain of global fashion retailers like Zara, Gap, Nike or Marks & Spencer, have been reported. In this connection, four main categories of mechanisms aiming to encourage supplier to assume socially responsible practices can be defined: international standards, extended frameworks, supplier codes of conducts and supplier social audits (Awaysheh and Klassen 2010).

2.3 Green Washing

> The World's leading Ozone destroyer takes credit for leadership in ozone protection. A mammoth greenhouse gas emitter professes the precautionary approach to global warming. A major agrichemical manufacturer trades in a pesticide so hazardous it has been banned in many countries, while implying it is helping feed the hungry. A petro- chemical firm uses the waste from one polluting process as raw materials for another hazardous process, and boasts of an important recycling initiative. Another giant multinational cuts timber from virgin rainforest, replaces it with monoculture plantations and calls the project "sustainable forest development." (Kelly 2008, p. 1)

The upcoming theme of environmental sustainability puts pressure on fashion retailer as they have been portrayed as one of the key causer of the climate change and environmental problems. In general company statements are not required by law and publishing environmental policy statements is voluntary. There is no third-party which proves if the company statements are implemented. Green washing is a relatively new concept but there are several definitions of the topic. One definition of greenwashing is stated by Ramus and Montiel (2005) "disinformation disseminated by an organization so as to present an environmentally responsible public image." Another definition comes from Walker and Wan (2012, p. 357), in their point of view green washing is defined "as a symbolic information emanating from within an organization without substantive actions. Or, in other words, discrepancy between the green talk and green walk." A third one is published by Laufer (2003, p. 253): "green-washing, a strategy that companies adopt to engage in symbolic communications of environmental issues without substantially addressing them in actions, has been identified by both academia and the mainstream media."

Underwriters Laboratories has identified seven different categories of green washing claims, the so-called "The Seven Sins" (The Seven Sins|The Sins of Greenwashing 2015):

- Sin of the hidden trade-off: a product is perceived as being "green" because it is just based on a narrowly defined set of attributes
- Sin of no proof: the environmental claim is unsubstantiated
- Sin of vagueness: a claim that is vague or ambiguous ("All-natural" is not necessarily "green")
- Sin of worshiping false labels: the company uses words or images that give the impression of a third-party endorsement where no endorsement exists (fake labels)
- Sin of irrelevance: irrelevant claims are employed ("CFC-free"- already banned by law)
- Sin of lesser of two evils: using descriptions for a product professing a green attribute when the overall product is commonly regarded as environmentally unfriendly (organic cigarettes)
- Sin of fibbing: the claims are false, so the unauthorized use of stamps, symbols and labels

As mentioned in the seven sins, product labelling or symbols can be false or misunderstood. Today we have a flood of labels and certifications which seem to mark a product as being "green". Fashion retailers can use their own labels to signal an ethical focus. Consumers often cannot judge the legitimacy and credibility of such labels.

But why do fashion retailers do green washing? Beside the mentioned upcoming pressure, companies have an incentive of publishing environmentally policy statements. Statements can positively influence public perceptions of the retailer commitment to environmental protection and sustainable development. Possibly this results in increased market share and improved stakeholder relations. All these positive effects could probably be annulled if an increased number of companies make claims to sustainable development through these policy statements because consumer will wonder if these policies are just a form of green washing. It is difficult for stakeholder outside a company to know if the published policy commitments results in internal organizational greening activities (Ramus and Montiel 2005).

2.4 Sustainability in Fashion Retail Store

Not only the implementation of sustainability aspect in the main part of the supply chain of fashion retailers is from great importance, also to mention the fashion retail store itself. Despite the increased interest in sustainable products and production processes there is not much information about the environmental friendliness of a fashion retail store (Thompson 2007). Figure 2.3 shows an exemplary process for a fashion retailer, who wants to implement a sustainability and energy management system in his retail stores. There is no standard solution for each retailer. The sustainability management system has to be customized for each business requirements.

The first step is to "Develop a Sustainability Strategy". Retailers should define a clear long-term oriented strategy and develop a plan of action to meet the strived goal. Commitments to sustainability and energy have to be formalized, core metric will be established and baselines have to be evaluated. Furthermore, a database of successes is compiled, performance is validated versus market benchmark data and an internal workgroup is formed to review sustainability issues gained. Up to this moment it is necessary to review the performance in periodic intervals and internal stakeholders are engaged and educated as well as a communication plan for progress is established. The result of step one is a sustainability roadmap document with detailed goals, benefits and risks (Jamieson and Hughes 2013).

The second step is to "Implement Strategic Energy Sourcing". This step contains the monitoring and analyzing of data to prioritize efficiency initiatives by return on investment. Market information across multiple geographies have to be analyzed to identify saving possibilities and reduce operating expenses. This step is also very important to negotiate with business partners about improving contract terms as well as rate structures have to be conducted. Additionally, rates and tariffs have to

Fig. 2.3 Sustainability and energy management system implementation. Adapted from Jamieson and Hughes (2013)

be analyzed, energy market research will be performed and risk tolerance is determined. Ongoing reports on contracts, market changes and opportunities are produced which include clear recommendations on how to move forward (Jamieson and Hughes 2013).

Step number three is to "Deploy energy assessments in a sample of stores". Fashion retailers have to use this step to examine audit results by store cluster and identify improvement. Stores have to be organized in categories such a building type and age or by equipment which is installed (Jamieson and Hughes 2013). Not only the equipment is important but also facts like using green energy and low energy bulbs for lightening. Green energy means to use power sourced from renewable or non-polluting energy sources. With the decision to source energy in a green way fashion retailers can support the general development of new and cleaner technologies that will reduce the environmental impacts (Thompson 2007). Not all stores, only a selection of stores undergoes in depth assessments to find inefficiencies because this procedure is much more cost effective than auditing all stores. Audit results can be used to find inefficiencies in similar categories of stores with similar characteristics (Jamieson and Hughes 2013).

The fourth step is to "Collect energy and resource data, report on key performance indicators". During this step, a report will be passed to internal and external stakeholders on progress toward goals. "A metering infrastructure which uses existing data collection infrastructure is designed and commissioned for new and current stores. Monitoring systems are defined to meet business-specific requirements." To be able to prepare a wide visibility into disparate data sources to internal and external stakeholders, energy and resource consumption data has to be collected and reports generated (Jamieson and Hughes 2013).

Step number five is to "Remotely monitor energy and resource consumption". This is necessary to make improvement recommendations. Therefore, energy and resource data has to be aggregated and centralized in a software and reviewed by remote energy experts to discover inefficiencies and recommend initiatives to optimize process or equipment. After that a ranking of the expert recommendations has to be done to improve efficiency based on the organizational goals. If there is a need, fashion retailer can also make site visits to implement initiatives in selected stores (Jamieson and Hughes 2013).

The last step is to "Identify Supply Chain Risks". This includes the examination of the supply chain, using a platform to view supplier characteristics as well as compliance and ratings. Furthermore, supplier information has to be analyzed to identify potential risks. This could vary from one retailer to another based on what is most relevant for the business. If retailers use an internet based platform for a supplier database, the supplier itself can add information. This assuaged the ranking according to supplier characteristics, survey response and compliance. It is more easy to compare and evaluate prospective supplier's sustainability profiles in an effort to minimize risk and promote compliance with identified sustainability targets (Jamieson and Hughes 2013).

Jamieson and Hughes (2013) say that a well-planned energy and sustainability management programme can reduce energy costs by 10–30 %. From their point of view, the triple bottom line is a good tool to support sustainability initiatives with harder to measure metrics such as organizational efficiency, customer loyalty, reduce environmental impact and productivity.

There are already fashion examples from retailers involving the theme sustainability in their supply chain and especially in their retail stores. One is VF Corporations (VF Corporations 2015). It is a global leader in branded lifestyle apparel, footwear and accessories with more than 30 brands. In their owned retail stores VF take care of reducing the environmental footprint by increased energy efficiency, reduced waste and sustainable materials sourced locally. One brand of VF which became a leadership in energy efficient retail, is "Timberland". With their energy guideline they achieved to consume 30 % less energy than in a traditional retail store. This results out of the implementation of LED spotlights and programmable internet protocol thermostats to control heating and air conditions.

To reduce the waste, VF uses paper shopping bags that are certified by the Forest Stewardship Council (FSC) or the Sustainable Forestry Initiative (SFI). By 2017, VF plans to have for all brands only paper shopping bags with a minimum content of recycled material of 30 %. Many brands (The North Face, Timberland, Nautica and Kipling) of VF are already using paper shopping bags with recycled content and some brands like "Vans" are also using reusable bags.

In addition to the use of recycled materials in paper shopping bags, Vans launched new hang tags that are made with 100 % post-consumer paper and soy-based inks and Timberland uses hang tags which contain 100 % recycled materials, 80 % of which is post-consumer paper and water-based inks. VF shows that it is possible to reduce costs through sustainable actions and guidelines. By consolidating all hangers purchasing with one supplier, they could reduce their costs for Europe by 20 %. All wooden hangers of VF owned retail stores in Europe have to be certified. Plastic hangers are also made from 100 % recycled materials and metal hangers contain 25 % recycled content. Another point in fashion retail is the furniture. VF brands like Timberland, Ella Moss and 7 For All Mankind only use FSC certified wood furniture and low emitting materials for the floors and floor furniture (VF Corporations 2015).

2.5 Conclusion

To give a conclusion, the findings of this literature review are briefly summarized. The theoretic basis of fashion retail and sustainability were presented in Sects. 2.1.2 and 2.1.3. The term fashion retail in this paper is equal to the definitions of the whole textile and clothing industry. It includes companies that are involved with fibres, primary textile manufacturers, who produce yarns and fabrics, as well as add coloration, apparel manufacturers and retailing enterprises. Sustainability in literature has many different definitions but one is the most cited and accepted. It is the definition used in the Brundtland report. It compromises to pay attention to the current needs of the generation but furthermore to keep in mind the needs of the following generations. The interaction of this definition and fashion retailers which implement a whole culture sustainability can be found in "triple bottom line".

In Sect. 2.2.1, findings about the reasons to "go green" were presented. The eco movement of people since the 1960s plays an important role and influences retailers still nowadays to have a look on the topic sustainability. Due to the increasing demand for sustainable products and transparent supply chains the decision of a fashion retailer to "go green" can arise from topics like social well-being, environmental stewardship, economic prosperity and governance. Afterwards the necessity and content of a sustainable supply chain management is presented in Sect. 2.2.2. To give a review of the whole theme, sustainable supply chain management was explained by the theoretical Five-R model which consists of recycle, reuse, reduce, re-design and re-imagine.

Sustainable supply chain management offers numerous chances, but it also implicates challenges, fashion retailers have to deal with. These challenges were presented in Sect. 2.2.3. Main challenges which were explained are fashion logistics, overproduction caused by forecasting errors, irresponsible consumption and uninformed customers and fulfilling corporate responsibility. Fashion retailers are often capable to meet these challenges, so they try to compass a real change to a cradle-to-grave sustainable business plan. Nevertheless, they communicate to be a sustainable fashion retailer without acting so. This is called “Green Washing” and can be defined by the Seven Sins that are seven different categories of green washing claims. This topic was presented in Sect. 2.3. The presented findings end with possibilities of sustainability in a fashion retail store in Sect. 2.4. An exemplary six steps process was given if fashion retailers are willing to implement a sustainability and energy management system. Furthermore, an example of a fashion company using a sustainability and energy management system was given.

2.6 Research Limitations and Future Research

The main limitations in literature are the little results concerning the demanded information content of the different understandings of sustainability in different regions of the world. In most literature there is a unification of the topic and cultural differences are not mentioned. The results are limited in their structure and quality. Most studies concerning the added value through sustainability consist of qualitative research methods with an explorative character. Moreover, there is a surplus of information from the United States and China and low level of information from other countries. Despite the increased interest in sustainable products and production processes there is not much information about the environmental friendliness in a fashion retail store and the topic of green washing. Sustainability in form of business frameworks or business plans is in most literature seen as a necessity which causes very low literature about the critical side of a sustainability programme for companies. The literature review of this paper itself is limited in the depth of each topic because it elaborated an overview of the whole topic of sustainability in fashion retail and concentrated on the fashion supply chain. It did not analyze topics like second-hand shopping and the social media part of sustainability.

In order to overcome the presented limitations, the qualitative results concerning the added value through sustainability should be tested with the help of quantitative studies. Especially for fashion retailers operating in international structures, intercultural similarities and differences should be examined as well in further research. Furthermore, sustainability in a fashion retail store as well as topics like second-hand shopping, the social media part and the critical parts of sustainability could be part of further research.

References

Allwood JM (2006) Well dressed? The present and future sustainability of clothing and textiles in the United Kingdom. University of Cambridge, Institute for Manufacturing, Cambridge

Awaysheh A, Klassen RD (2010) The impact of supply chain structure on the use of supplier socially responsible practices. Int J Oper Prod Manag 30(12):1246–1268. doi:10.1108/01443571011094253

Blackburn RS (ed) (2009) Sustainable textiles: life cycle and environmental impact. Cambridge, New York, Woodhead, CRC Press
Cachon GP, Swinney R (2011) The value of fast fashion: quick response, enhanced design, and strategic consumer behavior. Manag Sci 57(4):778–795. doi:10.1287/mnsc.1100.1303
Carson TL (2013) Free exchange for mutual benefit: Sweatshops and Maitland's 'Classical Liberal Standard'. J Bus Ethics 112(1):127–135. doi:10.1007/s10551-012-1236-3
Choi TM (2013) Carbon footprint tax on fashion supply chain systems. Int J Adv Manuf Technol 68(1–4):835–847. doi:10.1007/s00170-013-4947-4
Choi T-M (2014) Fashion retail supply chain management: a systems optimizati-on approach, 1st edn. CRC Press/Balkema, Boca Raton
Christopher M, Lowson R, Peck H (2004) Creating agile supply chains in the fashion industry. Int J Retail Distrib Manag 32(8):367–376. doi:10.1108/09590550410546188
Cooper MC, Lambert DM, Pagh JD (1997) Supply chain management: more than a new name for logistics. Int J Logist Manag 8(1):1–14. doi:10.1108/09574099710805556
Daneshvary N, Daneshvary R, Schwer RK (1998) Solid-waste recycling behavior and support for curbside textile recycling. Environ Behav 30(2):144–161. doi:10.1177/0013916598302002
de Geus A (1988) Planning as learning. https://hbr.org/1988/03/planning-as-learning. Accessed 26 Nov 2015
Diekamp K, Koch W (2010) Eco Fashion: Top-Labels entdecken die Grüne Mode. Stiebner, München
Durieu X (2003) How Europe's retail sector helps promote sustainable production and consumption. Ind Environ 26(1). Retrieved from http://www.uneptie.org/division/media/review/vol26no1/UNEP0103.PDF
Elkington J (2002) Cannibals with forks: the triple bottom line of 21st century business. Capstone, Oxford (Reprint)
Esty DC, Winston AS (2006) Green to gold: how smart companies use environmental strategy to innovate, create value, and build competitive advantage. Yale University Press, New Haven
Europäische Kommission (ed) (2009) Smarter and cleaner: consuming and producing sustainability. Office for Official Publications of the European Communities, Luxembourg
Fernie J, Sparks L (eds) (2004) Logistics and retail management: insights into current practice and trends from leading experts, 2nd edn. Kogan Page, Sterling
Foster R (2012). Creative destruction whips through corporate America. http://www.innosight.com/innovation-resources/strategy-innovation/upload/creative-destruction-whips-through-corporate-america_final2015.pdf
Global Compact Office (ed) (2011) Supply chain sustainability: a practical guide for continuous improvement. United Nations Global Compact Office, New York
Grooten M, Almond R, McLellan R, World Wide Fund for Nature (2012) Living planet report 2012 biodiversity, biocapacity and better choices. World Wide Fund for Nature, Gland. http://wwf.panda.org/about_our_earth/all_publications/living_planet_report/2012_lpr/
Gwilt A (2014) A practical guide to sustainable fashion. Fairchild Books, an imprint of Bloomsbury Publishing Plc, London
Hariembrundtland G (1985) World commission on environment and development. Environ Policy Law 14(1):26–30. doi:10.1016/S0378-777X(85)80040-8
Jamieson M, Hughes D (2013) A practical guide to sustainability and energy management in retail environmennts. Schneider Electric. Retrieved from http://www.climateactionprogramme.org/images/uploads/documents/creating-competitive-advantage-in-Retail.pdf
Kelly B (2008) The Greenpeace book of Greenwash. http://research.greenpeaceusa.org/?a=view&d=4588
Ko E, Kincade DH (1997) The impact of quick response technologies on retail store attributes. Int J Retail Distrib Manag 25(2):90–98. doi:10.1108/09590559710160382
Lang A, Murphy H (eds) (2014) Business and sustainability. Springer, Cham. doi:10.1007/978-3-319-07239-5
Laufer WS (2003) Social accountability and corporate greenwashing. J Bus Ethics 43:253
Mahler D (2007) Spotlight on supply chain. Manag Sustain Supply Chain 11:59–60

Maignan I, Ferrell OC, Ferrell L (2005) A stakeholder model for implemen-ting social responsibility in marketing. Eur J Mark 39(9/10):956–977. doi:10.1108/03090560510610662
Mattila H, King R, Ojala N (2002) Retail performance measures for seasonal fashion. J Fash Mark Manag Int J 6(4):340–351. doi:10.1108/13612020210448637
Muthu SS (ed) (2014) Roadmap to sustainable textiles and clothing. Springer, Singapore. doi:10.1007/978-981-287-110-7
Perry P, Towers N (2013) Conceptual framework development: CSR implementation in fashion supply chains. Int J Phys Distrib Logist Manag 43(5/6):478–501. doi:10.1108/IJPDLM-03-2012-0107
Pearce DW, Markandya A, Barbier E (1989) Blueprint for a green economy. Earthscan, London
Peters GM, Rowley HV, Wiedemann S, Tucker R, Short MD, Schulz M (2010) Red meat production in Australia: life cycle assessment and comparison with overseas studies. Environ Sci Technol 44(4):1327–1332. doi:10.1021/es901131e
Porter ME (2004) Competitive advantage: creating and sustaining superior per-formance (1. Free Press export ed), export edn. Free Press, New York
Powell I (2010) Leveraging EHS data and tools for a safe and greener supply chain. http://costello.pub.hb.se/login?url, http://search.proquest.com.lib.costello.pub.hb.se/docview/224603264?accountid=9670
pre-waste (2015) What is waste prevention? http://www.prewaste.eu/index.php?option=com_k2&view=item&id=55:what-is-waste-prevention?&Itemid=41
Pui-Yan Ho H, Choi T (2012) A Five-R analysis for sustainable fashion supply chain management in Hong Kong: a case analysis. J Fash Mark Manag Int J 16(2):161–175. doi:10.1108/13612021211222815
Ramus CA, Montiel I (2005) When are corporate environmental policies a form of greenwashing? Bus Soc 44(4):377–414. http://doi.org/10.1177/0007650305278120
Salzmann O, Ionescu-somers A, Steger U (2005) The business case for corporate sustainability. Eur Manag J 23(1):27–36. doi:10.1016/j.emj.2004.12.007
Savitz AW, Weber K (2006) The triple bottom line: how today's best-run companies are achieving economic, social, and environmental success-and how you can too, 1st edn. Jossey-Bass, San Francisco
Shedroff N (2009) Design is the problem. In: Brooklyn NY, Rosenfeld Media. Retrieved from http://public.eblib.com/choice/publicfullrecord.aspx?p=680182
Shishoo R (ed) (2012) The global textile and clothing industry: technological advances and future challenges. Textile Institute, Manchester
Sustainable Brands Insights (2012) http://www.sustainablebrands.com/insights. Accessed 29 Dec 2015
The Seven Sins|The Sins of Greenwashing: Home and Family Edition (2015) http://sinsofgreenwashing.com/findings/the-seven-sins/index.html
Thompson B (2007) Green retail: retailer strategies for surviving the sustainability storm. J Retail Leis Prop 6(4):281–286. doi:10.1057/palgrave.rlp.5100079
Trent RJ (2004) What everyone needs to know about SCM.(Basics). Supply Chain Manag Rev. https://www.highbeam.com/doc/1G1-114486224.html
University of Northernlowa (2015) The textile and apparel industry. http://www.uni.edu/csbs/sahs/tapp/textile-and-apparel-industry
Vezzoli C, Manzini E (2008) Design for environmental sustainability. Springer, Berlin
VF Corporations (2015) Growing our retail footprint while shrinking our environmental footprint. http://sustainability.vfc.com/products/retail/
Walker K, Wan F (2012) The harm of symbolic actions and green-washing: corporate actions and communications on environmental performance and their financial implications. J Bus Ethics 109(2):227–242. doi:10.1007/s10551-011-1122-4
Williams D, Baldwin N, Fletcher K (2009) Volume 3.0: Centre for Sustainable Fashion: tactics for change. Centre for Sustainable Fashion, London

Wilson JP (2015) The triple bottom line: undertaking an economic, social, and environmental retail sustainability strategy. Int J Retail Distrib Manag 43(4/5):432–447. doi:10.1108/IJRDM-11-2013-0210

Xie M, Qiao Q, Sun Q, Zhang L (2013) Life cycle assessment of composite packaging waste management—a Chinese case study on aseptic packaging. Int J Life Cycle Assess 18(3): 626–635. doi:10.1007/s11367-012-0516-6

Yang C-L, Lin S-P, Chan Y, Sheu C (2010) Mediated effect of environmental management on manufacturing competitiveness: an empirical study. Int J Prod Econ 123(1):210–220. doi:10.1016/j.ijpe.2009.08.017

Chapter 3
Closed-Loop Production: A Literature Review

Jochen Strähle and Franziska Philipsen

Abstract The purpose of this paper is to investigate how the practice of closed-loop production systems (CLPS) is implemented in the fashion industry. This paper offers a critical literature review to present a thorough understanding of the actual status of literature. Subsequently, the paper reveals that CLPS are of great importance. Generally, such systems include different activities that have to be integrated. Critical points are the product acquisition, the recovering process itself and the remarketing to the customer. A lack of reliable data concerning CLPS in the specific case of fashion industry can be identified. Important research fields could be marketing strategies, controlling the acquisition process, evolvement of return technologies and strategies, adaption of recovered products to the mass market, and the development of new technologies concerning recovering processes.

Keywords Closed-loop supply chain · Fashion supply chain · Sustainability · Product acquisition · Recovery options

3.1 Introduction

Environmental sustainability is an important subject today. Especially in the fashion industry, companies are facing many problems regarding sustainable procedures. Reasons for that development can be found in many sectors. A major problem regarding sustainability is the supply chain structure. In general, the fashion and apparel supply chain consists of six supply chain levels. First, fiber production, followed by spinning, fabric production, dyeing and finishing, clothing production, and finally clothing retailing (Goldbach et al. 2003). Because of the large number of involved parties and the huge number of processes, the fashion supply chain gets

J. Strähle (✉) · F. Philipsen
School of Textiles and Design, Reutlingen University, Reutlingen, Germany
e-mail: jochen.straehle@reutlingen-university.de

F. Philipsen
e-mail: franziska.philipsen@gmail.com

J. Strähle (ed.), *Green Fashion Retail*, Springer Series in Fashion Business,
DOI 10.1007/978-981-10-2440-5_3

very lengthy and complicated, which is a reason for the significant amount of carbon footprint creation and emissions of greenhouse gases (Turker and Altuntas 2014). Particularly, the production phase generates a high environmental impact. On the one hand, the cotton and wool production requires large quantities of water and pesticides. On the other hand regarding the use of synthetic fibers, a lot of energy is required. Besides a high amount of chemical products, natural resources are needed for the phases of dyeing, drying, and finishing of products (de Brito et al. 2008). Beyond that also the other phases of the supply chain pose a threat to the size of the carbon footprint and the environmental burden (Hu et al. 2014). According to actual developments in fashion industry, also the use phase generates environmental impact. Because of the evolution of fast fashion in recent years and the resulting shortened lifecycles of the products, each piece of clothing is only used for a short time and is then discarded while it has still a potential lifetime (Hu et al. 2014). The possibility to buy frequently at lower prices results in a throwaway attitude of consumers, which leads to a higher amount of waste and pollution (Dissanayake and Sinha 2015).

Contrariwise, there is also a high amount of people that becomes more aware of sustainability at the moment and demanding more efficient production methods to protect the environment. To face and solve the increasing environmental problems, companies have to implement sustainable strategies and practices and follow environmental standards and regulations. To reach this, the subject of environmental sustainability has to be considered as a key managerial issue (Caniato et al. 2012; Eryuruk 2015; Hu et al. 2014).

In the fashion industry there exists many methods to pursue environmental sustainability like using organic fibers, reusing and recycling of materials, vintage practices and second-hand clean technologies, green certifications, or green product and process design (Caniato et al. 2012). In the following, the focus is set on closed-loop production systems, a process to trim down the carbon footprint and the environmental burden. This paper presents a literature review on the mentioned closed-loop production systems. The first section gives an insight into the general idea of closed-loop production systems and their benefits. The following section describes important processes within closed-loop production systems. Then that challenges companies have to face are presented. Finally, concluding the limitations of research is carved out to evaluate further research options.

3.2 Closed-Loop Production Systems

3.2.1 Definition of Closed-Loop Production Systems

In recent years, the term "closed-loop production systems" has received increasing interest not only among companies and professionals, but also among researchers. In the linguistic usage, there exist various terms for this process. In addition to the used "closed-loop production system", the term "closed-loop supply chain" (CLSC) is also found in the literature. CLSC management can be defined as "the design, control and operations of a system to maximize value creation over the entire lifecycle of a product with dynamic recovery of value from different types and volumes of returns over time" (Guide and Van Wassenhove 2009, p. 10). Since that is a business definition, closed-loop supply chains can also be defined as the "taking back of products from customers and recovering added value by reusing the entire product, and/or some of its modules, components and parts" (Guide and Van Wassenhove 2009, p. 10).

The closed-loop construct consists of the common **forward supply chain** and the so-called **reverse supply chain** which closes the loop. In summary, there exist three different options to close the loop: Reusing the product as a whole, reusing the components or reusing the materials. In most cases, there will be a mix of reuse options where the various returns are processed through the most profitable way (Krikke et al. 2014).

Furthermore, it can be differentiated between different activities in the forward and reverse supply chain. Figure 3.1 shows the different activities in both supply

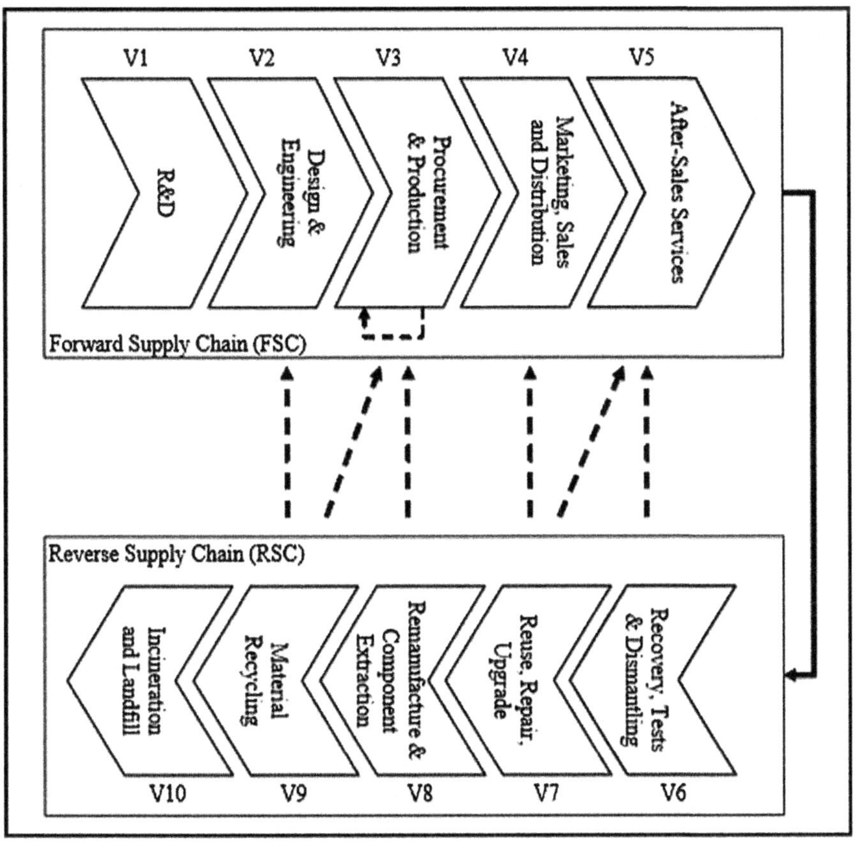

Fig. 3.1 Forward and reverse supply chain. Adapted from Talbot et al. (2007)

chains and the opportunities of connections between them. In the reverse supply chain, the recycled materials, components, or products can be reintroduced into the downstream production and distribution systems of the forward supply chain. The generated products can be reused either by the manufacturer itself, the suppliers, or by another company that contributes to other value chains benefits (Talbot et al. 2007). Generally, all supply chain participants such as suppliers, importers, manufacturers, and customers can contribute to a successful implementation of a CLSC (Pappis et al. 2004).

A closed-loop supply chain comprises five core business processes. The first process includes the **product acquisition** and aims the collection and retrieving of the products from the market. To achieve the right timing of quality, quantity, and composition, a close cooperation with the supply chain partners is needed. The next process called **reverse logistics** concerns the testing and inspection of the products as well as their transportation to the final location of recovery. In order to determine the right recovery option and the final route in the reverse supply chain, the products then need to be sorted on quality and composition. The following step includes the elected recovery option itself, which means the process of retrieving, reconditioning and regaining products, components, and materials. The final step is the redistribution and sales of the new gained products. The process mainly coincides with the distribution process of the forward supply chain, but additional marketing investments are needed to convince the customer of the product quality (Krikke et al. 2014).

These five processes can also be summarized in three subprocesses. First, the **product returns management** that concerns the timing, quantity of the used products, as well as the product acquisitions management. Second, the **remanufacturing operational issues** that deals with the process of reverse logistics as well as the testing, sorting, disassembling, repairing, and the remanufacturing of the products. Third, the process of **remanufactured products market development** can be stated. This process concentrates on the developing of channels, the remarketing and the occurring cannibalization (Guide and Van Wassenhove 2009). In order to achieve an effective and fully realized closed-loop production system, it is important that all processes are managed in a coordinated manner. At the best the processes should be smoothly integrated. Factors like the nonexistent availability of used products, technical remanufacturing issues or marketing and sales fear can lead to a failure of a CLSC (Guide and Van Wassenhove 2009). Regarding the fashion industry, the link between forward and reverse supply chain can be described as follows: The primary supply chain involves the processes from raw material extraction to the use phase of the products and their final disposal. On the other side, the individual "supply loops collect end-of-live products and reprocess them into secondary resources which replace primary resources in the forward supply chain" (Kumar and Malegeant 2006, p. 1129). This correlation is shown in Fig. 3.2.

The forward supply chain includes actors like raw material suppliers, yarn/fabric/apparel manufacturers and customers, whereas the reverse supply chain needs to involve collectors who obtain the products, reprocess them, and then provide the recovered products to the manufacturer or customer. Often, the

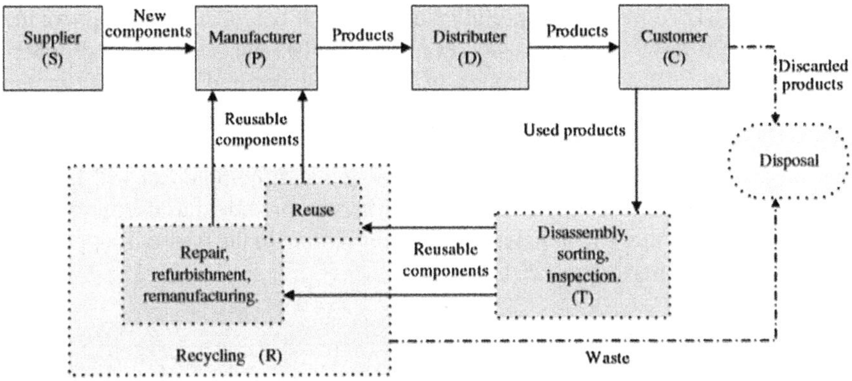

Fig. 3.2 Closed-loop supply chain. Adapted from Li (2013)

collectors are classified by the various recovery options (Oh and Jeong 2014). The diverse recovery options in the fashion industry are identified in Sect 2.1.3.

In the case study of ECOLOG, an ideal example of a fashion CLSC is presented that includes all stages of the forward and reverse supply chain. The ECOLOG Recycling Network GmbH is a textile company operating in Germany in the field of sports and outdoor wear. Summarizing, the aim of the ECOLOG network was the establishment of a closed-loop supply chain which focuses on the development and supply of homogenous polyester apparel products, but the system also offers a take-back option to the end-consumers. The post-consumer products are then transformed into virgin material. Figure 3.3 shows the CLSC of ECOLOG (Morana and Seuring 2007).

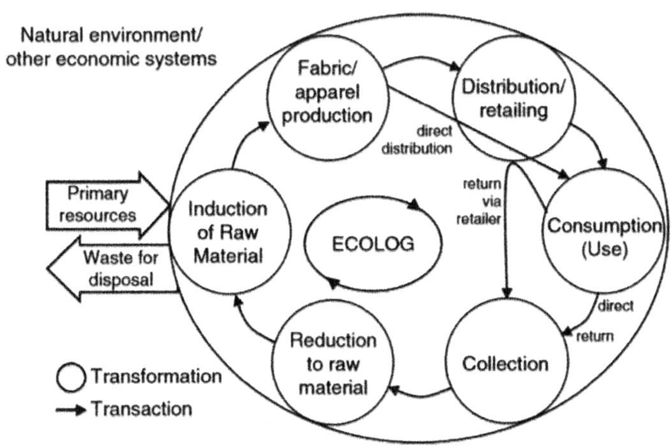

Fig. 3.3 ECOLOG closed-loop supply chain. Adapted from Morana and Seuring (2007)

Hu et al. (2014) developed a special system of a rent-based CLSC to improve the sustainability of fashion products (Hu et al. 2014). The consumer has the opportunity to receive the products through two channels, the traditional retail or the rental services. Through the rental service provider, the consumer books the products online or offline. When the use phase of the rented products is over, the consumer can send them back to the rental service and the products will be integrated in the "loop" again. Moreover, the rental service provider has the opportunity to dispose products that are no longer suitable for renting to the material supplier for further reprocessing (Hu et al. 2014).

3.2.2 Benefits of Closed-Loop Production Systems

In literature, many advantages of a closed-loop supply chain structure can be found. The most apparent benefit is the reduction of the environmental impact through mitigating environmental pollutions (Oh and Jeong 2014). Moreover, a CLSC can not only improve the environmental performance of a company, but also can increase firms' competiveness and product differentiation. On the one hand, the recovery options can raise the recycled content of a firm's products and consequently reduce factors like the intensity of raw materials extraction, energy consumption, and landfill use. On the other hand, toughening the environmental standards and raising the productivity of material resources at different product life cycle stages can lead to an enhanced competiveness (Talbot et al. 2007).

Additional benefits can be the improvement of customer satisfaction through the offered return service, the improved product quality through re-engineering, and the resulting green image of the company that can increase sales and the value of the company. Moreover, a CLSC results in a higher reliability of recalls for defect products and a decreasing necessity of a new production of spare parts. Regarding the value creation for the company, the most important benefits are the regain of value of materials and components and the avoiding of disposal costs (Kumar and Malegeant 2006).

Considering strategic alliances in a CLSC, there exist few specific benefits that arise only when the company cooperates with an eco non-profit organization (ENPO) and outsources the collection activities. These benefits can be illustrated through the presentation of the reuse-a-shoe program of Nike. Nike and the non-profit organization NRC (National Recycling Coalition) created an alliance in order to reprocess used tennis-shoes regardless of the brand. The NRC is responsible for the collection of the shoes whereas Nike focuses on the end uses for the resulting Nike Grind material. Through the alliance with ENPO's, the appearance as an environmental friendly, ecological active and community involved company can be increased. Moreover, by outsourcing the collection activity the company avoids investments in collection networks, storage or transportation costs and can save money in that way. The last benefit is the opportunity for the companies to concentrate on their core business and competencies because of the empowerment of their eco-partners (Kumar and Malegeant 2006).

Furthermore, regarding the increasing shortage of some raw materials like cotton and the increasing price of these materials in near future, searching for new technologies and practices in using materials and reusing them is becoming not only a sustainability concern, but also a business concern. A CLSC can help to guarantee the market survival of many firms in long term by remanufacturing or recycling the used products. Especially in times of increasing costs and decreasing availability of landfill space, an effective waste management can depict an advantage for a company. In the sector of revenues and strategic benefits, with the process of remanufacturing the company can save up to 60 % of the estimated costs of a completely new product. Moreover, CLSC management can increase the revenues by achieving green customers that set value on sustainable practices through the mentioned generated green image (Bouzon and Govindan 2015).

3.3 Important Processes in Closed-Loop Production Systems in Fashion Industry

3.3.1 Product Acquisition

The term product acquisition was already defined in one of the previous sections. Summarizing, it is the obtaining of the products from the end-users. The process of product acquisition can be seen as a key process as it is the initial trigger for all subsequent processes (Morana and Seuring 2007). Despite the importance of the right handling of product acquisition activities, there is "no published research [which] explicitly considers the problems of managing or controlling product returns" (Guide and Wassenhove 2009, p. 144). A reason for the low attention of this subject, also among firms, could be that the process is usually seen as an exogenous process and therefore out of the direct control of the firm (Guide and Wassenhove 2009).

Product returns can occur for a variety of reasons. In literature, many different types of returns can be identified. Many do not only focus on the return of the end-consumer, but also take other options into account. Generally, four main types of returns can be defined. The first ones are the **End-of-life** (EoL) returns that are taken back from the customer in order to avoid environmental or commercial damage and to process the collected items to formal prescriptions. The **End-of-use** (EoU) returns include products and components that are returned after some time of operations. These products can be remanufactured or traded on other markets. The most familiar type of return is the commercial returns. These are linked to the sales process and arise when the customer returns the products shortly after sales. Especially through the development of mail order and e-commerce the rates of commercial returns strongly increased. The last type is the return of reusable Items that are not part of the product itself, but contain or carry the actual product (Krikke et al. 2014).

Focusing on the product lifecycle stage of the products, three main return options can be identified. The manufacturing returns include raw material surplus, quality test returns, or production scrap. In contrast to that the distribution returns comprise product recalls, commercial returns, inventory adjustments, or redistribution of goods. The third option is the customer returns that include warranty returns, service returns, and end-of-use returns (van Nunen and Zuidwijk 2004).

Regarding the product acquisition from the end-consumer, some aspects have to be considered. As the first step, the products need to be classified. Two major criteria can help on the classification. The first one is the so-called **marginal value of time** that describes the remaining value of a product after a certain period of use. The second criteria is the **lifetime of a product**. These two criteria can be related to each other and can be build up in a matrix. This matrix is illustrated in Table 3.1. Regarding the fashion industry, situation 3 describes the case for long-life high quality apparel products such as outdoor jackets that have a long product lifetime (about 5–10 years), but a very low remaining value after use (Morana and Seuring 2007).

For the product acquisition, this case is extreme and also "the worst case". This may be the reason for the low attention for the subject among researchers so far. But it is important to mention that "if product acquisition from end customers were manageable in this situation, it should be possible to implement solutions for almost any other case" (Morana and Seuring 2007, p. 4427).

Furthermore, three important product characteristics that are of importance concerning the acquisition are identified. First, the predictability of acquisition is heavily influenced by the length of usage phase of the product. The longer the usage phase, the more difficult it gets to predict the product acquisition. Furthermore it gets more difficult to evaluate the last owner and the actual condition of the product. Second, the economic motivation of all included actors and therefore their motivation to support the CLSC is influenced by the marginal value of time. The higher the marginal value of time compared to the original selling price, the higher will be the will of the selling company to take back the product and offer an incentive to the dealer involved or the final customer (Morana and Seuring 2007). "The higher this incentive, the more willing the dealers are to advise the final customer, and he more willing the final customers are to return the old products" (Morana and Seuring

Table 3.1 Classification of products

	Low marginal value of time	High marginal value of time
Short use phase	1. Immediate collection Example: packaging	2. Frequent collection Example: electronic products
Long use phase	3. Rare returns Example: (high quality) apparel	4. Specific return routes Examples: cars

Own table based on Morana and Seuring (2007)

2007, p. 4427). Third, the mentioned remaining value also determines how worthwhile it is to take further periods of usage of the product into account, e.g., concerning the product design (Morana and Seuring 2007).

For a successful CLSC, the right understanding of customer behavior is of mayor importance. Therefore, a look at the costs customers incur for returning or disposing products is useful (Morana and Seuring 2007). For the analysis, the so-called transaction costs are utilized. Transaction costs "arise from the need for information and coordination between companies or economic actors to prepare and control contracts for the exchange of products" (Morana and Seuring 2007, p. 4428). The definition of transaction costs can also be transferred to the returning or disposal behavior of customers. Moreover these costs limit the will of customers to take part in such close-loop systems. The transaction costs can be split up into six main categories as illustrated in Table 3.2.

Regarding the return activities in a CLSC, four different sources of product acquisition can be defined (Dissanayake and Sinha 2015):

- Supply of the customer: as a disadvantage the different levels of quality and quantity as well as the required extra time for checking and accepting the products can be named.
- Collection from charity shops: leads on the one hand to increasing costs and a time-consuming selection of products but on the other hand to a better control of quality and quantity for the company.
- Sourcing the products from wholesalers: offers the companies better control of quality and quantity and a less time-consuming selection but also leads to higher purchasing prices.
- Sourcing from fabric merchants and from waste collection of sorting facilities.

As examples for take-back systems for the consumer the cooperation of the fashion retailer **Marks and Spencer** with Oxfam, a globally renowned aid and development charity can be stated. This cooperation shall motivate the consumers

Table 3.2 Transaction costs for product disposal

Cost element	Description
Information costs	The customer has to search for the location and the method to dispose the product when the information is not provided on the product itself
Planning costs	The customer has to plan the disposal
Inventory costs	Old products are stored for some time before disposal
Travel and transportation costs	The customer has to reach the disposal site (e.g., costs for using a car) or has to pay postage costs
Time costs	Describes the time needed to reach the disposal site (opportunity costs)
Psychological costs	Describes the level of how bothersome the disposal is seen by the costumer or the good or bad feeling arising by choosing a specific disposal method

Own table based on Morana and Seuring (2007)

to recycle unwanted clothes. Second, the system of the **SEOX group**, which consists of a network of retail organizations with collection boxes for discarded clothing can be mentioned. These boxes are then sent to SEOX for the next activities like sorting and processing (Dissanayake and Sinha 2015).

In the following, three case studies of closed-loop systems are introduced with the focus of their diverse product acquisition activities:

In the previously mentioned case study of **ECOLOG,** the returns occur at the EOL and EOU phase of the apparel products. The customers have the opportunity to return their products either to various retailers or to send them directly to ECOLOG. In spite of good sales volumes of the products over few years the system was not successful. The major problem was that the expected product return of customers did not take place. Neither the private customer nor institutional buyers did return the products. One reason for that may be the transaction costs. In this special case, these were—except in the field of psychological costs—high up to very high. Therefore, these transaction costs should not be underestimated while planning a CLSC. Moreover, the retailers had low interest in the promotion of the product return because they were just interested in selling. This failure of a CLSC leads to the assumption that before a company sets up such return channels for its products, a clear analysis of its economic viability has to be made (Morana and Seuring 2007).

In the case study of **Nike** that was also presented previously, the company offers the customer three different drop-off locations on its website. The customer can either send the shoes by mail, give his shoes to a local Nike store, or to an ENPO that is a member of the NRC. The specific process of collection depends on the location of the customer as well as on the number of shoes returning. Figure 3.4 illustrates the different collection options. For both of the participants, the different return options have various benefits and disadvantages. For Nike, the first option is a win–lose strategy. The company has no costs for transport and storage and can focus on the remanufacturing issues. On the other side, the donor has to pay for packaging costs choosing the first option. The second option, returning the products to the Nike store offers quasi no benefits to Nike due to the increasing product acquisition costs and the more complicated management. Therefore, the limitation of 10 shoes/donor is reasonable. For the customer, the second option represents a benefit because there are no further costs arising and the convenience is high. The third option of returning the products through an ENPO offers benefits for both parties. On the one hand, Nike has no product acquisition costs, on the other hand also the donor has the same benefits as in the second option. It only has to be added that for Nike incur few transportation costs for transporting the shoes from the ENPO to the recycling facility. Since its start in 1993, more than 15 million pairs of shoes have processed this way (Kumar and Malegeant 2006).

Another business model to acquire products from costumers is the case of **GETEX**, a franchise system for collection and redistribution of used apparel products. GETEX is one of 800 acquisition companies in Germany and buys the

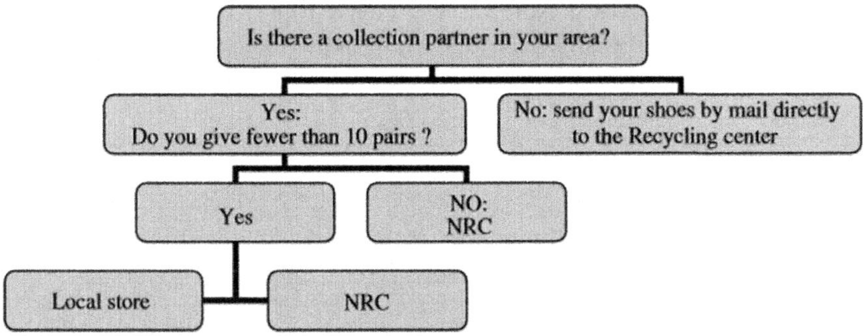

Fig. 3.4 Collection options Nike. Adapted from Kumar and Malegeant (2006)

apparel pieces from private customers. In that sense, the system is a specific closed-loop supply chain operating independent of the producers of the original products. The requirement for the system is the return of the products by the customers themselves to so-called acceptance companies. As an incentive, GETEX offers them an immediate voucher which is usable in these companies. As a next step, the apparel products are cleaned, graded, and sorted in the cleaning facilities. Then, GETEX collects the products and resale them either to second-hand shops in Germany or on international markets. Regarding the transaction costs for the customer, the information costs are reduced because a higher number of return and acceptance places are available. Furthermore, the psychological costs are higher than in the ECOLOG case, because it is not clear what happens to the products in the next stage (Morana and Seuring 2011).

Another factor that has to be considered regarding the product acquisition activities and opportunities in a CLSC are alternative competing disposal modes that pose a threat to the product acquisition process. Three alternative ways to dispose used apparel for the customer can be identified. The simplest option is the disposal through the municipal waste. For the customer, it is the easiest way because almost no transaction costs are incurred. The second alternative is the street and container collection that is common in many countries. The consumers can place their waste on the street at a certain time or dispose it in specific containers. The transaction costs are rather limited, the time for disposal is rather short for example. On the psychological cost side, the mode is rather critical because on the one hand it is a much better option than municipal waste disposal, but on the other hand it is not sure what happens to the products in detail. The third option named second-hand sales can either occur by a second-hand trader or directly by the consumer. Here, the transaction costs are considerably high because the products have to be stored, information about suitable dealers has to be collected and the products have to be transported to the second-hand shops or to the buyer itself. In contrast to that the psychological costs are usually rather low, because—in case of sale—the products are reused. Nevertheless, it might happen that the apparel is not sold, then the other options might be chosen (Morana 2006; Morana and Seuring 2007).

3.3.2 Reverse Logistics

In literature, different meanings of the term "reverse logistics" (RL) can be found. On the one hand reverse logistics can be seen as the whole CLSC process with all activities included (Bouzon and Govindan 2015). This view sees RL as "the process of moving products from their typical final destination for the purpose of capturing value or proper disposal", which "comprises all the activities involved in processing, managing, reducing and disposing of hazardous or nonhazardous waste from production, packaging, and use of products" (Bouzon and Govindan 2015, pp. 86–87). Therefore, all recovery options for the products are seen as reverse logistics activities. These options are presented in detail in Sect. 2.1.3.

On the other hand, one can just focus on the physical movement and the logistical processes when defining RL. Doing so, RL is defined as the "movement of product or materials in the opposite direction for the purpose of creating or recapturing value, or for proper disposal" (Tibben-Lembke and Rogers 2002, p. 271). It is the "process of planning, implementing and controlling the efficient, cost-effective flow of raw materials, in-process inventory, finished goods and related information from the point of consumption to the point of origin for the purpose of recapturing value or creating value or for proper disposal" (Venkatesh 2010).

Regarding the last definition, three main activities of RL have to be considered. Firstly, the distribution planning which includes the physical transportation of used products from the first consumer back to the re-producer. Secondly, the inventory management that deals with the sorting and the management of used products for conversion. Third, the production planning that includes the actual conversion of the old products into marketable new products (Abraham 2011).

Generally, the main characteristics of a reverse logistic are (Tibben-Lembke and Rogers 2002):

- Difficult to forecast: the forecasts are not just needed for the customer demand which is also challenging, but also for the availability of the products to be remanufactured. Because of this, the planning of RL is difficult and "predominantly reactive" (Venkatesh 2010)
- Many to one transportation: in contrast to the forward logistics where products are moved from one origin to many destinations, the reverse movement is from different places to one destination.
- Product Quality: in the reverse logistics, the variety of products is high. Therefore, a uniform quality cannot be guaranteed.
- Transparency in costs: the cost structure is not transparent because the processes are not standardized and difficult to monitor.

3.3.3 Evaluating and Identifying Recovery Options

In literature, there are various sources that deal with different options to close the loop by recapturing the value of used products. About the phase of evaluate what option is the right thing to do, there is no specific study available.

The phase of evaluation is named in literature as **inspection/separation phase** or as the phase of testing, sorting and disposition. The aim of this phase is to determine the condition of the products and their most economically attractive reuse option. Kumar and Malegeant name this phase of evaluation "inspection or separation" but are not going into detail what happens in this phase (2006; Morana and Seuring 2007).

Generally, the collected items are sorted by fabric type, color, and product category. As this process is manual, time-consuming and labor-intensive, the time taken depends on the quantity of products. Usually, the sorting process starts at the same time as the collection process in order to avoid unproductive time. As a next step the clothes may have to be cleaned. Finally, the sorted and cleaned fabrics are stored according the defined product categories (Dissanayake and Sinha 2015).

Regarding the different opportunities of recapturing value, five different recovery options that are appropriate in recovering value of used textiles can be identified (Dissanayake and Sinha 2006, 2015; Sinha et al. 2009).

- **Repair and reuse**: return the used products in working order, often the quality is less than the one of new products
- **Refurbishing**: disassembly to the module level, inspection and replacement of broken modules, the quality can be increased to a specified level
- **Remanufacturing**: complete disassembly to the component level, extensive inspection, and replacement of broken or outdated parts, should bring used products up to quality standards
- **Cannibalization**: recover a relatively small number of reusable parts from the used product and use it in one of the options mentioned above
- **Recycling**: reuse materials from used products in the production of the original or any other product

In contrast to the named five opportunities, Oh and Jeong (2014) consider the options refurbishing and cannibalization as not applicable to the fashion industry. They just define the three mentioned options **repair, remanufacturing, and recycling as practicable options in fashion CLSC's**. Depending of the chosen recovery option, the collectors and the levels of disassembly differ. The level of disassembly is defined as "the level of components that are produced by disassembling the end-of-life products" (Oh and Jeong 2014, p. 9030). Table 3.3 summarizes the three different recovery options, the respective collector, and the level of disassembly.

Table 3.3 Recovery options

Repairing	Remanufacturing	Recycling
Collector: repairer	Collector: remanufacturer	Collector: recycler
To fabric product	To yarn product	To raw material
To semi-apparel product	To semi-fabric product	To semi-yarn product

Own table based on Oh and Jeong (2014)

3.3.3.1 Repair/Reuse

The first option of repairing the used products requires only limited product disassembly and reassembly, because only broken parts of the products are fixed or getting replaced. Usually, these products are downgraded in quality. Regarding the process, the repairer collects the apparel products from the customers and fixes the defects. Then, the repairer supplies the finished parts or products to apparel manufacturers. Therefore, the repairer competes with the fabric supplier. For a perfect sustainable approach, the use of these repaired fabrics should have higher priority (Oh and Jeong 2014).

In literature, only one explicit example for a repair CLSC can be found. The previous mentioned system of a rent-based CLSC deals with repairing the received clothes from the customers and the subsequent rent of these products. The rental service provider collects the products periodically and is responsible for a sustainable laundry and repair process. This process includes the washing, disinfection, and repair of products. Out-of-date or worn-out products are sent to material suppliers for further reuse engineering processes (Hu et al. 2014).

Another option to close the loop and to reduce the environmental burden of clothes is the reuse option. In case that the textile products are still in good condition, they either can be given to commonly used second-hand shops, shops or wholesalers as third category clothes or to organizations that organize the supply to developing countries (Farrant et al. 2010).

Also famous fashion brands such as H&M started a global garment collection program in cooperation with I:CO and sell the collected items that can be worn again as second-hand clothes or reuse the clothes as cleaning cloths (H&M 2015; Wang et al. 2014).

3.3.3.2 Remanufacturing

The option of remanufacturing includes the complete disassembly of used apparel products, the sub-assembly of approved parts, and the assembly of remanufactured products. The purpose of remanufacturing is to bring used products to the same quality level than new products. In order to receive these products of equal value, technological upgrades can be used. On the process side, the remanufacturer collects the products from customers and remanufactures semi-fabric or yarn products. Therefore, the remanufacturer competes with yarn manufacturers or yarn supply. In case of a perfect sustainable supply chain, the remanufactured fabrics should also be used with a higher priority (Oh and Jeong 2014).

Remanufactured fashion can be defined as “fashion clothing that is constructed by using reclaimed fabrics, which can either be post-industrial or post-consumer waste, or a combination of both” (Dissanayake and Sinha 2015, p. 2). The quality of the emerging fashion products should be equal or even better than brand new fashion clothing. Whereas post-industrial waste includes waste that is generated in the manufacturing process of textiles or apparels, post-consumer waste consists of the discarded garments by the end-consumers. The process of remanufacturing includes in detail “the disassembly of used products, inspection, cleaning and reworking of components parts, and [the] use in a manufacturing process to create a product as new quality” (Dissanayake and Sinha 2015, p. 2).

There persist five key steps in the product development process for remanufactured fashion products. They seem to be fairly similar to the general product development, however the approach offers many differences. Figure 3.5 summarizes the five key steps that are described below (Dissanayake and Sinha 2015).

Whereas in the conventional design process gathering trend information is the first step, in the **research and analysis** phase of the remanufacturing process the designers only gather trend information in order to evaluate general design directions. Trend information like seasonal colors, fabrics, or silhouettes is not taken into account. The designers receive the inspiration mainly by the fabric itself through intense analyzing of the available apparels and fabrics. By analyzing the products, their adequacy and suitability for new designs can be identified. Because of many constraints like large variation of colors or quality differences, the success of this material analysis is highly dependent on the creativity of the designers and their ability to evaluate the suitability of materials (Dissanayake and Sinha 2015).

The next step is the **concept development** and aims the generation of design ideas. In conventional design processes this is realized through sketches, whereas in the remanufacturing process the design ideas are generated through experiments with diverse possible shapes and color combinations that can be realized with the available material. In order to maintain a workable and flat piece for redesigning, the apparel pieces are disassembled. Disassembly is the unpicking of the seam threads or the cutting along the seams of a garment. It is a manual and time-consuming operation and above all an unproductive activity. Therefore, the use of low skilled workers for this activity could be an option. As mentioned before, in order to develop creative design ideas, a high level of design thinking and creativity is necessary. To explore unusual and unique design ideas, the draping technique is a willingly used method. Moreover, the disassembled pieces can be mixed with

Fig. 3.5 Product development process in remanufacturing. Own figure based on Dissanayake and Sinha (2015)

remnant fabrics in order to overcome material restrictions and to produce products that are commercially viable and repeatable. In the end, the most suitable designs that are feasible with the available materials are chosen as final designs. Depending on the available amount of fabrics, there is the opportunity of repeating specific designs (Dissanayake and Sinha 2015). Moreover, most of the styles "are created for multi-functional purpose, i.e. one design could be worn in few different ways" (Dissanayake and Sinha 2015, p. 5).

The phase of the **sample preparation** passes off as in the conventional design process. The samples are made of working patterns and the appropriate fabrics. Altogether, all samples form a sample collection with about 20–50 pieces that can be presented on catwalk events and fashion shows. After the sample preparation and the presentation for the retailers, the **pattern development and cutting** phase begins. In order to create the production patterns, the working patterns that were designed in the design development phase are used and—if the retailers request any modifications—developed. The process of cutting is the most time-consuming and critical one. Because of the high dimension of various used materials, colors and prints, each piece has to be hand-cut individually. The last phase of **manufacturing** includes garment construction and testing. The cut pieces are finally stitched together and have to be checked for quality and fit. Depending on the availability of fabrics, the output of the manufacturing could be either a single garment from each design called one-off pieces, or few repeats of a specific design. Moreover, through the standardization of the basic design and fabric type it is possible to create a collection of garments that seem to be similar (Dissanayake and Sinha 2015).

3.3.3.3 Recycling

The option of recycling deals with the reuse of materials from used products and components with the aim of making new parts or products. This recovery option leads to the loss of identity and functionality of original products. Regarding the process, the recycler collects the products from the customers and produces raw materials for yarn manufacturers, therefore the recycler competes with raw material suppliers. As mentioned before, the yarn manufacturer should use the recycled materials with first priority in order to realize a sustainable supply chain (Oh and Jeong 2014).

The recycling process of post-consumer textile waste is very lengthy process, because it contains many operations such as sorting, separation, and processing. Recyclers are confronted with issues such as the right separating of the blended components and the quality of the separated and recycled materials. These barriers have to be considered in the future and solutions have to be found (Muthu et al. 2012).

The previously mentioned case study of Nike focus on its recycling CLSC, which aims the hundred percent recycling of post-consumer and defective athletic shoes. Nike's cooperation partner NRC separates three main materials out of these shoes through a chemical process and grinds them up. The three main materials consist of upper fabric, midsole foam and outsole rubber. Each of these materials is then used for new products in the field of sports surfaces like grind rubber, grind foam, and grind fluff (Kumar and Malegeant 2006).

3.3.4 Remarketing

In literature, only little research can be found regarding the right marketing strategy in a CLSC. In the specific case of remanufactured fashion, it is stated that this is not yet applicable to the mass fashion market. One problem is that it cannot be guaranteed that the fabrics are equal and of equal quality. Retailers can only guarantee a standard design. Nevertheless, offering remanufactured fashion to customers could be a successful marketing point for fashion retailers because they could inform their customers about their sustainable initiatives and approaches. As a strategy, the products should be promoted as trans-seasonal and sustainable and should be offered at a higher price. The higher prices can be justified through the expected higher willingness to pay of the target costumer (Dissanayake and Sinha 2015).

Regarding the channel choice of a remanufacturer, two options can be considered. Either the remanufacturer competes with the manufacturer or cooperates with him. Both options are illustrated in Fig. 3.6. In the case of competition, the remanufacturer sells the remanufactured products directly to the customers, whereas in the cooperation mode the remanufacturer sells his products to the manufacturer at a wholesale price (Wang et al. 2014). Moreover, online could be a short-term solution. But in the long term, the remanufactured products should be offered in store parallel to the normal collection. This could contribute to the green awareness of the remanufactured collection (Dissanayake and Sinha 2015).

Another point to consider is the design that should be marketable. A solution to reach the mass market could be simple standard designs rather than complex and

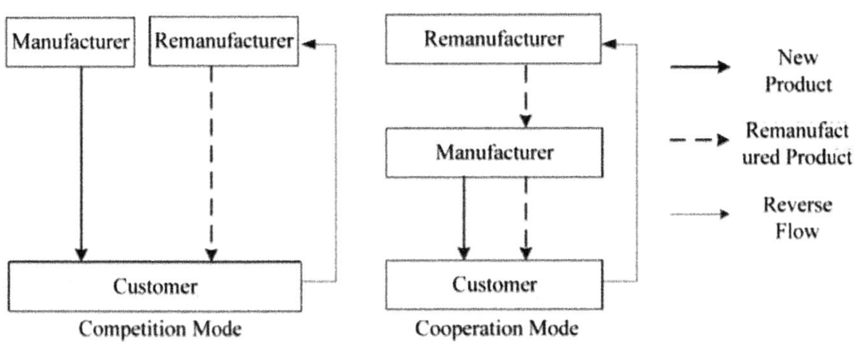

Fig. 3.6 Competition versus cooperation mode. Adapted from Wang et al. (2014)

unique products. For a successful selling, the marketing should be very clear and has to state the remanufacturing aspects. Moreover, in order to avoid cannibalization among remanufactured and normal products, different price ranges, and marketing approaches should be pursued and the prices and markets have to be understood in a widely manner (Dissanayake and Sinha 2015; Guide and Van Wassenhove 2009). One assumed approach is either the perfect substitution with complete cannibalization or the sales to secondary markets without cannibalization (Guide and Van Wassenhove 2009).

3.4 Challenges of Closed-Loop Production Systems

There are many challenges companies have to face considering a CLSC implementation and that hinder the general acceptance of such systems. First of all, many firms do not see the importance of such systems yet and instead want to focus on their core business. Moreover, the implementation requires a significant initial investment that detains many firms from implementing a CLSC. There exists also fear that a CLSC approach could affect the brand image of a firm because of the expected lower quality and lower standards of the products (Kumar and Malegeant 2006). This expectation could also lead to limited market demand for the secondary products and thus to a loss of sales for the firm (Geyer and Van Wassenhove 2005).

Another factor that hinders the implementation is the uncertainty in timing and quantity of the returns (Kumar and Malegeant 2006). Because of the little control over the CLSC, the firms are dependent of unpredictable sources of products, and thus it arises a high variability of quality and quantity of materials and finished products (Dissanayake and Sinha 2015). The unpredictably processing times lead also to a higher complexity in production planning as well as an increasing complexity of the whole supply chain management (Dissanayake and Sinha 2015; Geyer and Van Wassenhove 2005). Furthermore, through the higher complexity and the resulting additional costs for additional space and specialized labor the overall production and operational costs are increasing (Dissanayake and Sinha 2015; Larney and van Aardt 2010).

Moreover, for the firms exist, the danger of failing in the product acquisition process. This problem is visible in the case study of ECOLOG, where the success of the designed CLSC failed because of the low return rates of products (Morana and Seuring 2007). The lack of awareness of consumers or the high transaction costs that incur for the consumer may be the reasons for that failure (Larney and van Aardt 2010; Morana and Seuring 2007). Another factor that has to be considered is the lack of equipment or technology for recycling or remanufacturing programs, especially for textile and apparel manufacturers, which hinders the implementation of successful CLSC's (Larney and van Aardt 2010). Figure 3.7 summarizes the three main challenging factors of a CLSC.

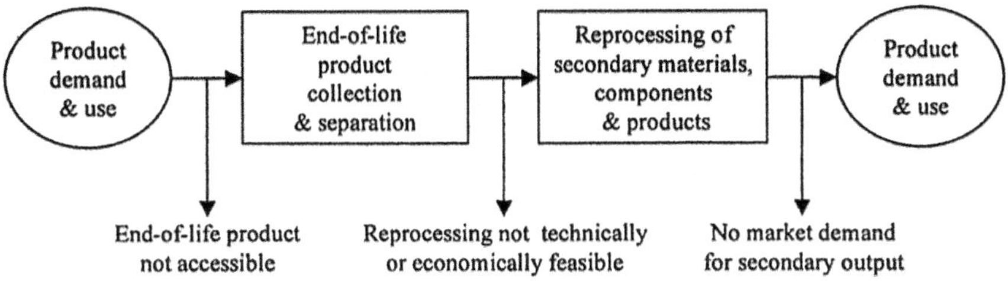

Fig. 3.7 Three types of constraints. Adapted from Geyer and Van Wassenhove (2005)

3.5 Conclusion

This paper provides an insight on the literature about closed-loop production systems and current practices in the fashion industry. Main characteristics of a CLSC are the forward and reverse supply chain as well as the five core processes namely product acquisition, reverse logistics, sorting, process of recovery, and redistribution. In order to realize a successful CLSC, these five processes need to be smoothly integrated.

Regarding the product acquisition, factors such as the marginal value of time, the remaining value of the product and most importantly the behavior of the customer have to be analyzed in a detailed manner. For the fashion industry, five recovery options can be identified that are appropriate: Repair/Reuse, Refurbishing, Remanufacturing, Cannibalization and Recycling. In case of remanufacturing, a specific product development process can be stated.

In the future, it will be difficult to overcome challenges such as the uncertainty of return quality and quantity, the technical or economic feasibility of reprocessing the textile products, or the lack of market demand for these recovered products. Nevertheless, companies have to recognize the importance of such systems in order to decrease the environmental burden of textile clothes. Therefore, solutions for the named challenges have to be found in order to realize a successful implementation.

3.6 Research Limitations and Future Research

Regarding the conducted literature review, some limitations can be outlined. Firstly, there is a lack of reliable data concerning closed-loop production systems in the specific case of the fashion industry. Little research is taken on this specific case, instead the focus is mainly set on other sectors like electronics or automotive. Moreover, many sources describe in fact many aspects of a CLSC but do not focus on a specific part. Therefore, the sources seem to be alike. Some aspects such as the right marketing strategy are rarely discussed in literature.

With the aid of the mentioned limitations some aspects for further research can be identified. These aspects are not only of importance for researchers, but also for companies who concern themselves with this subject. In the field of the product acquisition process, it would be interesting to develop strategies for the right managing and controlling of this process. Moreover, technologies should be evolved that focus on the improvement of the return process. This should also include strategies to convince customers to return their textile products. Moreover, especially in the fashion industry, research about the sorting process as well as the evaluation of the right recovery option should be conducted in a detailed manner. Another interesting point to consider is the adaption of remanufactured or recovered products to the mass market. Different strategies and approaches have to be developed. In this context, also the right marketing strategy is of great importance. The evolvement of specific advertisement approaches and pricing strategies of recovered products should be conducted. In order to evaluate the right marketing strategy, research should regard and analyze the target group and evolve forecasting systems for the demand of recovered products. In addition, the development of new technologies concerning the remanufacturing or recycling process of textile products could be an interesting field of research.

References

Abraham N (2011) The apparel aftermarket in India—a case study focusing on reverse logistics. J Fash Market Manag: Int J 15(2):211–227. doi:10.1108/13612021111132645

Bouzon M, Govindan K (2015) Reverse logistics as a sustainable supply chain practice for the fashion industry: an analysis of drivers and the Brazilian Case. In: Choi T-M, Cheng TCE (eds) Sustainable fashion supply chain management. Springer International Publishing, Cham, pp 85–104. http://link.springer.com/10.1007/978-3-319-12703-3_5

Caniato F, Caridi M, Crippa L, Moretto A (2012) Environmental sustainability in fashion supply chains: an exploratory case based research. Int J Prod Econ 135(2):659–670. doi:10.1016/j.ijpe.2011.06.001

de Brito MP, Carbone V, Blanquart CM (2008) Towards a sustainable fashion retail supply chain in Europe: organisation and performance. Int J Prod Econ 114(2):534–553. doi:10.1016/j.ijpe.2007.06.012

Dissanayake G, Sinha P (2015) An examination of the product development process for fashion remanufacturing. Resour Conserv Recycl 104:94–102. doi:10.1016/j.resconrec.2015.09.008

Eryuruk SH (2015) Life cycle assessment method for environmental impact evaluation and certification systems for textiles and clothing. In: Handbook of life cycle assessment (LCA) of textiles and clothing. Elsevier, pp 125–148. http://linkinghub.elsevier.com/retrieve/pii/B9780081001691000071

Farrant L, Olsen SI, Wangel A (2010) Environmental benefits from reusing clothes. Int J Life Cycle Assess 15(7):726–736. doi:10.1007/s11367-010-0197-y

Geyer R, Van Wassenhove LN (2005) The impact of constraints in closed-loop supply chains: the case of reusing components in product manufacturing. In: Fleischmann B, Klose A (eds) Distribution logistics, vol 544. Springer, Berlin, pp 203–219. http://link.springer.com/10.1007/978-3-642-17020-1_11

Goldbach M, Back S, Seuring S (2003) Co-ordinating sustainable cotton chains for the mass market. Greener Manag Int 43:65–78

Guide VDR, Van Wassenhove LN (2009) OR FORUM—the evolution of closed-loop supply chain research. Oper Res 57(1):10–18. doi:10.1287/opre.1080.0628
Guide VDR, Wassenhove LN (2009) Managing product returns for remanufacturing. Prod Oper Manag 10(2):142–155. doi:10.1111/j.1937-5956.2001.tb00075.x
Hu Z-H, Li Q, Chen X-J, Wang Y-F (2014) Sustainable rent-based closed-loop supply chain for fashion products. Sustainability 6(10):7063–7088. doi:10.3390/su6107063
Krikke H, Le Blanc I, Van de Velde S (2014) Product modularity and the design of closed-loop supply chains. Calif Manag Rev 46(2):23–39
Kumar S, Malegeant P (2006) Strategic alliance in a closed-loop supply chain, a case of manufacturer and eco-non-profit organization. Technovation 26(10):1127–1135. doi:10.1016/j.technovation.2005.08.002
Larney M, van Aardt AM (2010) Case study: Apparel industry waste management: a focus on recycling in South Africa. Waste Manag Res 28(1):36–43. doi:10.1177/0734242X09338729
Li C (2013) An integrated approach to evaluating the production system in closed-loop supply chains. Int J Prod Res 51(13):4045–4069. doi:10.1080/00207543.2013.774467
Morana R (2006) Management von closed-loop supply chains. Deutscher Universitätsverlag, Wiesbaden. http://link.springer.com/10.1007/978-3-8350-9061-3
Morana R, Seuring S (2007) End-of-life returns of long-lived products from end customer—insights from an ideally set up closed-loop supply chain. Int J Prod Res 45(18–19):4423–4437. doi:10.1080/00207540701472736
Morana R, Seuring S (2011) A three level framework for closed-loop supply chain management—linking society, chain and actor level. Sustainability 3(12):678–691. doi:10.3390/su3040678
Muthu SS, Li Y, Hu JY, Ze L (2012) Carbon footprint reduction in the textile process chain: recycling of textile materials. Fibers Polym 13(8):1065–1070. doi:10.1007/s12221-012-1065-0
Oh J, Jeong B (2014) Profit analysis and supply chain planning model for closed-loop supply chain in fashion industry. Sustainability 6(12):9027–9056. doi:10.3390/su6129027
Pappis CP, Daniel SE, Tsoulfas GT (2004) LCA as a tool for the evaluation of end-of-life options of spent products. In: Dekker R, Fleischmann M, Inder-furth K, Van Wassenhove LN (eds) Reverse logistics. Springer, Berlin, pp. 333–356. http://link.springer.com/10.1007/978-3-540-24803-3_14
Sinha P, Dissanayake DG, Hussey C, Bartlett C (2009) Recycled fashion. http://eprints.hud.ac.uk/17252/
Talbot S, Lefebvre É, Lefebvre L (2007) Closed-loop supply chain activities and derived benefits in manufacturing SMEs. J Manuf Technol Manag 18(6):627–658. doi:10.1108/17410380710763831
Tibben-Lembke RS, Rogers DS (2002) Differences between forward and reverse logistics in a retail environment. Supply Chain Manag: Int J 7(5):271–282. doi:10.1108/13598540210447719
Turker D, Altuntas C (2014) Sustainable supply chain management in the fast fashion industry: an analysis of corporate reports. Eur Manag J 32(5):837–849. doi:10.1016/j.emj.2014.02.001
van Nunen J, Zuidwijk R (2004) E-enabled closed-loop supply chains. Calif Manag Rev 46(2):40–54
Venkatesh VG (2010) Reverse logistics: an imperative area of research for fashion supply chain 7 (1/2):77–89
Wang K, Zhao Y, Cheng Y, Choi T-M (2014) Cooperation or competition? Channel choice for a remanufacturing fashion supply chain with government subsidy. Sustainability 6(10):7292–7310. http://doi.org/10.3390/su6107292

Chapter 4
Impact on Sustainability: Production Versus Consumption

Jochen Strähle and Katharina Hauk

Abstract The purpose of this paper is to examine the impact on sustainability of fashion production and consumption in order to discuss what the main lever is to reduce the negative impact. The research methodology applied is a literature review examining academic references. Key findings suggest that fashion production and consumption have a single comparable impact on sustainability. Moreover, as the fashion production follows the demand, the consumer steers the production in a certain direction. Therefore, consumers take over responsibility and need to be informed. To reach a long-term change in the fashion industry, the consumer has to be the focus of the sustainable efforts. Most results in literature were conducted by qualitative research methods, so that further quantitative testing of the results is recommended. Furthermore, most surveys were conducted with young fashion consumers in the EU or UK which does not represent the fashion consumer in general.

Keywords Fashion environmental impacts · Fashion production · Fashion consumption · Consumer behaviour · Sustainability

4.1 Introduction

With the establishing of the 1987 Brundlandt Report, the concept of sustainability became globally (Grunwald and Kopfmüller 2012). By facing global challenges such as poverty, the climate change and the increasing consumption of resources, sustainability is becoming a more and more important topic for governments, society and companies. These sustainability issues are also affecting the fashion industry which is highly addressed for sustainability impacts particularly following

J. Strähle (✉) · K. Hauk
School of Textiles and Design, Reutlingen University, Reutlingen, Germany
e-mail: jochen.straehle@reutlingen-university.de

K. Hauk
e-mail: katharina_hauk@live.de

J. Strähle (ed.), *Green Fashion Retail*, Springer Series in Fashion Business,
DOI 10.1007/978-981-10-2440-5_4

high-profile events such as the April 2013 collapse of Rana Plaza in Bangladesh (Kozlowski et al. 2015). Due to these media reports the awareness regarding social and environmental impacts in the fashion industry has been raising (Beard 2008). Thereby mainly the fashion production as the main contributor of negative sustainability impacts is addressed. The part of fashion consumption as a contribution, too, is mainly not focus of the public view (Gwilt and Rissanen 2010). This paper draws on this point, examining the impact of both sides: fashion production and consumption to get a more decidedly picture of the sustainability impacts in the fashion lifecycle. Therefore, the paper addresses the specific research question: *Where is the main lever when it comes to the impact on sustainability in the fashion industry: production vs. consumption?*

To answer this question a literature review examining academic references was conducted to first clarify the single impacts and related problems of each side. In the production phase, besides the environmental and social impacts, the production structure itself is subject of discussion. Regarding the consumption, the paper is focusing on the purchasing behaviour as well as the usage and disposal of fashion. Thereby the knowledge of the consumer concerning these impacts and the discrepancy between attitude and behaviour is outlined. Subsequently both sides are compared to get an overview of the levels of impact. Aside the single impacts, the coherence of fashion production and consumption is discussed as well as the question of responsibility along the fashion lifecycle. Based on this discussion, leverage points are given to implement a long-term change concerning the current sustainability impacts in the fashion lifecycle.

4.2 Literature Review

4.2.1 The Concept of Sustainability

In literature, numerous definitions of sustainability are attainable, but, in principle, they all remain similar to one definition from the World Commission on Environment and Development in the 1987 Brundlandt Report (Glavič and Lukman 2007). Within this report, sustainability is defined as the ability to "meet the needs of the present without compromising the ability of future generations to meet their needs" (WCED 1987, p. 41). With regard to this definition, Baumgärtner and Quaas (2010) define that sustainability is based on justice within human–nature relationships which includes justice between different generations, justice between human beings of the same generation and justice between humans and nature (Baumgärtner and Quaas 2010). Based on this understanding, John Elkington developed a framework—the so-called "Triple Bottom Line" in 1994 (Slaper and Hall 2011). It constitutes a concept which consists of three dimensions regarding social, environmental and economic aspects (Shen 2014; Lampikoski et al. 2014). These dimensions are commonly called the 3P's: people, planet and profit. This concept should prevent companies from only using traditional measures of economic and

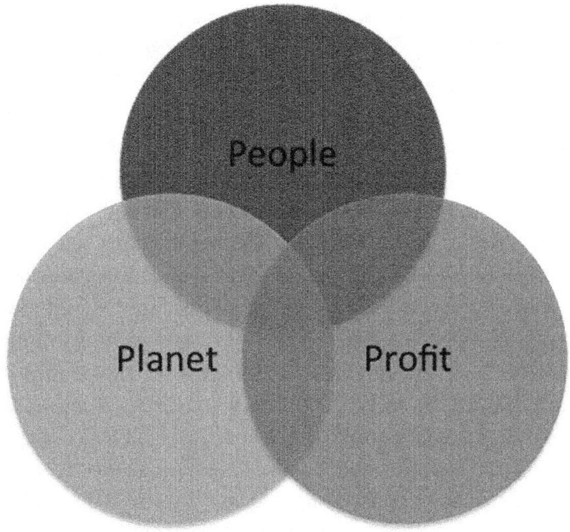

Fig. 4.1 3P's: people, planet and profit. Based on Slaper and Hall (2011)

monetary indicators. It should rather encourage companies to install integrated objectives serving sustainability. Thereby, companies should strive for a complementary of objectives between all three dimensions (Slaper and Hall 2011; Fig. 4.1).

4.2.2 Impact of Fashion Production on Sustainability

4.2.2.1 Fashion Production Structure

The fashion supply chain is highly complex as it involves many different stages as well as several individual companies (Karthik and Gopalakrishnan 2014; Taplin 2014). According to Kogg (2009), big Western fashion companies nowadays have up to 500 suppliers in the first tier (apparel manufactures). These in turn have sub-suppliers in the second tier (e.g. dyeing and finishing) and so on. This drastically increases the number of involved actors in the whole manufacturing process of garments. Moreover, there is the element of change and dynamism due to the fact that individual suppliers are added and dropped according to the evolving needs in the supply chain. In the end it is difficult to define how many individual companies are involved (Kogg 2009). Thereby the focal company, the fashion brand, is in direct contact with the consumers and puts the product on the market (Gold et al. 2010). As the focal company can set requirements for suppliers and sub-suppliers, it is responsible for the activities in the whole supply chain including all companies and organizations that the focal company interacts with, directly or indirectly. Therefore, it is common that the focal company is the first counterpart for other stakeholder groups such as authorities, NGOs and the media (Fransson et al. 2013). The fashion supply chain gets even more complex as it has evolved following the trend of global purchasing (Fransson et al. 2013). Caused by international competition, firms of industrialized nations make use of manufacturing plants in developing countries as they offer lower production costs (Nagurney and Yu 2012). This is possible due to lower wages (Allwood et al. 2006) as well as less strict environmental regulatory systems (Nagurney and Yu 2012). Consequently, fashion companies can produce in Asian countries under more favourable economical conditions by exploiting social and environmental regulations (Piegsa 2010). Thereby, these countries like, for instance, Bangladesh are dependent on western firms, as they are labour-rich, but capital-poor countries (Taplin 2014). The high number of involved sub-suppliers as well as the global structure including different legal guidelines makes it difficult for the focal company to monitor and control the whole production process (Draper et al. 2007). Therefore, fashion brands are criticized that they do not recognize responsibility regarding environmental and social impacts in the production process (Taplin 2014; Fig. 4.2).

Aside from the complexity of the supply chain, it is important to consider how the production is driven: The fashion production is driven by the exigencies of the dynamic patterns of fashion demand (Bruce et al. 2004). In practice, this means the consumer has enough supply to choose from, and therefore brands react with their supply on current consumption patterns to gain more customers (Chen 2008). In order to meet the fast moving and demanding fashion consumer needs, the fashion industry developed the current concept of fast fashion (Bruce et al. 2004; Cachon and Swinney 2011). This includes two essential aspects: Trendy, fashionable

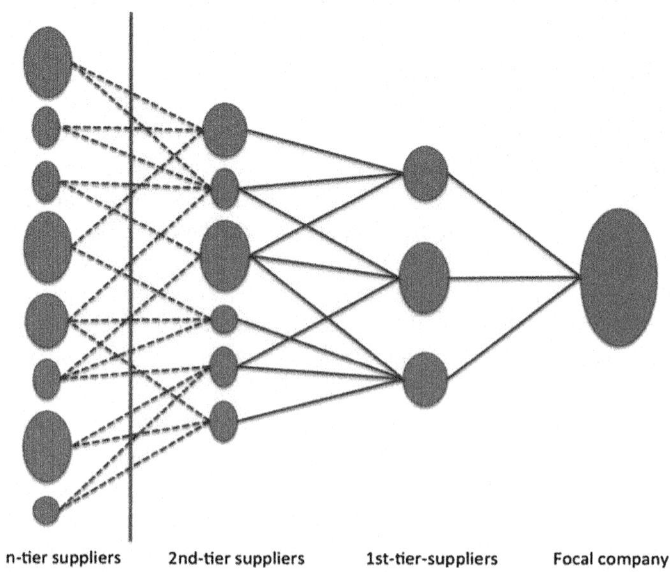

Fig. 4.2 Complexity of the fashion supply chain. Adapted from Kogg (2009)

products that are universally affordable and continually changing, as well as a short time gap between production and distribution (Cachon and Swinney 2011). This means, the current supply chains are at the forefront of this new business model. As part of changes in overall production strategies, fashion brands like H&M or Inditex embraced a quick response model of production, which was designed to reduce inventory and dramatically reduce the time spent between initial design of garments and their arrival in fashion retail stores. This puts pressure on suppliers to meet strict delivery deadlines as well as strong cost mandates (Taplin 2014). Different authors agree that the concept of fast fashion has highly intensified the overall negative impacts of fashion on sustainability (Morgan and Birtwistle 2009; Taplin 2014), which will be explained in the following.

4.2.2.2 Fashion Materials and Processing

Fashion Materials

Regarding the measurable impact of production on sustainability the production phase accounts for the highest impact when it comes to eutrophication, agricultural land occupation and natural land transformation in the fashion lifecycle, mostly due to the use of natural fibres. This is because natural fibres require land and fertilisers during the cultivation (Beton et al. 2014).

As cotton is the most important apparel fibre with an annually produced amount of 24 tons (Draper et al. 2007) and an extensive need of resources, it has the highest impact on sustainability (Beton et al. 2014). During the whole cultivation process cotton demands heavy use of pesticides which account for 24 % of the world's pesticides (Chen and Burns 2006). This amount even tends to rise, as there comes a pest resistance in shorter gaps, which leads to salinization of soil, desertification and water pollution (Piegsa 2010). The majority of these chemical pesticides are used by farmers in the developing world, where access to safety equipment, training and understanding of the dangers posed by hazardous chemicals is often lacking. This results in acute pesticide poisoning and therewith a massive health risk. Genetically modified cotton plants reduce these effects on environment and humans, but long-term effects are not explored yet, which has a high environmental risk. Moreover, the cotton cultivation has an enormous need for water: In some cases, over 10 tons of water are used to grow enough cotton to make 1 pair of jeans (Draper et al. 2007). Besides the washing in the usage phase, the cotton cultivation has the highest impact on water depletion in the fashion lifecycle (Beton et al. 2014). This leads to a decreasing groundwater level and social problems due to water shortages in the developing countries (Adamcyk 2015). Besides plant fibres like cotton, the fashion industry uses various animal fibres. Thereby, wool is the most important animal fibre used in textiles (Chen and Burns 2006). The most significant factor is the use of land: Despite the fact that wool only constitutes 2 % of all fibres used, it totals 69 % of all land used in the fibre production. Furthermore, in the animal farming as well as in the conditioning pesticides are used. These burden the animals and end up untreated in the sewage by the wool washing process (Piegsa 2010). Moreover, many animal fibres are not by-products of the food industry. There are many concerns about the animal welfare especially when it comes to leather, fur and silk (Draper et al. 2007). In the last years, the fibre production in general has been grown steadily which intensifies the impacts. However, this is mainly attributed to an increased use of man-made (chemical) fibres, especially of polyester (Allwood et al. 2006). In contrast to the use of land, water and pesticides in the production of natural fibres, the production of man-made fibres requires high amounts of oil, chemicals and energy (Piegsa 2010): Thereby, the main problem is that these fibres like polyester are generally made from non-renewable by-products of the oil industry, which take a long time to degrade and are therefore difficult to dispose of (Draper et al. 2007). Second, the implemented chemicals in the man-made fibre production are carcinogenic and therewith harmful to health and at the same time hazardous for the environment (Piegsa 2010; Fig. 4.3).

Fashion Processing

The negative environmental impacts in the fashion production are also linked to the processing procedure (Beton et al. 2014). These impacts are mainly caused by the significant use of chemicals, water and energy during the dye, print and finishing processes (Draper et al. 2007). According to a study from Orhon et al. (2003), up to 230 m^3 of water is needed to produce 1 ton of textile fabric in Turkish factories (Orhon et al. 2003). Thereby, the main problem besides the amount of used water is that during the processes several different chemicals such as acids or dyes are used which contaminate the water. Moreover, in some cases heavy metals are discovered in the wastewater (Visa et al. 2011). As many as 2000 different chemicals are used in textile processing, especially in textile wet processing, and many of these are known to be harmful to human health (Choudhury 2014) as they are carcinogenic and genetically harmful (Adamcyk 2015). In dyeing factories, especially in India

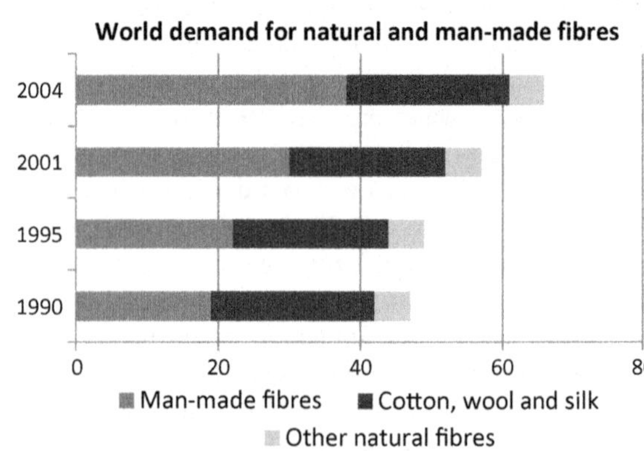

Fig. 4.3 World demand for natural and man-made fibres. Adopted from Allwood et al. (2006)

and China, this untreated wastewater is often discharged directly into the environment such as local streams and rivers (Kaye 2013). This has an enormous impact on the environment as the chemicals in the water do not readily degrade, they can accumulate in organisms and are simply toxic (Choudhury 2014). Finally, this leads to freshwater and marine eutrophication (Beton et al. 2014). This constitutes not only a big impact on the environment, as humans are also suffering from it. The accessibility to clean drinking water is an increasingly important issue in many countries where clothing dye houses operate and pollute the drinking water (Draper et al. 2007). Moreover, as the chemically treated textiles are dried in special tenter frames, the pollutants also accumulate in the exhaust air (Piegsa 2010). These impacts are intensified by the fact that most dye houses are working inefficiently: There is a significant amount of resource wastage during the dyeing processes, particularly as a lot of fabric is dyed twice (Draper et al. 2007). In Europe countries certain institutions control these impacts, but these legal guidelines like in Europe do not exist in Asian countries where the main production of textiles and fashion is conducted. Therefore, chemicals that are forbidden in Germany are still used in these countries. Especially, chemicals for finishing that could not be eroded by water can evoke allergies and can have cancer-causing impacts (Piegsa 2010). In the pursuit of low production costs, apparel firms took advantage of lower environmental awareness and looser environmental regulatory systems in these developing countries (Nagurney and Yu 2012).

4.2.2.3 Fashion Social Workforce

Besides these environmental impacts of the production there are also social issues to be considered in the fashion industry (Strähle et al. 2015a, b). As already explained in Sect. 4.2.2.1, due to the increasing globalization and therewith the increasing price war, firms of industrialized nations make use of manufacturing plants in developing countries that offer much lower production costs (Nagurney and Yu 2012). Thereby firms argue that they pay the legally defined minimum wage in each country, but social groups assert that this “minimum legal wage” is not to be mistaken for a “minimum living wage”. It may not be possible to escape from a cycle of poverty with only the minimum legal wage (Allwood et al. 2006). Moreover, due to the development of fast fashion, what requires quick production time and strict delivery deadlines, it results in work intensification (Taplin 2014) and therewith excessive hours or unpaid overtime (Draper et al. 2007). The use of repeated temporary contracts or the absence of any employment contracts combined with delayed payment is common practice in these countries (Allwood et al. 2006). According to Draper et al. (2007) these payment and contract conditions are largely due to workers being prevented from forming and joining trade unions (Draper et al. 2007). Thereby, the industry workforce is largely made up by young, low-skilled woman, who are vulnerable to various forms of abuse and may not know how to claim their rights as employees (Allwood et al. 2006; Fig. 4.4).

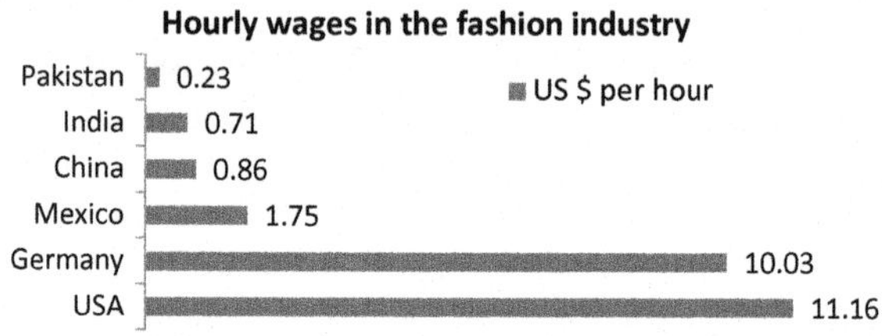

Fig. 4.4 Hourly wages in the fashion industry. Adopted from Allwood et al. (2006)

Furthermore, the textile and fashion industry has been heavily criticized for bad working conditions and violation of human rights, particularly following high-profile events such as the April 2013 collapse of Rana Plaza in Bangladesh (Kozlowski et al. 2015). In this disaster 1127 people were killed, which illustrates the dangerous nature of working in Asian garment factories (Taplin 2014). Included in these dangerous working conditions are, besides the insecurity of the buildings and equipment, the health issues caused by hazardous chemicals used in the production process, fibre dust and noise. Moreover, cases are noted in which women are threatened by their superiors and unable to complain, without risk of losing their jobs. A further problem is the child labour as it is mostly difficult to monitor subcontractors, indirect workers and home workers (Allwood et al. 2006).

4.2.3 Impact of Fashion Consumption on Sustainability

4.2.3.1 Fashion Purchasing Behaviour

Fashion Consumption Level

In the first row, consumers buy clothes to fulfil an emotional need, to fit in, impress others and to satisfy basic needs (Cao et al. 2014). Especially for young people wearing the newest trends in fashion is a way of socializing (Morgan and Birtwistle 2009). According to Cachon and Swinney (2011) there is a socio-cultural change in the lifestyles of the fashion consumer, who is constantly knowledgeable about the latest fashion trends and feels the need to adapt to the reality around him in an affordable, dynamic manner (Cachon and Swinney 2011). Hence, consumers frequently ask for new fashion items to buy as there is a peer pressure to constantly wear new clothes (Jørgensen and Jensen 2012). This approach is not confined on age, gender or social class. It is shared by young people who wish to follow the trends by buying inexpensive clothes and at the same time by more mature consumers who are willing to relinquish their high-quality standards to renew their wardrobes more often (Bhardwaj and Fairhurst 2010). This fact has required the fashion industry to react (Christopher et al. 2004). As already explained in Sect. 4.2.2.1 the industry acts on the growing demand of affordable and trendy fashion with the concept of "fast fashion". Consequently, this introduction of low price fast fashion has fuelled rapid consumption. It has become common practice to buy clothes one may never wear and consume excess garments to satisfy the mentioned psychological needs. As an example in the UK, in the years from 2001 to 2005, spending on women's clothing grew by 21 % and that on men's by 14 %. During the same time, price dropped by 14 %, so sales by volume increased by 37 % (Allwood et al. 2006). Furthermore, a survey from Morgan and Birtwistle (2009) confirms an excessive fashion purchasing behaviour from young fashion consumers. About one in ten respondents purchased more than one fashion item weekly, and one in five purchased something every week. Overall, more than half the respondents purchased fashion garments at least every 2 weeks. In terms of expenditure, it is evident that many of these items are purchased from the fast fashion retailers such as Primark, Top Shop and H&M. Consequently, young fashion consumers prefer to buy several cheaper items of fashion than one more expensive piece. Thereby many stated that they purchased actually more than they really needed (Morgan and Birtwistle 2009). Thereby most fashion purchases are independent from a real necessity as most fashion items are rarely or never used after buying them (Jørgensen and Jensen 2012; Wahnbaeck et al. 2015).

These surveys exhibit clearly a development to an overconsumption of fashion in the last years provided by fast fashion (Morgan and Birtwistle 2009). According to Chen and Burns (2006) global textile consumption is estimated to amount to more than 30 million tons a year, which causes heavy environmental impact during production (Chen and Burns 2006): While in the last 30 years industrial development has achieved environmental improvements and has moved to a smaller environmental impact, consumption has increased by the same level and therewith production. This erodes the environmental benefit of the technological advances (Niinimäki and Hassi 2011). This described phenomenon is called "the rebound effect" (Throne-Holst et al. 2007). According to Sheth et al. (2011) the environmental concerns arising from rapid growth in consumption are twofold: First, the eco-system of the earth is constrained and therewith it cannot support unlimited growth in consumption. It is estimated that humanity's current collective consumption level already needs the resources of 1.4 earths. If the whole world consumed at the level of consumption of America, five planets would be needed. Second, it leads to environmental degradation risks such as biodiversity loss, deforestation and pollution of water systems and land (Sheth et al. 2011).

Fashion Buying Criteria

It has not only an impact how much consumers buy, but also what kinds of products they buy (Niinimäki and Hassi 2011). According to a study from Beckmann (2014), which concentrates on fashion buying criteria of young fashion consumers, the most important buying criteria are the fitting, quality and the visual appearance (Beckmann 2014; Table 4.1).

This shows that consumers' general textile decision process follows a specific order: Only products are taken into consideration which convince by their style;

Table 4.1 Fashion buying criteria

Buying criteria	Midpoint	Level of measurement	Meaning
Fitting	2.38	1	Most important
Quality	2.61	2	Important
Visual appearance	2.65	3	Rather important
Price	3.48	4	Rather unimportant
Fair trade	4.75	5	Unimportant
Environmental impact	5.04	6	Most unimportant

Adopted from Beckmann (2014)

second, the consumer examines the products' touch and fitting; and finally he checks the price in relation to performance (Meyer 2001). Products are assessed as "good" if they include the mentioned three criteria of style, quality and perceived low price. Thereby the quality is judged by the brand of the retailer or manufacturer and by the "feel" of items, particular by the strength of seems (Fisher et al. 2008). As the quality is mainly judged by the image of a brand, the country of origin as well as fair trade and ecological textile labels are no relevant buying criteria (Beckmann 2014). When it comes to the price, consumers prefer a cheaper product, as the most trends are short. At the same time they are aware of the fact that these cheap, fashionable clothes from low budget retailers will not last long as they are made to be worn less than 10 times (Birtwistle and Moore 2007). Generally, different studies reveal that in this decision process, the least important considerations are fair trade, health and environmental considerations (Beckmann 2014; Laitala and Klepp 2013). According to Hiller Connell and Kozar (2012), 70 % of the fashion consumers have never even considered the environmental impacts of garments when making clothing purchases (Hiller Connell and Kozar 2012).

4.2.3.2 Fashion Usage and Disposal

Fashion Usage

After a fashion item has been bought, the usage phase continues the environmental impacts. The usage phase in this context means the washing, tumble drying and ironing of clothing by a consumer (Strähle et al. 2015a, b), which is in many sources combined to the laundering process. Laundering involves exposing the textile article to the combined effects of water, energy and detergent (Slater 2003).

The washing of clothing is a highly water-consuming process and therewith responsible for the largest contribution to water depletion in the whole fashion lifecycle. In comparison to the production and processing it accounts for 67 % of the impact on water depletion (Beton et al. 2014).

Besides the high use of water, the usage phase accounts for the most significant energy use in the clothing lifecycle. Allwood et al. (2006) made explicit findings in this context: They examine the energy consumption of a garment in the different lifecycle phases using the example of a T-Shirt. With 25 washes at 60 °C, followed by tumble drying and ironing, 60 % of the total energy in the fashion lifecycle is required (Allwood et al. 2006). This means, depending on which materials the clothes are made from, as much as 80 % of the carbon footprint of clothing can be made caused in its washing and care (Draper et al. 2007). Thereby, the most significant impact is contributed by tumble drying as it uses electricity to generate heat. A tumble dried load of laundry produces three times that much CO_2 emissions as compared to drying on the line (Berners-Lee and Clark 2010). As the energy consumption is clearly dominated by the usage phase, changes in production structure, including recycling, have only a little effect on energy use as the same use requirements remain (Allwood et al. 2008).

In addition to the impact of energy and water usage in laundering, chemical additives in detergents act as pollutants (Goworek et al. 2013). The detergent use has a very high impact on toxicity, metal depletion and freshwater eutrophication. In the overall consideration of the fashion lifecycle, most of the toxic emissions affecting human beings and aquatic ecosystems are associated to the usage phase, mainly caused by laundry detergents (Beton et al. 2014). Besides the homecare of clothing, Hiller Connell (2011) underlines the massive use of chemicals in dry cleaning processes that are environmentally hazardous (Hiller Connell 2011). These used chemicals do not only have a massive impact on the environment, but even more they pose a health risk to industry workers and consumers (Gwilt and Rissanen 2010). All in all the usage phase scores the highest contribution in the fashion lifecycle to the damage to ecosystems (Beton et al. 2014), like burden to wastewater systems and sewage treatment plants (Paloviita and Järvi 2008), as a potential consequence of the significant contribution to freshwater and marine toxicity. And at the same time the usage phase has a high negative impact on human health (Beton et al. 2014).

As the consumer decides the method and frequency of laundering, he mainly controls the mentioned sustainability impacts in the usage phase (Laitala et al. 2011). According to a study from Fisher et al. (2008), the consumer is only partly aware of these impacts of usage activities. However, even if he is informed, he has a pragmatic approach. Perceived cleanliness is important to consumers and most of the respondents wash their clothes after every single use to restore the freshness. This was associated with the social development to a peer pressure to change clothes daily (Fisher et al. 2008). Moreover, this is supported by the increased fashion consumption: Consumers wash in summary more, which counteracts the benefits of technological improvements in laundry concerning the length of time of washing (Laitala et al. 2011). To this, a research by Shove (2002) indicates that clothes may be washed more frequently from habit than out of necessity (Shove 2002). This extensive laundry behaviour is the main reason for the intensity of the environmental impact in the usage phase (Laitala et al. 2015).

Fashion Disposal

The environmental impact stretches even further once an item has been worn out. Thereby clothes are not only culled when they are no longer wearable. Two-thirds of consumers cull clothes because they do not like them anymore or the clothes are out of style. Only 20 % of consumers cull clothes only because they are broken or do not fit anymore (Wahnbaeck et al. 2015). Furthermore, consumers are reluctant to repair their clothing to extend its life, because of insufficient time or skills and the low price of replacing it (Gibson and Stanes 2011). Especially, the second aspect is the main reason for growing fashion waste (Morgan and Birtwistle 2009). According to a study from Wahnbaeck et al. (2015) the wearing time of clothes gets shorter. Trousers and tops are worn less than 3 years in average (Wahnbaeck et al. 2015). This means there is a tendency for consumers to buy increasing amounts of fashion and to keep it for a relatively short time until disposing of it, as the low price and the poor quality (Achabou and Dekhili 2015) of fashion products equates with disposability in the minds of consumers (Birtwistle and Moore 2007). This also includes that consumers are more likely to throw away cheap clothes than expensive ones as they would have feelings of guilt (Morgan and Birtwistle 2009). This had lead to the coining of the term "disposal fashion" (Birtwistle and Moore 2007). Consequently, the relationship between fast fashion and increasing textile waste is unmistakable (Morgan and Birtwistle 2009). Thereby most consumers never did clothes-swapping, borrowing or second hand as an alternative for disposing fashion (Wahnbaeck et al. 2015). Therefore, the clothing waste is growing. As an example, in Britain, clothing and textile waste is estimated to be the fastest growing waste stream between 2005 and 2010 and amounting to 1.5–2 million tons annually (Niinimäki and Hassi 2011). It is estimated that in the UK, the average consumer disposes 30 kg of clothing and textiles waste to landfill per year (Allwood et al. 2006). In the landfills this waste causes particular environmental problems: synthetic products do not decompose, while woollen garments decompose and produce methane, which contributes to global warming. Moreover, the toxic chemicals pollute the groundwater (Fletcher 2014). According to the study from Birtwistle and Moore (2007), the consumer has only little awareness of this impact of disposing of high volumes of textile waste (Birtwistle and Moore 2007).

4.2.3.3 Fashion Sustainability Knowledge

Regarding the mentioned environmental impacts there is a lack of knowledge along the fashion consumers (Morgan and Birtwistle 2009; Shen et al. 2013). According to a study from Morgan and Birtwistle (2009) about young fashion consumers' consumption patterns, there is not only insufficient knowledge when it comes to the impacts of fashion disposal and the need for recycling. In general, they are not aware of how and where clothing is disposed of, how the items have been made and what environmental consequences the production process entails (Morgan and Birtwistle 2009). Especially, the lack of knowledge about the impact of the used materials is significant. Different studies reveal that there is a common misperception that natural fibres like cotton have little environmental impact than synthetic fibres (Austgulen 2013; Laitala and Klepp 2013). Moreover, consumers are not aware about their own contribution to the sustainability impact in the form of purchasing behaviour as well as the usage activities (Fisher et al. 2008). A study from Shen et al. (2013) confirms these findings and emphasizes that this lack of knowledge applies not only for young fashion consumers. There is a general lack of knowledge in terms of sustainable fashion and demographic traits are no determining factors when it comes to this (Shen et al. 2013). Besides this lack of environmental knowledge in several areas of fashion consumption (Laitala et al. 2015) there is also little knowledge about the holistic concept of sustainability itself (Hill and Lee 2012). While most fashion consumers consider environmental issues as part of sustainability, only few consider social and economic aspects as part of this concept (Hill and Lee 2012; Kagawa 2007). This leads to an uni-structural understanding demonstrated by a single broad element of sustainability (Carew and Mitchell 2002).

Moreover, most of the consumers who claim to be aware of sustainable development could not really explain the term. This shows that also consumers who have awareness of sustainability are not really informed (Kagawa 2007). As in recent years awareness of sustainability in fashion has been raised (Gam 2011; Morgan and Birtwistle 2009) the consumer desires more information to make informed choices (Morgan and Birtwistle 2009). However, more information about sustainable fashion provided by companies does not necessary lead to a higher knowledge of consumers. Around 30 % of fashion consumers do not even read the information provided on product labels (D'Souza et al. 2006; van der Merwe et al. 2014). And if they do so, the information is not completely correctly understood and interpreted (van der Merwe et al. 2014). The same problem is associated with ecological and fair trade textile labels. Most consumers are only aware of one or two existing labels and do not know what they explicit mean for the production process and the product itself (Austgulen 2013; Beckmann 2014).

According to different studies the described lack of knowledge of sustainability impacts of the fashion industry (Morgan and Birtwistle 2009) may act as a barrier to sustainable consumption (D'Souza et al. 2006; Gam 2011; Shen et al. 2013). Before consumers make any decision, they need to be knowledgeable about their choices so that they are able to make evaluation of competing alternatives. Not choosing one particular alternative might be by the lack of awareness and knowledge (Shen et al. 2013). As consumers are not informed about certain issues like the positive use of organic cotton or fair trade, they do not realize the need for purchasing sustainable fashion items and therewith accepting a higher price (Bhaduri and Ha-Brookshire 2011). In addition, there is lack of knowledge concerning currently available sustainable fashion products and the majority of consumers have trouble making the link between fashion and sustainable consumption (Sisco and Morris 2012). Moreover, in general, they do not know how to change their behaviour in the whole consumption phase, including usage and disposal, in order to be more environmentally sustainable (Hiller Connell and Kozar 2014).

4.2.3.4 Fashion Attitude–Behaviour Gap

Yet it is important to mention that even a higher knowledge and therewith awareness of the sustainable impacts does not necessary lead to a change in behaviour (Strähle et al. 2015a, b): As sustainability is a topic with increasingly importance in the fashion industry, different studies reveal that consumers state to consider environmental and social issues in their consumption behaviour and the majority has a positive attitude towards sustainable fashion (D'Souza et al. 2006; Shen et al. 2013). However, in spite of consumers' expressed concern for the environment and the growing prevalence of green product shelves, the demand for sustainable products and services is consequently not as high as expected (Gleim et al. 2013; Kaufmann et al. 2012). In fact, sustainable clothes only have a small share of the clothing market (Jørgensen and Jensen 2012). As Sect. 4.2.3.1 already showed, in actuality there are other more important criteria when buying fashion (Laitala and Klepp 2013). This discrepancy between the stated attitude by consumers and the actual behaviour is named in literature as the "attitude-behaviour-gap" (Antonetti and Maklan 2015; Laitala et al. 2015). This means that there is "[…] a significant difference between what consumers say about the importance of consumption-related ethical issues and their actual behaviour […]" (D'Astous and Legendre 2009, p. 255). These discrepancies between attitudes and behaviour are mainly explained by the fact that shopping for clothes can be a complicated process where several factors must be taken into account simultaneously (Laitala et al. 2015). These different factors can operate as barriers to sustainable fashion consumption (Sisco and Morris 2012).

According to Hiller Connell and Kozar (2014) the relationship between knowledge and behaviour is slight, when there are considerable external constraints. Thereby the most decisive factor is the price (Gleim et al. 2013; Sisco and Morris 2012). Consumers perceive ethical clothing to be too expensive (Sisco and Morris 2012). Especially in comparison to traditional ("non-sustainable") fashion products, sustainable fashion products are shown to be considerably more expensive which deters many non-green consumers (Gleim et al. 2013). According to Gleim et al. (2013) not only the price is considered as costs for the consumer, but also the time and effort needed to evaluate and search for sustainable products are part of the cost of consumption (Gleim et al. 2013). As the today's fashion consumer is used to an anytime and easy availability of new fashion products (Billeson and Klasander 2015) he is in general reluctant to conduct extensive information search and elaborate cognitive processing (Gleim et al. 2013). Sustainable product alternatives are yet not mainstream (Sisco and Morris 2012) which requires the consumer to actively search and ask for these products (Gleim et al. 2013). Moreover, when deciding for sustainable fashion items, it reduces the range of fashion items. As different studies in this context show, the consumer does not want his choice limited by a small assortment of eco-labelled clothes (Jørgensen and Jensen 2012; Meyer 2001). This explained constrain of convenience is also the main reason for the unsustainable behaviour in the usage phase (Hiller Connell and Kozar 2014).

Moreover, there is a common consumer perception that sustainable clothes are not stylish or fashionable and eco-clothing is unattractive (Achabou and Dekhili 2015; Sisco and Morris 2012). As the today's fashion consumer feels the peer pressure to constantly wear fashionable clothes, he is not willing to make constraints when it comes to the style of a fashion item (Billeson and Klasander 2015; Meyer 2001). As an example, consumers accept sustainable aspects of clothes like organic cotton only as there would not be any compromise on the design (Billeson and Klasander 2015).

According to different studies, social influences can act as a crucial constraint in the green consumption process (Gleim et al. 2013; Kim and Choi 2005). First of all, the consumer's decision-making is largely influenced by the attitudes of his fellows such as family, friends and other groups. Individual consumers are likely to act in a manner that is consistent with beliefs expressed by such groups. Therefore, if an individual associates with people not concerned with sustainability, his consumption is likely to be non-green (Gleim et al. 2013). This is especially important when it comes to a broader perspective in form of society: Ecological consumption choices are, unlike general product purchase decisions, future oriented. This is because using green products often provides benefits for the entire society in the long term (Kim and Choi 2005). Therefore, the perceived efficacy or the extent to which one believes that he can make a difference towards achieving a goal impacts green behaviour (Gleim et al. 2013; Hiller Connell and Kozar 2014; Kaufmann et al. 2012). In literature this is called the perceived consumer effectiveness (PCE; Kaufmann et al. 2012). When it comes to purchasing sustainable fashion, the perceived consumer effectiveness is low. The same was revealed for the behaviour in the usage phase (Fisher et al. 2008; Fig. 4.5).

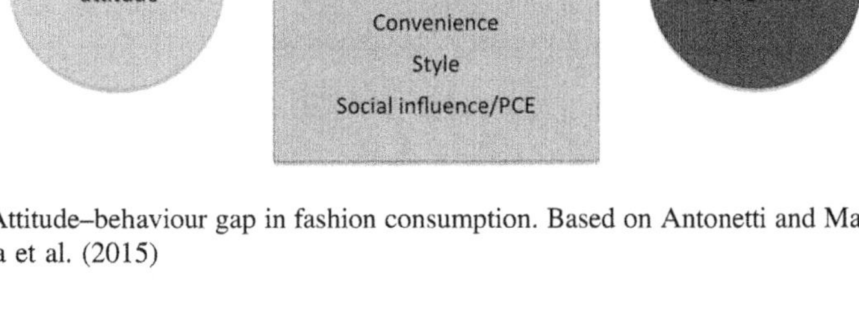

Fig. 4.5 Attitude–behaviour gap in fashion consumption. Based on Antonetti and Maklan (2015) and Laitala et al. (2015)

4.3 Discussion

4.3.1 Impact of Fashion Production and Consumption in Comparison

Coming back to the research question posed at the beginning: *Where is the main lever when it comes to the impact on sustainability in the fashion industry: production vs. consumption?* It is first important to summarize the findings of the single sustainability impacts in each phase and to contrast them. As the previous literature research shows the fashion production as well as the fashion consumption has different impacts on sustainability. Figure 4.6 summarizes the environmental

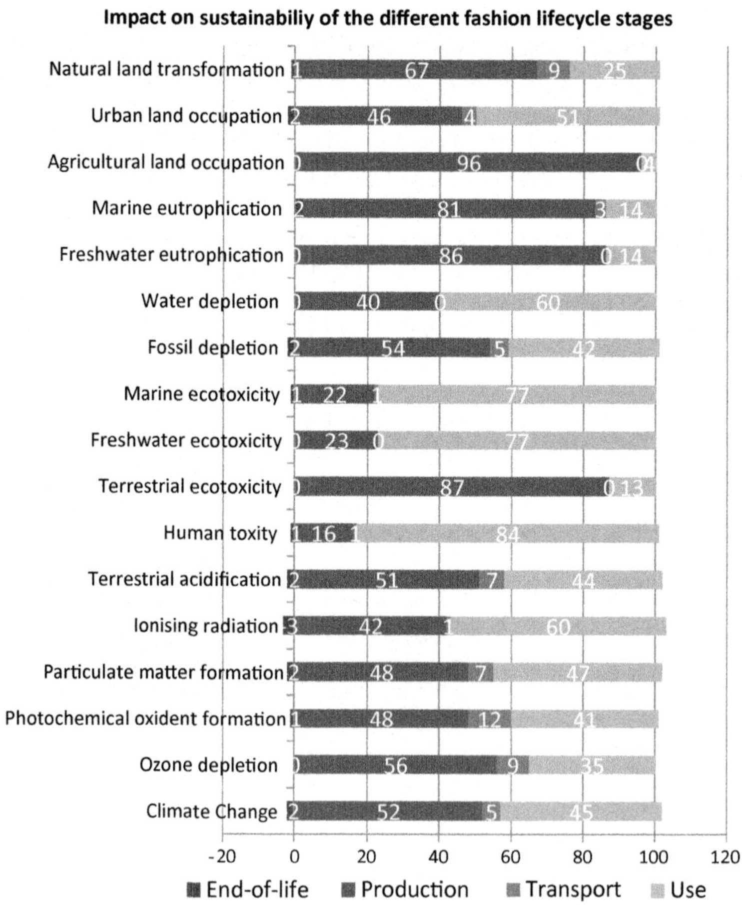

Fig. 4.6 Impact on sustainability in percentages of the different fashion lifecycle stages. Adapted from Beton et al. (2014)

impacts of the different fashion lifecycle stages and shows therewith a comparison between fashion production and fashion consumption.

In an overall consideration, every single side, the fashion production as well as the fashion consumption, has a comparable negative environmental impact on sustainability as shown in Fig. 4.6. The production phase accounts for the most significant impact when it comes to eutrophication, agricultural land occupation and natural land transformation in the fashion lifecycle as explained in Sect. 4.2.2.2. Besides the environmental impact, Sect. 4.2.2.3 showed that companies are also criticized for social issues in the production phase which is not shown in this figure. On the other side the usage in the consumption phase has a main impact on water depletion, marine and freshwater ecotoxicity and human toxicity mainly due to the high amount of water used as well as the detergent in the waste water. Furthermore, the enormous use of energy and the increasing fashion disposal from consumers contributes to global warming, as explained in Sect. 4.2.3.2.

Therewith it becomes clear that a negative impact on sustainability in the fashion industry cannot be allocated only to the lifecycle stage of production like it is done nowadays. It has to be allocated to both sides to the same extent, but the focus on today's sustainable improvements in fashion is almost entirely placed on the side of production (Gwilt and Rissanen 2010). Thereby, the significant impact from the usage phase in the consumption should not be underestimated. Different studies state that the consumption due to the consumer care has even a higher impact on sustainability in the overall fashion lifecycle as compared to the production (Kogg 2009). This emphasizes the consumer itself as a main contributor to a negative sustainability impact in the fashion lifecycle. This means consequently, current positive developments in the production have only little effect when the usage behaviour of the consumer is not changing. Moreover, these positive developments are eroded by an intensified usage behaviour and increasing waste of fashion by the consumer. Regarding this role as a main contributor of sustainability impacts in the fashion lifecycle, the consumer is not aware of. He does not see his own actions contributing in this extent to the overall sustainability impact due to less knowledge. Thereby, the problem is that even NGO's and the government do not really address these negative sustainability impacts of the consumer, but set a focus on the production side (Gwilt and Rissanen 2010). Therefore, there is no learning for the consumer to a changing behaviour in the usage of clothes.

4.3.2 Coherence of Fashion Production and Consumption

Moreover, the fashion lifecycle stages of production and consumption do not only have an impact on their own, and they are mutually dependent. Therefore, besides the single measurable impacts of each side, it is important to consider the coherence of fashion production and consumption. As explained in Sect. 4.2.2.1, the fashion production is driven by the demand of the consumers. This means fashion companies

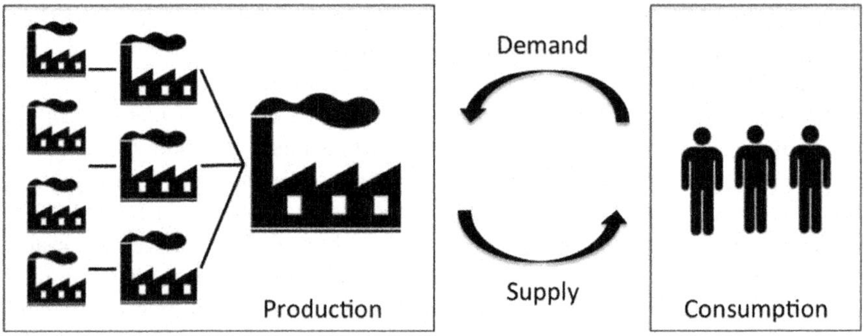

Fig. 4.7 Coherence of fashion production and consumption. Own illustration

act on what and how much the consumer is demanding for to maximize profit. Consequently, the consumer steers the production in a certain direction (Fig. 4.7).

As the consumer states that he regards increasingly sustainability aspects in the fashion purchasing, the producing fashion companies reacted with a broader supply of sustainable fashion. As in Sect. 4.2.3.4 explained, this did not lead to the expected demand from the consumer's side yet. Actually other buying criteria like style, fitting and price are predominant and environmental as well as social issues are only few or not considered by the consumer. Moreover, the consumer is not really informed about what sustainable fashion actually means. Therefore, companies react with the phenomenon of "greenwashing" (Strähle et al. 2015a, b) which means companies position theirselves to be more sustainable as they really are to gain a better image from the consumers. This means they do not produce in a more sustainable way, but they give the consumer the feeling of it. Due to insufficient knowledge and convenience of the consumer this strategy is successful. As for the consumer's purchasing behaviour the sustainable image of a brand as well as the general brand image is more important than the actual sustainable behaviour (Strähle and Köksal 2015), there is no incentive for companies to actually produce more sustainable as required. Furthermore, as the aim of companies is to make profit, they have no interest in informing the consumer correctly about negative sustainability impacts as the consumer might rethink his consumption and need for fashion (Strähle et al. 2015a, b).

Much more the production "learns" from the consumer as the consumer's demand decides which fashion products and therewith which fashion companies are successful. As the consumer is focused in his purchasing behaviour to constantly wear new and fashionable clothes to satisfy a certain peer pressure, fashion has to be stylish and affordable in the first row and regularly changing. The today's fashion market shows that the fast fashion retailers like Inditex or H&M that satisfy exactly these needs are the most successful fashion retailers and have an increasing market share. Accordingly, these companies have the strongest purchasing power regarding the production and therewith dictate the way of production as well as the prices. Due to the increasing demand, the consumer supports this supremacy of the fast fashion retailers.

At the same time the fast changing fashion supply by the producing fashion retailers supports the overconsumption of fashion. This shows that the production also steers the consumption. This finally ends in a circulatory of more demand of cheap and not sustainable fashion and therewith an intensification of sustainability impacts in the production. In this circulation of demand and supply, both sides pursue their own goals, the producing fashion company strives for maximizing profit and the consumer wants to satisfy his emotional needs. In this deadlocked system the main question is where responsibility is to seek and in which extent.

4.3.3 Responsibility Among Fashion Production and Consumption

As explained in Sect. 4.2.2.1 in a general understanding the producing company or also called focal company is responsible for all sustainability impacts in the fashion supply chain till the transition of the product to the consumer. Coming to the responsibility of the consumer regarding sustainability impacts in the fashion lifecycle, it is not that clear defined. While the responsibility of the producing company is undeniable, the literature research clarified that the consumer sees himself in his consumption, which includes purchasing as well as the usage and disposal behaviour, not responsible for sustainable impacts (Fig. 4.8).

Primarily, this is due to the lacking knowledge of sustainability. He is simply not informed about the consequences of his behaviour and even if he is, he has mostly the feeling of that a change of behaviour has no effect on the general sustainability impact made in the overall fashion lifecycle. This clarifies a disinterest from

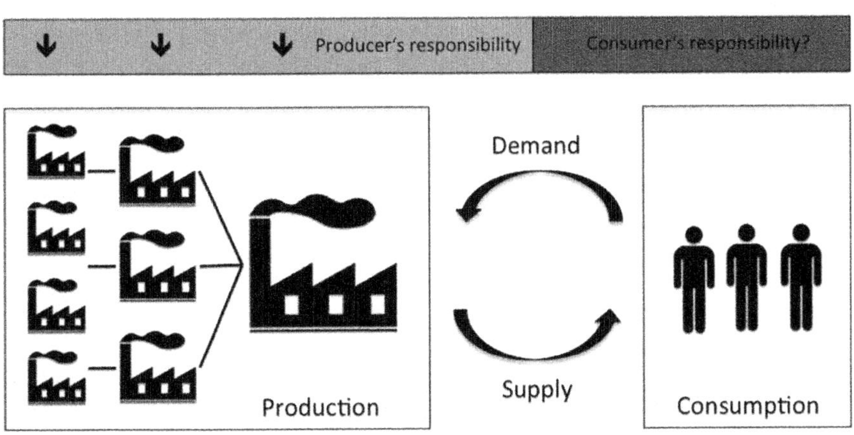

Fig. 4.8 Responsibility among fashion production and consumption. Own illustration

consumer's side (Achabou and Dekhili 2015) as he is not addressed for responsibility. Second, there seems to be no initiative from organizations and the government to make the consumer rethink his behaviour. As green consumption behaviour is future oriented and therewith influenced by the whole society, the consumer will not accept responsibility when it is not common practice (Kim and Choi 2005). Therewith the main question is *When the consumer in the consumption phase has a similar negative impact on sustainability like production and moreover he steers the production in a certain direction, why is he not made responsible for that? Thereby the question can be pursued: When the producing company, the fashion brand, is made responsible for every single step till the finished product, why is the consumer as the buyer of a certain product not responsible for every step made before?*

It becomes clear in this discussion that the consumer has to be made responsible for his behaviour; otherwise there will be no long-term changes in the fashion industry. Changes in production act positively on a short-term basis, but this will be eroded by the negative impacts of consumption in the long-term as explained. As the production is driven by the demand only a coincident changing purchasing behaviour will actually lead to a long-term change in the overall system.

4.3.4 Leverage Points Regarding the Sustainability Impact in the Fashion Industry

As fashion production and consumption is mutually dependent and should not be considered detached from each other, leverage points have to be found that do not only focus on production. As today's sustainability developments in fashion only take place on the production side, they are not successful due to a lack of acceptance from demand side (Niinimäki and Hassi 2011). Therefore, it is important to incorporate the consumer in these efforts to achieve a positive development.

In the first row, as discussed, the consumer has to be educated in a way that he feels responsible for his actions regarding the purchasing and especially the usage of fashion. All efforts will be in vain when the consumer does not see the necessity of his contribution (Gwilt and Rissanen 2010). Only when the consumer is interested in sustainability and receptive for information, there is the chance to make a positive change. As sustainable fashion consumption is only a partial aspect of a general sustainable lifestyle, it requires a general rethinking in society. This can only be reached when all parties work together: the government, NGO's, companies and the media. As companies have no interest in educating the consumer, there have to be clear regulations by the governance.

This is especially important when it comes to information. Today, the consumer is confused by the variety of different sustainable products and wordings and he does not know what is best. Therefore, it has to be easier and clearer for the consumer to obtain relevant information. This includes first an easier access to relevant product information by the company and second a transparency of the production processes of the fashion item and therewith a better insight for the consumer. Moreover, general regulations need to be found such as one worldwide accepted and well-known textile label to help the consumer in the purchasing of sustainable fashion. Regarding the usage behaviour it is even more important to enlighten the consumer about his impact and especially how he can change his behaviour. Thereby, care instructions of fashion products should be clearer and the government as well as NGO's should put a focus on encouraging to changing usage behaviour. An overall higher knowledge and responsibility of the consumer should lead to a more reflected consumption behaviour which requires the fashion companies to react in their production.

Aside the aspect of responsibility, there has to be an added value for the consumer when it comes to sustainability in fashion. Today, the consumer has barriers when purchasing sustainable fashion which includes for instance price and style. Thereby, the main focus has to be on reducing these barriers and encouraging the consumer to buy sustainable fashion. As a solution different studies claim that the consumer should already be integrated in the design process of sustainable fashion to make it more accepted (Niinimäki and Hassi 2011). This requires making the consumer the focus of the sustainable efforts like it is done for example in the prosumer concept. Moreover, in general it should be easier for the consumer to find sustainable options. Studies revealed that if the consumer can make a contribution to sustainability without huge effort, he is more likely to do it (Morgan and Birtwistle 2009). Accordingly, this means when the consumer has easy access to options with a positive sustainability impact such as sustainable fashion products, recycling or second hand, he is more likely to use it. Thereby it is always important to offer the consumer an added value which can be for instance the price and the same time good quality in branded second hand stores.

Regarding the mentioned aspects of more transparent and clearer information by the fashion companies as well as a changing supply, there have to be incentives respectively punishments by the government. As it is the main goal to make profit, companies only contribute to this development when it is profitable in the long-term. Without the contribution of the fashion companies there is no change in the system possible. Therefore, the government should first make stricter regulations regarding information to publish as mentioned before. There have to be financial penalties for companies that do not follow, which makes it indispensable for companies. Moreover, the government should support sustainable concepts like branded second hand stores with subsidies or grants.

Finally, with a more targeted and transparent sustainable supply by companies and at the same time a more reflected demand and usage behaviour by the consumers due to a higher perception of responsibility, the circulatory of production and consumption can be steered in a more sustainable direction.

4.4 Conclusion

This paper focused under the topic: Impact on sustainability: production vs. consumption the question: *Where is the main lever when it comes to the impact on sustainability in the fashion industry: production vs. consumption?* Thereby first of all the single impacts of fashion production as well as fashion consumption were determined: Sect. 4.2.2 clarified that main impacts in production are in the fibre production and the processing regarding the steps of dyeing and finishing. Moreover, the production faces several social issues such as poor wages and bad working conditions. Due to a complex supply chain the fashion company has in many cases no oversee of these impacts. On the other hand the consumer contributes to a negative impact on sustainability, too, as described in Sect. 4.2.3. In his purchasing behaviour he buys steadily more fashion products which leads to an increasing overconsumption of fashion. At the same time he does not regard sustainable issues in his buying decision what supports the current supply of non-sustainable clothes. Moreover, in the usage of clothes a main impact on sustainability in the fashion lifecycle is made due to the high use of water, energy and detergent. Thereby, the consumer is not aware of his own contribution and has a general lack of knowledge regarding sustainability. Besides the lacking knowledge other barriers such as costs, convenience, style, social influences and especially the perceived consumer effectiveness deter the consumer to a more sustainable consumption.

In the discussion a comparison between these single impacts of both sides clarified that they have a comparable negative impact on sustainability when regarded detached from each other. As the production and consumption are mutually dependent, one should not only regard these single impacts. The production adjusts its supply according to the demand of the consumer. As the consumer is increasingly asking for stylish and cheap clothes, the fast fashion retailers are most successful. Sustainable clothes are not requested and the image of a brand is more important than the actual sustainable behaviour. This gives companies no incentive to a more sustainable supply and current positive developments in the production are eroded by the overconsumption of cheap clothes. Accordingly, responsibility should not only be searched at the producing fashion companies as done today, but considerable more at the consumer's side. With a changing consumption in a more sustainable direction, the production has to follow to satisfy this demand. Only with a changing consumer behaviour a positive development of the sustainability impact in the long-term can be reached. This means in practice, the consumer has to be made responsible for his behaviour and he has to be educated. Thereby, he needs an easier access to information and sustainable options such as sustainable clothes, second hand and recycling. Moreover, current barriers of sustainable consumption have to be reduced by focusing more on the needs and requirements of the consumer. Only when the consumer sees the necessity and at the same time the opportunity of sustainable consumption, a shift in the whole system of demand and supply can be reached. Of course the contribution of the fashion companies at the same time should not be forgotten. Companies have to work together with consumers and make him the focus of the sustainable supply. The government should support this development with stricter regulations for companies and educational work for the consumers. In conclusion the question raised can be answered with the consumption. As not only a similar negative sustainable impact is made in the usage of clothes compared to production, but moreover the consumption steers the production in a certain way, only with a changing consumption the current sustainability impact can be changed in the long-term.

4.5 Research Limitations and Future Research

The main limitations of this paper are that most studies concerning fashion consumption are from the UK and the EU. Big fashion markets like China or the US are mostly not considered which therefore only gives insight in the European consumption patterns. Moreover, most studies are focused on young fashion consumers in the age between 18 and 35 and their behaviour which does not represent the fashion consumer in general concerning all demographic levels. Furthermore, most studies consist of qualitative research methods and were conducted with a small sample, so that the results are not transferable to the population. In general, most studies are focused on the single impacts of fashion production and consumption and only few surveys were found which consider the coherence of fashion production and consumption. The paper itself is limited in its selection of topic concerning the sustainability impact. It is solely focused on the fashion production and consumption not considering factors such as transport or retail. Furthermore, it does not give concrete solutions, but represents general leverage points.

In order to overcome the presented limitations the qualitative results of the studies concerning the fashion consumption should be tested with the help of quantitative studies. Moreover, studies should involve testing in different parts of the world with people from different demographic levels to reach a higher validity concerning the general fashion consumption. Furthermore, the discussed coherence of fashion production and consumption should be more subject of further research.

References

Achabou M, Dekhili S (2015) Constraints and drivers of growth in the ethical fashion sector: the case of France. In: Choi T-M, Cheng TCE (eds) Sustainable fashion supply chain management: from sourcing to retailing. Springer, Cham, pp 167–181

Adamcyk A (2015) Clean fashion: water footprint von Textilien. PL Academic Research, Frankfurt am Main

Allwood J, Laursen S, Rodriguez C, Bocken N (2006) Well dressed? The present and future sustainability of clothing and textiles in the United Kingdom. University of Cambridge, Cambridge

Allwood J, Laursen S, Russell S, Malvido de Rodriguez C, Bocken N (2008) An approach to scenario analysis of the sustainability of an industrial sector applied to clothing and textiles in the UK. J Clean Prod 12:1234–1246. doi:10.1016/j.jclepro.2007.06.014

Antonetti P, Maklan S (2015) How categorisation shapes the attitude-behaviour gap in responsible consumption. Int J Mark Res 57(1):51–72. doi:10.2501/IJMR-2015-005

Austgulen M (2013) Consumer perspectives on eco-labelling of textiles—results from five European countries. National Institute for Consumer Research, Oslo

Baumgärtner S, Quaas M (2010) Sustainability economics: general versus specific, and conceptual versus practical. Ecol Econ 69(11):2056–2059

Beard ND (2008) The branding of ethical fashion and the consumer: a luxury niche or mass-market reality? Fash Theory J Dress Body Cult 12(4):447–467

Beckmann F (2014) Bachelorthesis: Wirkung von textilen Labels auf den Konsu-menten. Hochschule Reutlingen, Reutlingen

Berners-Lee M, Clark D (2010) What's the carbon footprint of … a load of laundry? Retrieved 16 September 2016, from http://www.theguardian.com/environment/green-living-blog/2010/nov/25/carbon-footprint-load-laundry

Beton A, Dias D, Farrant L, Gibon T, Le Guern Y (2014) Environmental im-provement of textiles (IMPRO Textiles). Publications Office, Luxembourg

Bhaduri G, Ha-Brookshire JE (2011) Do transparent business practices pay? Exploration of transparency and consumer purchase intention. Cloth Text Res J 29(2):135–150

Bhardwaj V, Fairhurst A (2010) Fast fashion: response to changes in the fashion industry. Int Rev Retail Distrib Consum Res 20(1):165–173

Billeson K, Klasander K (2015) Dress code: sustainable fashion: bridging the attitude-behaviour gap. Jönköping University, Jönköping

Birtwistle G, Moore CM (2007) Fashion clothing—where does it all end up? Int J Retail Distrib Manag 35(3):210–216. doi:10.1108/09590550710735068

Bruce M, Daly L, Towers N (2004) Lean or agile: a solution for supply chain management in the textiles and clothing industry? Int J Oper Prod Manag 24(2):151–170

Cachon G, Swinney R (2011) The value of fast fashion: quick response, enhanced design, and strategic consumer behavior. Manag Sci 4:778

Cao H, Chang R, Kallal J, Manalo G, McCord J, Shaw J, Starner H (2014) Adaptable apparel: a sustainable design solution for excess apparel consumption problem. J Fash Mark Manag 18 (1):52–69. doi:10.1108/JFMM-08-2012-0046

Carew AL, Mitchell CA (2002) Characterizing undergraduate engineering students' understanding of sustainability. Eur J Eng Educ 27(4):349

Chen Z (2008) Defining buyer power. Carleton University, Ottawa

Chen H-L, Burns LD (2006) Environmental analysis of textile products. Cloth Text Res J 24 (3):248–261. doi:0.1177/0887302X06293065

Choudhury AKR (2014) Environmental impacts of the textile industry and its assessment through life cycle assessment. In: Subramanian M (ed) Roadmap to sustainable textiles and clothing. Springer Singapore, Singapore, pp 1–40

Christopher M, Lowson R, Peck H (2004) Creating agile supply chains in the fashion industry. Int J Retail Distrib Manag 32(8):367–376. doi:10.1108/09590550410546188

D'Astous A, Legendre A (2009) Understanding consumers' ethical justifications: a scale for appraising consumers' reason for not behaving ethically. J Bus Ethics 87(2):255–268. doi:10.1007/s10551-008-9883-0

D'Souza C, Taghian M, Lamb P (2006) An empirical study on the influence of environmental labels on consumers. Corp Commun Int J 11(2):162–173

Draper S, Weissbrod I, Murray V (2007) Sustainability: a review of the sustainability impacts of the clothing industry. Retrieved 16 September 2016, from http://www.caseplace.org/d.asp?d=5450

Fisher T, Cooper T, Woodward S, Hiller A, Goworek H (2008) Public understanding of sustainable clothing—a research report completed for the Department for Environment, Food and Rural Affairs by Nottingham Trent University and Sheffield Hallam University. Defra, London

Fletcher K (2014) Sustainable fashion and textiles: design journeys. Earthscan, London [ua]

Fransson K, Brunklaus B, Molander S, Zhang Y (eds) (2013) Managing che-mical risk information: the case of Swedish retailers and Chinese suppliers in textile supply chains. In Sustainability in fashion and textiles: values, design, production and consumption. Greenleaf Publ, Sheffield

Gam H (2011) Are fashion-conscious consumers more likely to adopt eco-friendly clothing? J Fash Mark Manag 15(2):178–193

Gibson C, Stanes E (2011) Is green the new black? Exploring ethical fashion consumption. In: Lewis T, Potter E (eds) Ethical consumption: a critical introduction. Routledge, London, pp 169–185

Glavič P, Lukman R (2007) Review of sustainability terms and their definitions. J Clean Prod 15 (18):1875–1885. doi:10.1016/j.jclepro.2006.12.006

Gleim MR, Smith JS, Andrews D, Cronin J, Joseph J (2013) Against the green: a multi-method examination of the barriers to green consumption. J Retail 89:44–61. doi:10.1016/j.jretai.2012.10.001

Gold S, Seuring S, Beske P (2010) Sustainable supply chain management and inter-organizational resources: a literature review. Corp Soc Responsib Environ Manag 17(4):230–245. doi:10.1002/csr.207

Goworek H, Hiller A, Fisher T, Cooper T, Woodward S (2013) Consumers' attitude towards sustainable fashion. In: Sustainability in fashion and textiles—value, design, production and consumption. Greenleaf Publishing, Sheffield

Grunwald A, Kopfmüller J (2012) Nachhaltigkeit: eine Einführung. Campus-Verlag, Frankfurt am Main

Gwilt A, Rissanen T (2010) Shaping sustainable fashion: changing the way we make and use clothes. Earthscan, Washington

Hill J, Lee H-H (2012) Young generation Y consumers' perceptions of sustainability in the apparel industry. J Fash Mark Manag 16(4):477–491. doi:10.1108/13612021211265863

Hiller Connell K (2011) Exploring consumers' perceptions of eco-conscious apparel acquisition behaviors. Soc Responsib J 7(1):61–73. http://doi.org/http://dx.doi.org/10.1108/17471111111114549

Hiller Connell KY, Kozar JM (2012) Social normative influence: an exploratory study investigating its effectiveness in increasing engagement in sustainable apparel-purchasing behaviors. J Glob Fash Mark 3(4):172–179. doi:10.1080/20932685.2012.10600847

Hiller Connell K, Kozar J (2014) Environmentally sustainable clothing consumption: knowledge, attitudes, behaviour. In: Subramanian M (ed) Roadmap to sustainable textiles and clothing environmental and social aspects of textiles and clothing supply chain. Springer Singapore, Singapore, pp 41–63

Jørgensen MS, Jensen CL (2012) The shaping of environmental impacts from Danish production and consumption of clothing. Ecol Econ 83:164–173. doi:10.1016/j.ecolecon.2012.04.002

Kagawa F (2007) Dissonance in students' perceptions of sustainable development and sustainability: implications for curriculum change. Int J Sustain High Educ 8(3):317–338

Karthik T, Gopalakrishnan D (2014) Environmental analysis of textile value chain: an overview. In: Subramanian M (ed) Roadmap to sustainable textiles and clothing: environmental and social aspects of textiles and clothing supply chain. Springer, Singapore, pp 153–158

Kaufmann H, Panni M, Orphanidou Y (2012) Factors affecting consumers' green purchasing behavior: an integrated conceptual framework. University of Nicosia, Cyprus

Kaye L (2013) Clothing to dye for: the textile sector must confront water risks. The Guardian

Kim Y, Choi SM (2005) Antecedents of green purchase behavior: an examination of collectivism, environmental concern, and PCE. Adv Consum Res 32(1):592–599

Kogg B (2009) Responsibility in the supply chain: interorganisational management of environmental and social aspects in the supply chain; case studies from the textile sector. International Institute for Industrial Environmental Economics, Lund

Kozlowski A, Searcy C, Bardecki M (2015) Corporate sustainability reporting in the apparel industry: an analysis of indicators disclosed. Int J Prod Perform Manag 64(3):377–397

Laitala K, Klepp I (2013) Environmental and ethical perceptions related to clothing labels among Norwegian consumers. Res J Text Appar 17(1):50–58

Laitala K, Boks C, Klepp IG (2011) Potential for environmental improvements in laundering. Int J Consum Stud 35(2):254–264. doi:10.1111/j.1470-6431.2010.00968.x

Laitala K, Harvik M, Klepp I (2015) Responsibility without means. In: Roadmap to sustainable textiles and clothing: regulatory aspects and sustainability standards of textiles and the clothing supply chain. Springer, Singapore, pp 125–152

Lampikoski T, Westerlund M, Rajala R, Möller K (2014) Green innovation games: value-creation strategies for corporate sustainability. Calif Manag Rev 57(1):88–116. doi:10.1525/cmr.2014.57.1.88

Meyer A (2001) What's in it for the customers? Successfully marketing green clothes. Bus Strategy Environ 10(5):317–330. doi:10.1002/bse.302

Morgan LR, Birtwistle G (2009) An investigation of young fashion consumers' disposal habits. Int J Consum Stud 33(2):190–198. doi:10.1111/j.1470-6431.2009.00756.x

Nagurney A, Yu M (2012) Sustainable fashion supply chain management un-der oligopolistic competition and brand differentiation. Int J Prod Econ 135:532–540. doi:10.1016/j.ijpe.2011.02.015

Niinimäki K, Hassi L (2011) Emerging design strategies in sustainable production and consumption of textiles and clothing. J Clean Prod 19:1876–1883. doi:10.1016/j.jclepro.2011.04.020

Orhon D, Kabdasli I, Germirli Babuna F, Sozen S, Dulkadiroglu H, Dogruel S et al (2003) Wastewater reuse for the minimization of fresh water demand in coastal areas—selected cases from the textile finishing industry. J Environ Sci Health A 38(8):1641–1657. doi:10.1081/ESE-120021486

Paloviita A, Järvi P (2008) Environmental value chain management of laundry detergents in the use phase. Int J Consum Stud 32(6):607–612. doi:10.1111/j.1470-6431.2008.00692.x

Piegsa E (2010) Green Fashion: ökologische Nachhaltigkeit in der Bekleidungsin-dustrie. Diplomica-Verl, Hamburg

Shen B (2014) Sustainable fashion supply chain: lessons from H&M. Sustainability 6(9):6236–6249. doi:10.3390/su6096236

Shen D, Richards J, Liu F (2013) Consumers' awareness of sustainable fashion. Mark Manag J 23 (2):134–147

Sheth J, Sethia N, Srinivas S (2011) Mindful consumption: a customer-centric approach to sustainability. J Acad Mark Sci 39(1):21–39. doi:10.1007/s11747-010-0216-3

Shove E (2002) Sustainability, system innovation and the laundy. Lancaster University, Department of Sociology, Lancaster

Sisco C, Morris J (2012) The nice consumer—toward a framework for sustainable fashion consumption in the EU. Danish Fashion Institute, Copenhagen

Slaper TF, Hall TJ (2011) The triple bottom line: what is it and how does it work? Indiana Bus Rev 86(1):4–8

Slater K (2003) Environmental impact of textiles: production, processes and protection. CRC Press, Boca Raton

Strähle J, Köksal D (2015) Impact of brand- and country image on the perception of sustainability in the fashion business. Int J Bus Commer 5(1):25–58

Strähle J, Shen B, Köksal D (2015a) Sustainable fashion supply chain: adaption of the SSI-index for profiling the sustainability of fashion companies. Int J Bus Manag 3(5):232–241

Strähle J, Will C, Freise M (2015) Communication of sustainability at European fashion online shops. Int J Econ Commer Manag 3(7):71–86
Taplin IM (2014) Who is to blame? A re-examination of fast fashion after the 2013 factory disaster in Bangladesh. Crit Perspect Int Bus 10(1):72–83
Throne-Holst H, Stø E, Strandbakken P (2007) The role of consumption and consumers in zero emission strategies. J Clean Prod 15:1328–1336. doi:10.1016/j.jclepro.2006.07.018
van der Merwe D, Bosman M, Ellis S, van der Colff N, Warnock M (2014) Consumers' knowledge of textile label information: an exploratory investigation. Int J Consum Stud 38 (1):18–24. doi:10.1111/ijcs.12053
Visa M, Pricop F, Duta A (2011) Sustainable treatment of wastewaters resulted in the textile dyeing industry. Clean Technol Environ Policy 13(6):855–861. doi:10.1007/s10098-011-0362-4
Wahnbaeck C, Brodde K, Groth H (2015) Greenpeace—usage & attitude mo-de/fast fashion. Nuggets - Market research & consulting GmbH, Hamburg
WCED (1987) Report of the World Commission on Environment and Development: our common future. Oslo, United Nations. Retrieved from http://www.un-documents.net/wced-ocf.htm

Chapter 5
Impact of Sustainable Manufacturing Standards for Retail Brands

Jochen Strähle and Sarah Kreuzhermes

Abstract The purpose of this paper is to define, what impacts sustainable manufacturing standards have for retail brands concerning the communication policy and to find possible solutions of how the companies can deal with them. Therefore, sustainable standards and the impacts on the internal and external communication are described. The enclosed discussion finds possible solutions for the negative impacts. A literature discussion has been conducted to investigate the purpose. Generally, there are many impacts fashion retails have to consider, if they want to transform their company to become more sustainable, because only the impacts on a defined part of the communication policy were huge. A limitation of this paper is that the proposals, how retailers could deal with the impacts of the transformation of the company toward more sustainability, need further research and tests until they are practicable.

Keywords Sustainable textile standards · Communication policy · Internal communication · External communication · Impacts of sustainability for fashion retail brands

5.1 Introduction

The current discussion about changing climate, lack of water, lack of arable land, and countless environmental problems lead to a growing interest of consumers toward the aspect of products that is mostly not visible during purchase, the production (Piegsa 2010). Where does the product come from? Who produced it and how was it produced are questions that most of the textile industry companies are not able to answer, because the clothing sector is organizationally complex and

J. Strähle (✉) · S. Kreuzhermes
School of Textiles and Design, Reutlingen University, Reutlingen, Germany
e-mail: jochen.straehle@reutlingen-university.de

S. Kreuzhermes
e-mail: s.kreuzhermes@googlemail.com

J. Strähle (ed.), *Green Fashion Retail*, Springer Series in Fashion Business,
DOI 10.1007/978-981-10-2440-5_5

supply chains can be very long with many different parties involved. Moreover, globalization influenced these supply chains, they became broader and more international (Gardetti 2013), which increased the complexity. However, the environmental impact of conventional production of textiles is enormous. The very cotton cultivation has a tremendous use of water and pesticide with negative effects also on human health. The Aral Sea is an example for the negative impacts of cotton cultivation. The once fourth largest lake of the world has shrunk to 20 % of its original size (Piegsa 2010), because the rivers feeding the Aral Sea were used for intensive irrigation of cotton fields. The area is now a desertified wasteland, people lost most of their fishing industry there and the rate of illness rises, because not only the water, but also fruits and other food are contaminated because of the polluted water and soil (*Uzbekistan—Aral Sea In Danger*, 1995). There are further similar developments with contaminated water through cotton cultivation in the world, e.g., the playa lakes of the southern high plains (Anderson et al. 2013).

For years, governments and non-governmental organizations have tried to influence the textile branch toward more sustainable processes in the textile chain (Piegsa 2010), but the effectiveness is not proved. Until now, there are no reliable numbers about the market share of sustainable clothes for one or for several markets, because the definition of sustainable clothes is not stated. Based on a consumer survey of the "Gfk", a non-profit-organization for market research, eco-textiles generate 3.5 % of the turnover from whole market, consumer defined the term eco-textiles in this survey for themselves ('GfK' 2015; Schaus 2013). As a result this number is discussed and not everywhere accepted, because what the consumer perceives as eco-textile is often just labeled by "Öko-Tex 100", and this label just states a residue analysis (Schaus 2013). However, the number shows that the textile chain still offers a huge potential for optimization in sustainability (Piegsa 2010).

The effort of institutions over years to make the textile chain more sustainable, compared with the current situation of manufacturing textiles, shows a huge gap. But the growing interest of consumers toward the manufacturing process of textiles may lead to a higher interest of companies to meet this interest. This leads to the research question of this paper: *What are the influences of sustainable manufacturing standards on a Retail Brands' communication policy and how can they deal with them?*

This paper commences with a definition of sustainable manufacturing standards, since this term can be defined differently. The second part of this section will shortly present the most important sustainable manufacturing standards, fashion retail brands should be aware of. Next embedded in the paper is a section with a closer view on the impacts of sustainable manufacturing standards for retail brands concerning the communication policy. It defines impacts on the internal and external communication. A subsequent discussion and evaluation give ideas, how fashion retails can deal with these impacts. Conclusion and foresights offer a review of all gained findings, and sum up the outcomes of this research paper.

This paper focuses on the steps of the textile chain until the product is finished, for example until a T-Shirt is ready for transportation to its sales market. Furthermore, this paper's extent is limited. Especially, the possible solutions for the impacts on communication policy need further research until they are practicable.

The term "retail brand" in this paper describes a fashion retail company that sources its collections globally from multiple suppliers and sells its products internationally through different sales channels online and offline (including wholesale and own retail). This paper focuses on the change of such a company from conventional to sustainable manufacturing standards. Moreover, it is assumed, that retail brands with a sustainable manufacturing incorporate the topic sustainability also in their company culture. Therefore, sustainability becomes an elementary part of the understanding of the brand and in the daily work of each employee. However, many aspects also affect other retail brands, for example start-ups or implemented brands which work already in a sustainable way and who want to find out more about the topic of this paper.

5.2 Definition of Sustainable Manufacturing Standards

The textile chain is very complex and has globally located production steps. Each step has its own problems concerning sustainability, e.g., use of pesticides for cotton cultivation (Piegsa 2010). The question is how sustainable manufacturing standards can be defined, because the word "sustainability" alone offers a broad variation in understanding. Today more than 200 definitions for it can be found (Tobler-Rohr 2011). In the USA, a university's faculty's, staff's and students definitions of sustainability were analyzed. Results showed that the definitions of the people did not match highly when compared to a list of the most frequent words from published definitions. But when the definitions were examined more closely, they showed nuances in understanding (Owens and Legere 2015). That shows the necessity to deal with this topic and to come to an understanding of this word in order to understand this paper.

The top ten of the most frequently used words in the mentioned analysis were environment, social, economic, life/live, system, nature, resource, human, develop, and need. The list involves over 100 previously published definitions of sustainability (White 2013). However, the definition in the Brundtland report of the World Commission on Environment and Development is still considered as one of the most accepted, and in fact indicates indirectly most of these words. The definition describes sustainable development as a development, which should meet the needs of the present without compromising the ability of future generations to meet their needs. This description is not very detailed and exact, but it indicates a long-term-strategy, including economic, human/social and environmental resources (Tobler-Rohr 2011). The Oxford Dictionary defines sustainable as "involving the use of natural products and energy in a way that does not harm the environment" and as "that can continue or be continued for a long time" (Oxford Dictionary 2015). These two parts combined also indicate the long-term-aspect of the word, but they specifically highlight, in contrast to the definition above, the aspect to not harm the environment. Generally, most definitions of sustainability refer to the following three pillars: economy, society and (ecological) environment (Tobler-Rohr 2011).

As a result of this comparison and in connection to the research question, for this paper sustainability can be defined as a long-term-strategy, which considers and respects economic, human/social, and environmental resources during usage in a way that does not harm one of these resources. In this paper, usage refers to all steps of the textile chain, until the product is finished. Therefore, the following sustainable standards aim to reach a situation that is sustainable in terms of the named definition.

The processes of the textile chains are very complex, therefore the following sentences should give just an overview about them. The chain begins with the fiber material. It can be divided up in natural fibers and chemical fibers (Piegsa 2010). Natural fibers come from animals or plants, and have to be harvested and prepared for yarn formation (spinning), while chemical fibers have to be chemically extracted from crude oil fractions, or other natural resources, before they can be spun. Therefore, the spinning process differs very much between these two categories. The next step is the weaving or knitting of the yarn, while weaving is not that important for the apparel industry, because it mostly requires knitted materials. Fibers, knitted or woven fabric, or the finished product can be dyed and finished in the different production steps (Tobler-Rohr 2011).

Each step in the textile chain has its specific negative impact on the resources of the earth, like natural and chemical fibers, for example. Natural fibers use the resources water, land, and partially pesticides while chemical fibers use energy, water, oil, and chemicals (Piegsa 2010). In order to give an idea of the negative impacts and the complexity of each process of the textile chain, the main problems with conventional production of synthetic fibers are shortly described in the following. The enormous consumption of energy, water, chemicals, and the pollution of waste water and exhaust air are the main problems. In addition, the raw material of synthetic fibers can also be harmful, e.g., polyamide can be made out of benzol, which can cause cancer. Moreover, the production of synthetic chemical fibers needs crude oil, where the availability in the future is also a big problem. Also the auxiliary materials can endanger health. Residues of heavy metal can, if the treatment plant is not able to recognize and filter it, get out of the production- and treatment plan-area and harm the environment, because this metal can cause cancer. The types of chemical fibers differ very much in their raw material composition, and therefore, also in their production process. Moreover, the environmental impact differs too, e.g., lyocell is less harmful than compared to viscose (Piegsa 2010). In general, the negative impacts of the textile chain mostly address the use of water, land, energy, chemicals, auxiliary materials, and pollution of waste water, exhaust air, and working conditions (Piegsa 2010).

To address these impacts, companies have different possibilities. For example, they can search for energy-saving and environmentally friendly production techniques, machines, and processes. Moreover, companies can examine their chemicals and auxiliary materials considering their ingredients, and make the processes more efficient, e.g., minimize the waste during cutting. Concerning the problem of the need for crude oil during synthetic fiber production, companies can use post-consumer recycling (PCR), for example. This method recycles waste from industry and private households to raw materials. According to a report from a Californian testing institute, PCR fibers could save 650,000 million barrel crude oil, 240,000 tons plastic and 375,000 tons emissions. Moreover, companies could recycle PET-bottles to generate new fibers. The quality of this recycled fiber does not differ from fibers, which are conventionally generated. At the moment, this recycling costs very much, but concerning the lack of crude oil in the future, it can become a valuable alternative (Piegsa 2010).

Another solution is a sustainable manufacturing standard, which defines criteria for sustainable manufacturing. Companies can choose between different sustainable standards, which differentiate in their criteria. Moreover, some standards have low requirements that many companies can fulfill, other have high standards which are more difficult to fulfill. The advantage for companies is that they do not have to think about how they can be sustainable. They can choose one standard and if they fulfill the requirements, they receive a credible certification. This proves their engagement to external parties. Otherwise, they can define sustainable standards for them independently, but do not get such a proof for their sustainability.

The criteria are, for example, restrictions or bans for chemical inputs due to environmental and/or toxicological reasons, limits for the substances in the exhaust air, regulations for the origin of raw materials or prohibition of methods that are considered to be harmful to the workers, e.g., sand blasting of denim (Global Organic Textile Standard International Working Group (IWG) 2014).

The following sustainable production standards consider the three eco-labels with the highest and most comprehensive requirements. The independent and international organization "Greenpeace" has the goal to protect the environment and compared eco-textile labels in 2014. They only recommended the labels global organic textile standard (GOTS), the IVN Best-labeled clothes and additionally for clothes out of chemical fibers (The Blue Angel). Generally, Greenpeace criticized that there is no perfect standard at the moment, but these two labels are the best for textiles so far (Sadik 2014). Also other institutions made similar analyses and came to a similar result. One of the results was contributed by Schaus (2013), in her report about sustainable clothing. She mentioned the GOTS standard as the best, as well. Moreover, she found out that GOTS has the most certified members compared to other labels, apart from labels without comprehensive requirements like "Öko-Tex Standard 100" (Schaus 2013).

GOTS is a standard for organic fibers, including ecological and social criteria, and includes an independent certification of the entire textile supply chain (Sadik 2014). According to this standard, textiles should be made from at least 70 % certified organic natural fibers. Moreover, it has provisions that include a ban of genetically modified organisms, a restricted substance list, and strict wastewater treatment practices (Textile Industry's Eco Efforts 2015). Other important advantages of the standard are independency, transparency, traceability of the ecological flow of goods, that it requires annual analysis of the products and annual on-the-spot inspections. Therefore, GOTS has reached a high credibility. But the standard also has disadvantages. GOTS considers no chemical fibers, while these fibers have a higher market share than natural fibers, so ecological requirements for chemical fibers would therefore make sense. The second most important disadvantage is that the GOTS does not consider the efficiency of usage of resources (Schaus 2013). Furthermore, Greenpeace criticizes that the GOTS has a few weaknesses in the chemical management (Sadik 2014). 2014 the GOTS rules were modified on permissible fiber materials that now may consist of up to 30 % of regenerated, respective synthetic fibers provided they are environmentally improved and certified. This was a loosening in regulations and therefore a point of criticism (Textile Industry's Eco Efforts 2015).

Naturtextil IVN certfified BEST from the International Association of Natural Textile Industry (IVN) as a quality standard with the strictest requirements for ecological textile production. It inspects the entire textile chain in terms of ecology and social accountability, but also considers only natural fibers. Greenpeace criticizes the label for the partly too high limits, or for partly missing limits. The outcome therefore is that some products are not able to be produced at the moment, because of too high standards. Moreover, for the production it allows only 100 % natural fibers, which is also a very strict requirement. Furthermore, Greenpeace criticizes that not all prohibitions have limits. Compared to GOTS, this label has a stricter chemical management, because it requests a stricter limit at one dye. However, both standards have less strict limits concerning chemical management compared to "Öko-tex Standard 100" or "Bluesign" (Sadik 2014). Further, the involvement of the IVN in the implementation of GOTS led to a high similarity of both standards (Piegsa 2010).

The blue Angel is an environmental label organized by the federal government of Germany and sets very exact standards. It is independent and has offered a standard for textiles since 2011 (Sadik 2014). Indeed, there was no textile company certified with this label until end of 2015 (Survey of all Basic Award Criteria 2015). The big advantage compared to the other mentioned standards is that it certifies natural and chemical fibers. The label receives criticism from Greenpeace concerning their limits, since they have not embedded them for all prohibitions. Furthermore, the level of the limits differs between the criteria, therefore some limits do not meet the actual state of analysis (Sadik 2014).

The author of the report about sustainable clothing Schaus (2013) concludes that it is not appropriate to implement a new sustainable standard, irrelevant whether it is from governmental or from non-governmental organizations, because the GOTS is an already accepted standard (Schaus 2013). According to Greenpeace, these three standards are not perfect, but it makes more sense to adopt one standard and modify it, than to implement a new one. First, companies can perform better than the minimum requirements of the standards, if they are able to do so, and if they think that the standards are too loose. Second, they can try to influence the standards toward higher or better requirements, so that the requirements of, e.g., Greenpeace are fulfilled, and a very high level of sustainability in the textile chain is reached.

How to realize a standard like GOTS is a question, which this paper is not able to answer, but 3,016 already certified companies prove, that it is possible (Schaus 2013).

5.3 Impact of Sustainable Manufacturing Standards on the Communication Policy

5.3.1 Definition of Communication Policy

If a retail brand wants to implement sustainable manufacturing standards successfully, they should be aware of the impacts of this change. On the one hand, the manufacturing process changes and influences departments like purchasing and logistics, but on the other hand also other departments of the company are affected, like human resources, controlling, or sales. For example, the sales department has to think about new possible sales channels, because they now have the opportunity to sell products in sustainable retail shops online and offline. However, they also have to think about negative impacts, like customers or retailers, who do not want to buy sustainable clothes, because they perceive them as unfashionable. Nearly everyone involved in the business of the company is affected by the transformation from a conventional to a sustainable retail brand. The main reason therefore is that not only the manufacturing process changes, it is also the company culture which is affected, and that means that each employee has to live and practice sustainability in the whole company.

This paper focuses on impacts on the communication policy of retail brands. Communication is a "Two-way process of reaching mutual understanding, in which participants not only exchange (encode–decode) information, news, ideas and feelings but also create and share meaning" (Business Dictionary 2015). Moreover, the function of communication in business is elementary, because it connects levels, departments, employees, business partners, and many more (Business Dictionary 2015). Communication policy belongs to the element "promotion" of the four instruments of marketing (four P's of marketing), the other instruments are product, price, and place. These instruments can build up the marketing strategy of a company and have multiple different connections among themselves (Bruhn 2015). The basic idea of marketing is to think about business in terms of needs of the market, and how the company can satisfy them (Kirchgeorg 2015). Communication policy comprises all actions that give information about a company to several different stakeholders, and internally to the employees, to influence the recipients in the sense of marketing. To achieve this communication, different instruments like sponsoring, public relations, advertising, personal selling, sales promotion, events, product placement, or internal communication (behavioral branding) can be used. The mix of several instruments is called communication mix. The target of communication policy is positioning the offer of the company to make it attractive for customers, and to differentiate from competitors (Esch 2015).

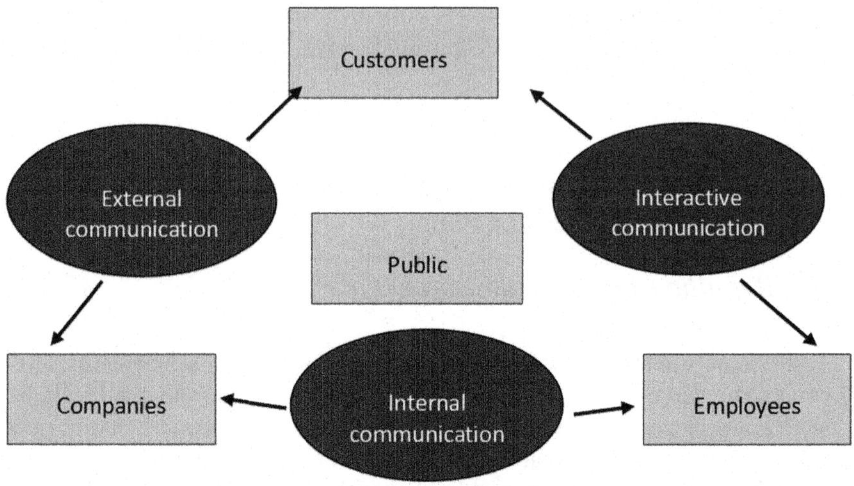

Fig. 5.1 Forms of communication. Adapted from Bruhn (2015)

There are three forms of communication of companies; external communication (e.g., advertisements), internal communication (e.g., intranet), and interactive communication between employees and customers (e.g., customer support) (Bruhn 2015). The enclosed image visualizes these forms of communication (Fig. 5.1).

The following two sections explain impacts of the change of a company toward being more sustainable, on the internal and external communication, because interactive communication is influenced by the information that employees and customers get. The section about internal communication describes how employees are affected by the transformation of the company in terms of communication policy. The section about external communication focuses on the purchasing behavior that appears for sustainable clothing, because is different compared to conventional products.

The enclosed discussion first deals with the topic, how the communication policy of the company can address the mentioned impacts of the transformation of the company on the employees. Second it is discussed, how communication policy can address the aspects of the sustainable purchasing behavior. Both parts of the discussion have the target to find possible solutions for the negative impacts, so that these parts contribute to a successful transformation of the company toward being more sustainable.

5.3.2 Internal Communication

The increasing importance of employees as convincing brand ambassadors inside and outside of the company increases the importance of internal communication, because employees are able to develop the perception of customers of the brand (Wittke-Kothe 2001). If companies want to influence the communication process of employees toward customers, they can strategically influence the personal and mass media communication inside their company (Tomczak et al. 2012). To prevent a wrong understanding of the brand, companies can develop an internal communication policy. To make this policy successful, it is necessary to be aware of the parts of communication processes. As a help, companies can use the communication formula of Lasswell (1967). The author states, that a convenient "[...] way to describe an act of communication is to answer the following questions (Lasswell 1967):

- Who
- Says What
- In Which Channel
- To Whom
- With What Effect?

One step out of this act of communication is more important for companies than others, because the success of internal communication depends especially on the chosen communication channel. Generally, companies can choose between personal and mass media communication. Both channels have their specific advantages and disadvantages, therefore it depends on the message that channel suits best (Bruhn 2015).

The process of change of an organization toward sustainable manufacturing standards and toward sustainability as a new element in the company culture, influences each individual employee. The people within the company are the primary source of change, because they are the ones who put the strategy into practice. However, changes have often to do with anxieties, for example fear of learning new ways of doing the work, or fear of dismissal. A company culture that appreciates change, crises, and making mistakes can help to decrease such anxieties (Dievernich 2015). Moreover, multilevel dialog and interaction involving the complete personnel is demanded between management and employees, because the new strategies must be communicated to employees, so that they understand what the strategy means for their everyday work. In sum, it is a challenge to implement communication strategies globally and reorganizing companies with multilevel structures (Balogun and Johnson 2005; Ikävalko 2005).

One positive impact of the change of the company toward sustainability is that this change is able to enhance the efficiency of employees. The ability of companies to attract and keep talented staff increases, and therefore a more innovative and efficient working environment can be reached (Duralia 2014).

One condition to reach the overall goal, to become a sustainable fashion retail brand is to create a team. This is necessary, because without the understanding and the believing in the strategy of the employees the company is not able to reach the target. Therefore, it is also necessary to find the right people (Danciu 2006; Duralia 2014).

5.3.3 External Communication

The communication policy toward customers in this paper should achieve, that regular customers stay and new customers buy. Retail brands have to find solutions that how they can use their change toward a sustainable retail brand, to maximize their sales volume. Therefore, strategies have to be found, which solve negative impacts of this change. One impact of the transformation of the company toward being more sustainable is that the purchasing behavior of the customers for sustainable products differs from the purchasing behavior of conventional products. In the following the aspects of purchasing sustainable products are described.

Generally, the transformation of the company toward being more sustainable is positive, because it creates a competitive advantage. The differentiation from competitors through the sustainable development of the whole company results more and more in a successful positioning in the market. One main reason therefore is that the retail brand not only serves the requirements of the consumer, but also serves the requirements of the environment, which in combination is more valuable for the consumer and can lead to a higher customer loyalty (Duralia 2014).

To communicate the advantage of the sustainability of products and the brand and thereby influence the buying behavior of customers in a positive way, specific colors of advertisement layouts can be used. Concerning an American research report, companies are perceived as sustainable from American consumers, if they use the color green and natural graphics as a dominant element in their advertisements and brand logo (Visser et al. 2015; The Green Revolution 2014). A research concerning the communication of sustainable shoes to mainstream consumers indicates that sustainability has a positive effect on buying intentions, regardless of what benefit it addresses (e.g., personal benefit and environmental benefit). Moreover, the combination of a personal benefit with a green layout led to the highest buying intention (Visser et al. 2015).

Moreover, the use of a sustainable standard like the mentioned standards in the previous sections, customers are more able to distinguish environmentally friendly products from those, which are not and from those, which promote that they are sustainable but in fact they are not (Ottman 1993). Consumers have to be convinced of the environmental benefit of the product and its superior sustainability image (Peattie 2001). Therefore, the use of such a label is an advantage, because it proves sustainability. However, the problem in this case is that the majority of people do not know the meaning of textile labels like GOTS. Only 5 % of consumers think that they have sufficient knowledge about textile labels (GfK Fashion Talk 2015). Therefore, consumers need more knowledge about sustainable clothing to be able to act in a more sustainable way (Joergens 2006), so to purchase sustainable clothes.

That leads to the next impact, because the change of the brand toward sustainability can also lead to revise the definition of the target group, hence which customers the company wants to address, because they buy sustainable clothes. According to Professor Ken Peattie from Cardiff Business School there is nothing like a "green consumer" who companies can address, it is more a green purchasing behavior (Peattie 2001). Another research done by students also found out that distinguishing of groups of consumers in terms of behavioral or attitudinal constructs is not so effective, than to type them according to the strategy they employ for making their lives more sustainable (McDonald et al. 2012).

Therefore, green purchasing behavior has to be examined. If all products were equal, the majority of consumers would prefer the environmental friendly alternative. But most consumers buy both, products with and without environmental benefits (McDonald et al. 2012). In line with this stand, findings of an experiment from Delft University of Technology in the Netherlands (Vogtländer et al. 2002) found out that customers first look for products which fit in terms of quality and price. Afterwards, they check the environmental issues of the product before making the purchase decision. Therefore, the last aspect finalizes the purchase decision: the product that meets and communicates the environmental belief of the customer best will be bought. But if there is another product that fits better than the sustainable product in terms of quality and price, the environmental aspects are not taken into consideration during purchasing. This is a layered decision-making process, or also called double filter theory (Vogtländer et al. 2002). The general steps of this process also appear for textile-related consumer preferences, but the steps differ in the special criteria of clothing. In the first step, consumers take products into consideration that they like because of their visual appearance (color, shape, and style). If they purchase in a store, the touch of the textile is also a criteria, and the following step is to try the clothes on. Important aspects are the fit and the comfort during wearing. At last, consumers check price in relation to performance. The most important criteria are appearance, functionality, and price (Meyer 2001). A quantitative research about eco-fashion found out that eco-aspects can only bring additional value to products, if they are otherwise attractive. This appears especially in the fashion and luxury sector. Quality and esthetic aspects are highly important aspects for all consumers, when purchasing clothes, and this applies to eco-fashion, too (Niinimäki 2010). In addition to that it is a problem that most consumers regard eco-clothing as less fashionable is also an issue. Consumers perceive these clothes as shapeless, colorless, overpriced, etc (GfK 1999; Meyer 2001; Otto Versand 2000). Moreover, consumers buy only if there are no disadvantages to the consumer in terms of higher price, loss of quality or discomfort in shopping, but sustainable clothes can mainly be bought from catalogs (Joergens 2006); therefore, the limited availability has negative effect on purchasing intentions. However, significant drivers toward purchasing of eco-clothes are ethical commitment in clothing purchasing and ethical values. Consumers with strong ethical commitment value their norms much more than their personal esthetic feeling. But these people are a minority group and remain a niche. Furthermore, the general attitude of the majority of people toward ethical consumption is positive, but the step from ethical interest to ethical purchasing is more complex (Niinimäki 2010).

Furthermore, another research about ethical fashion from Catrin Joergens points out that the knowledge of consumers about the ethical problems of clothes are not the reason why they do not buy sustainable apparel, even if they are aware of the problems, it does not significantly affect their purchasing decisions. Reasons therefore are that consumers feel like they do not have a real choice, because most of their garments are produced in developing countries. Another reason is that they feel they are not in the position to judge about unethical manufacturing in developing countries, because with Western norms in mind, they cannot judge about the culture of the manufacturing country. This does not indicate that consumers do not care about impacts of conventional manufacturing standards, but it does not directly affect the purchasing behavior, particularly for fashion items. Everything which directly affects the health of consumers influences their acting, e.g., if chemicals in their clothes would harm their skin they would care about it, but as long as clothes do not harm, they do not care about working conditions in manufacturing companies, and the environmental impact of their production. The same accounts for food, because consumers buy fair trade or organic food since it directly affects their health. Consequently, it is assumed that consumers only care about issues that influence them directly (Joergens 2006).

In addition to these aspects of the purchasing behavior of sustainable clothes, the research includes two case studies about the marketing of retailers who sell sustainable fashion. These companies are "coop Switzerland" and "Patagonia". They do not have high sustainable standards like GOTS, but the case studies show how these retail companies dealt with problems concerning sustainability with partly are solved with their communication policy (Meyer 2001).

5.4 Discussion

5.4.1 Internal Communication

The aspect that employees have the ability to develop the perception of customers of the brand appears in the fashion branch, especially for the sales staff, because they have the most contact with the customers. Therefore, the sales staff has the highest chance to create or develop the knowledge of consumers about the brand. Consequently, it is very important that they do understand the brand right. Also all other employees are important for internal branding, because they all contribute to the company and a wrong understanding of the brand from employees can cause mistakes in their work.

To prevent a wrong understanding of the brand, a communication policy can be developed. The formula of Lasswell states the parts of communication, which sound simple but in fact they can help to realize, how the internal communication can be improved, because it can affect a conscious use of the parts of communication, to achieve that the employees understand the message in the right way. Only the chosen communication channel can make a huge difference to this understanding. This is especially important if the company wants to transform. Because, the already mentioned importance of multilevel dialog and interaction involving management and all employees, depend also on the right understanding of the messages from the recipient. The implementation of an internal communication policy cannot be necessary, too. An already existing company can already use an internal communication policy. This implemented itself through the daily work, so through the way the employees communicate certain message, e.g., for the changed times of the canteen companies would write an E-mail with this information and send it to every employee. However, in these companies it can be useful to think about their way of internal communication, especially if they want to transform the company like in this example toward a more sustainable company.

This transformation also can have positive effects on the employees, like already mentioned the efficiency can be enhance. Moreover, the ability of companies to attract and keep talented staff increases, and therefore a more innovative and efficient working environment can be reached. On the one hand, a reason for this development can be that employees rather would like to work for a brand, which contributes to their own values and norms and therefore prefer a brand which works in a sustainable way. But on the other hand, it is also possible that the topic sustainability does not contribute to values and norms of persons. In that case, these persons are not attracted by the brand and already employed persons would maybe leave the company or they stay but are not that motivated. However, another reason why the change can cause the mentioned positive effects is that sustainability in fashion industry has a huge potential and the chance to influence this development and reach the goal of, e.g., a sustainable fashion brand which serves the mass market. This can be an interesting incentive for some job applicants and employees. To achieve a goal like this, it is necessary to create a team and to become a team it is important, concerning the mentioned literature that the employees understand the strategy and believe in it. As a result, it is important to find, attract and keep the right employees, which want to work for the company, because they understand and believe in the strategy. To create a team like this, out of an existing team, can be difficult, because the personal norms and values of each individual employee cannot fit to the new strategy of the company. Furthermore, it is also possible that the personal values and norms first do not fit, but change through the right internal communication. It is also possible that these persons just accept that the strategy of the company changed and that they have to adapt it to their daily work.

5.4.2 External Communication

Generally, it can create a competitive advantage to become a sustainable retail brand, because the products are more valuable through the aspect of sustainability. As a result it has a positive effect on buying intentions, but this aspect has to be communicated to customers to be able to use the positive effects. One possible opportunity is to change the design of the advertisements, so that the aspect of sustainability of the clothing and of the brand is underlined by the layout of the advertisement. The same appears to the brand name and logo. They can also underline the sustainability of the company with their design. However, the advertisement layout also has to be attractive toward the target group. Moreover, advertisements of already implemented retailer may be also remembered for their layout and for the brand name and logo. Therefore, the advantages and disadvantages of this change should be balanced out carefully. However, it is possible to make a compromise of these aspects, e.g., they can only adapt the color green to the brand name and the logo, but remain the established font of the brand name and the established style of the logo.

Another problem concerning the communication of the sustainability of the retail brand and their products is that the majority of customers do not have sufficient knowledge about sustainable textile standards. Due to the research, consumers have to be convinced of the environmental benefit of the product and its superior sustainable image, which are proved by these standards. Therefore, consumers need more knowledge about sustainable textile standards. One possible solution for this problem can be to use the marketing campaigns also to explain the chosen textile standard in a condensed format. That would increase the awareness about the label in the target group.

In addition to that the revision of the target group also has risks. According to the mentioned literature, it can be said that it is not recommendable for retails to create a target group with a definition of a "green consumer", because it is more a green purchasing behavior, for example. Instead of adding a target group, retail brands can try to address the purchasing behavior of their target group with their communication policy.

Furthermore, the purchasing behavior of sustainable products is also complex. Summarizing the mentioned aspects of it in the previous research, it is recommendable for retail brands to be aware of the aspects of green purchasing behavior. Because, the companies are able to influence the communication of the sustainability of their products, so that it matches the environmental beliefs of the target group. Moreover, the clothing still has to meet the requirements of the target group in terms of appearance, functionality and price, because if they do not meet this need the sustainable aspect of the products cannot develop to the competitive advantage, which it could be. Therefore, the conclusion could be drawn that the advertisement should focus on the promotion of the attractiveness of the product and additionally on the sustainability of the brand and the product.

To address this issue that eco-clothing is regarded as less fashionable, retail brands can convince their target group with the clothing itself. If it meets the requirements of the target group in terms of esthetic attractiveness and a reasonable price–value–relationship, the negative image of sustainable clothing may disappear; or at least it does not influence the purchasing decision. Moreover, companies can influence the negative perception through their communication policy, if they use their communication policy instruments in the right way with a convincing message. They can make reference to this issue with an extensive slogan, but this has also to be appealing for the target group.

Another mentioned problem concerning the purchasing behavior is the limited availability to purchase sustainable clothes, because they can be mainly bought from catalogs. If the retail brand is able to sell their products other multiple distribution channels, like own retail stores, shop-in-shops, leased departments or online-shops, the availability can be increased and the comfort of shopping enhanced. However, the consumers also have to be aware of the multiple options to buy the sustainable clothes. Therefore, the communication toward consumers maybe should also include messages about the availability of the products. They can, for example, promote the number of stores in the country, which sell their products and/or the address of the online-shop. Moreover, they can concretely name the closest store, if they use advertising of local mediums like radio channels or newspapers.

The fact that only issues which directly affect the health of consumers influence their acting, which is their purchasing behavior in this case, is also an impact witch the communication policy of retail brands can use. Out of this fact appears a chance for the communication policy. Companies can communicate, e.g., that their clothes are 100 % free of harmful substances, because this addresses the health of consumers. Their own health is a sensitive theme for consumers and offers the chance to promote the sustainable clothing of the retail brand. Moreover, this goes in line with the mentioned proposal to explain the chosen sustainable standard. The goal of this is to increase the awareness about the label in the target group, because the label proofs the sustainability of the company and communicates this competitive advantage.

5.5 Conclusion

The starting point of this paper was the growing interest of consumers toward the manufacturing process of textiles, which may lead to a higher interest of companies to meet this interest. Referring back to the research question it can be said that there are generally many impacts fashion retailers have to consider, if they want to transform their company to become more sustainable.

To sum up, companies first have to define how sustainable their company should be. Second, they can implement a given sustainable standard like GOTS, or they can define their own standards. Third, they have to define a strategy, how they want to become sustainable. Lastly, they need to plan how they want to make their strategy work and put it into practice. However, to transform from a conventional to a sustainable fashion retail brand is complex. But, as a result out of the research and the discussion part of this paper, it is possible. Especially, the discussion part indicates that there can be found solutions for problems appearing. Because the most negative impacts, could have been possibly solved by communication policy. Moreover, there are already sustainable fashion brands that are certified by GOTS.

5.6 Research Limitations and Future Research

A limitation of this paper is that the proposals how retailers could deal with the impacts of the transformation of the company toward more sustainability, need further research and tests until they are practicable. Assumed that the trend toward the purchasing of sustainable clothes expands, it could become easier for fashion retail brands to become sustainable.

References

Anderson TA, Salice CJ, Erickson RA, McMurry ST, Cox SB, Smith LM (2013) Effects of landuse and precipitation on pesticides and water quality in playa lakes of the southern high plains. Chemosphere 92(1):84–90. doi:10.1016/j.chemosphere.2013.02.054

Balogun J, Johnson G (2005) From intended strategies to unintended out-comes: the impact of change recipient sensemaking

Bruhn M (2015) Kommunikationspolitik: systematischer Einsatz der Kommunikation für Unternehmen (8., überarb. Aifl.). Franz Vahlen, München

Business Dictionary (2015) What is communication? Definition and meaning. http://www.businessdictionary.com/definition/communication.html. Accessed 29 Dec 2015

Danciu V (2006) Marketing ecologic (green marketing). Economica

Dievernich EP (2015) Change management and the human factor: advances, challenges and contradictions in organizational development. Springer, Berlin, Heidelberg, Cham

Dipl. Ing. Katharina Schaus. (2013) Gutachten: Der Weg zu nachhaltiger Klei-dung- Standards, Siegel und politische Rahmenbedingungen. https://www.gruene-bundes-tag.de/fileadmin/media/gruenebundestag_de/themen_az/verbraucherschutz/Gutachten_Nachhaltige_Kleidung.pdf. Accessed June 2013

Duralia O (2014) Applying sustainable marketing strategies—the key to obtaining competitive advantages on the industrial products market. Stud Bus Econ 9(3):21–28

Esch F-R (2015) Stichwort: Kommunikationspolitik. Gabler Wirtschaftslexikon. Springer Gabler. http://wirtschaftslexikon.gabler.de/Archiv/10971/kommunikationspolitik-v15.html

Gardetti MA (2013) Sustainability in fashion and textiles: values, design, production and consumption. Greenleaf Publications, Sheffield

Gesellschaft für Konsumforschung (GfK) (1999) Öko-Labels für Textilien [Eco-Labels for Textiles]

GfK Fashion Talk (2015). http://www.gfk-verein.org/compact/rahmendaten/gfk-brancheninfos/gfk-fashion-talk-informationen-fuer-fashion-lifestyle

Global Organic Textile Standard International Working Group (IWG) (2014) Global Organic Textile Standard (GOTS) Version 4.0. http://www.global-standard.org/de/information-centre.html. Accessed 3 Jan 2014

Ikävalko H (2005) Strategy process in practice: practices and logics of action of middle managers in strategy implementation

Joergens C (2006) Ethical fashion: myth or future trend? J Fash Mark Manag Int J 10(3):360–371. doi:10.1108/13612020610679321

Kirchgeorg M (2015) Stichwort: marketing. Gabler Wirtschaftslexikon. Springer Gabler. http://wirtschaftslexikon.gabler.de/Archiv/1286/marketing-v9.html

Lasswell HD (1967) The structure and function of communication in society. Read Public Opin Commun 2:216–228

McDonald S, Oates CJ, Alevizou PJ, Young CW, Hwang K (2012) Individual strategies for sustainable consumption. J Mark Manag 28(3/4):445–468. doi:10.1080/0267257X.2012.658839

Meyer A (2001) What's in it for the customers? Successfully marketing green clothes. Bus Strategy Environ (Wiley) 10(5):317–330. doi:10.1002/bse.302

Visser M, Gattol V, Helm R (2015) Communicating sustainable shoes to mainstream consumers: the impact of advertisement design on buying intention

Niinimäki K (2010) Eco-clothing, consumer identity and ideology. Sustain Dev 3:150

Ottman JA (1993) Green marketing: challenges and opportunities for the new marketing age, 1st edn. McGraw-Hill/Contemporary, Lincolnwood

Otto Versand (Ed) (2000) Qualitativ-psychologische Studie zum ökologisch orien-tierten Textilangebot [Qualitative Psychological Survey Concerning the Environmentally Benign Offerings]

Owens KA, Legere S (2015) What do we say when we talk about sustainability?: analyzing faculty, staff and student definitions of sustainability at one American university. Int J Sustain High Educ 16(3):367–384

Peattie K (2001) Golden goose or wild goose? The hunt for the green consumer. Bus Strategy Environ (Wiley) 10(4):187–199. doi:10.1002/bse.292

Piegsa E (2010) Green Fashion: ökologische Nachhaltigkeit in der Bekleidungsin-dustrie. Diplomica Verlag, Hamburg

Sadik O (2014) Einkaufsratgeber: Textil—Label. https://www.greenpeace.de/sites/www.greenpeace.de/files/information_material_guidebooks/greenpeace-ratgeber-textil-label-2014-e00972-es_0.pdf. Accessed 15 Oct 2014

Survey of all Basic Award Criteria (2015) https://www.blauer-engel.de/en/companies/basic-award-criteria Accessed 23 Dec 2015

Oxford Dictionary (2015) Definition: sustainable. Oxford University Press. http://www.oxforddictionaries.com/de/definition/learner/sustainable

Textile industry's eco efforts (2015) WWD: women's wear daily 209(89):31

The Green Revolution (2014) Grail research. http://www.grailresearch.com/pdf/ ContenPodsPdf/Grail-Research-Green-Evolution-Study.pdf

Tobler-Rohr MI (2011) Handbook of sustainable textile production. [Elektronische Ressource]. Oxford; Philadelphia; Cambridge: Woodhead

Tomczak T, Esch F-R, Kernstock J, Herrmann A (2012) Behavioral branding [Elektronische Ressource]: Wie Mitarbeiterverhalten die Marke stärkt. Wiesbaden: Gabler Verlag

Vogtländer JG, Bijma A, Brezet HC (2002) Communicating the eco-efficiency of products and services by means of the eco-costs/value model. J Clean Prod 10:57–67. doi:10.1016/S0959-6526(01)00013-0

White MA (2013) Sustainability: I know it when i see it. Ecol Econ 86:213–217. doi10.1016/j.ecolecon.2012.12.020

Wittke-Kothe C (2001) Interne Markenführung. Verankerung der Markeniden-tität im Mitarbeiterverhalten, 1st edn. Deutscher Universitäts-Verlag, Wiesbaden

Chapter 6
The Prosumer Concept in Fashion Retail: Potentials and Limitations

Jochen Strähle and Anna-Katharina Grünewald

Abstract The purpose of this paper is to highlight potentials and limitations of the prosumer concept in fashion retail. The paper illustrates the evolution of prosumption and in which directions the concept is being developed. The primary research is based on a literature review containing different sources of academic and non-academic references. Findings suggest that the prosumer concept is no new phenomenon. Recently, it has moved into the focus of companies that have noted that it is efficient when engaging with customers in order to strengthen their brand loyalty. An increasing number of companies offer innovative business models that underlie the concept. However, lately smart prosuming machines are changing the objectives of the concept. Even if the prosumer concept exists since many years and scholars investigate its potentials continuously, it is the fashion industry that has been researched comparatively little up to now.

Keywords Prosumption · Smart prosuming machines · Co-creation · Third Wave DIY · Web 2.0

6.1 Introduction

The term prosumption was used by Alvin Toffler in 1980 for the first time. The concept of prosumption refers to a combination of production and consumption. Scholars argue that the concept is not new but is actually primordial. Many scholars have dealt with the issue implicitly. Only recently, they have begun to deal with it explicitly. Prosumption has always existed, but various social changes, like the rise of Web 2.0 and of social networking on it, have significantly expanded both the practice of prosumption and scientifically attention to it.

Prosumption has its most obvious and direct relevance to the economy as the use of social media (e.g. blogs), for instance, has grown remarkably during recent years.

J. Strähle · A.-K. Grünewald (✉)
School of Textiles and Design, Reutlingen University, Reutlingen, Germany
e-mail: jochen.straehle@reutlingen-university.de

J. Strähle (ed.), *Green Fashion Retail*, Springer Series in Fashion Business,
DOI 10.1007/978-981-10-2440-5_6

Especially fashion retail benefits from the fact that the largest segment of blogs consists of fashion blogs, i.e. blogs that focus on fashion brands, fashion products, and fashion e-commerce. Hence, it is of growing importance how commercial actors operate in the "sphere" of the fashion blogging and how they interact with the users. Using new technologies, consumers/users have become able to create distribution channels for communicating fashion information which in parts compete with traditional fashion journalism. Through the practice of blogging, bloggers take part in the creation of fashion by influencing the spread of new trends throughout the fashion system. Thus, consumers are able to exert a high degree of influence and act as institutional entrepreneurs by engaging in prosumption activities. The social practices created by influential fashion bloggers have provided the key element from which the value creation processes of the organizational field take their starting point. Consumers and prosumers have thus played a key function in creating new channels for fashion marketing.

Apart from that tendency of the concept, there is also a new Do-It-Yourself model that involves ordinary people inventing, designing, making, and/or selling physical things of value, ranging from jewelry over furniture to off-road vehicles. In practice, two types of DIY may be distinguished—the web-based DIY on the one and the workshop-based DIY on the other hand. While the web-based approach involves combining the read–write functionality of Web 2.0 with computer-aided (CAD) design, the workshop-based concept contains people using handheld tools together with CAD and manufacturing machines. It has been argued that Third Wave DIY can be carried out by anybody at any location. However, much Third Wave DIY requires users with certain computer skills.

Another evolving form of prosumption is the so-called co-creation, which occurs between fashion brands and their customers. Fashion retailers applying such an approach will be able to strengthen long-term buying relations by turning towards central value connections. Compared to other industries, fashion retail has remained largely unexploited in this regard. The industry mostly relies on traditional marketing methods, instead of pushing strategies towards co-creation in order to integrate consumers into the value creation chain. Market performance in the fashion industry will become increasingly dependent on the knowledge base of consumers, who strongly identify with the company. Hence, companies will face new challenges in order to position their brand(s) in the market. A central task will then be the identification of these customer groups and to figure out how to engage them in efficient dialogs and to capture and use their knowledge. Co-creation in the fashion business means to address consumers as individually and personally as possible. Companies have to turn value exchange of the primordial manner into an emotional and individualized experience.

As economies are continuously developing rather than being static, a new dimension of prosumption has recently arose, replacing the human prosumer. Smart prosuming machines are a new phenomenon appearing in the context of prosumption, just as the human prosumer finally gets the attention he has deserved. While the impact of these machines on producers has long been obvious, the growing impact on consumers has changed the most. Human consumers/prosumers are often unaware of the prosumption being done by smart (prosuming) machines. While they offer many advantages, there are also threats, caused by the replacement of human by non-human technologies, as these smart (prosuming) machines work autonomous and control themselves. This is especially the case on the Internet, where many functions operate as self-organizing devices.

6.2 Theoretical Background

6.2.1 The Anatomy of Fashion Retail

The fashion retail industry is considerably growing due to an industrialized and developing world. The society becomes more affluent and is also continuously growing. This in combination with ever-accelerating communication tools forms the basis for the reality of the global fashion retail. Consumers changed from national customers to variety-loving, international shoppers, having clothes from brands spread all over the globe.

6.2.1.1 Dynamics of Fashion Retailing

Due to the fact that fashion gets increasingly faster, as trends are appearing and disappearing very rapidly, fashion retailers have to make sure to deliver quickly and to always carry the latest trends in their assortment. Therefore, this fast-moving industry requires fashion buyers who make precise and quick decisions on what products need to be ordered.

If a retailer does not offer the right product at the right place in the right time, he will not be competitive and will suffer. That is why buying and merchandising are two of the business' core management functions. Even for an excellent sales team it is almost impossible to generate high sales with a weak merchandise offer. Hence, merchandise planning and buying is of growing importance. Therefore, fashion companies invest heavily in the right employees and complex planning tools. However, fashion buying is about supplying what consumers are asking for rather than a buyer's right intuition and personal preferences. Thus the observation of stores (sales, tops/flops list, etc.) and collection of customer feedbacks is crucial.

The picture of the glamorous world of fashion created by the media is only an illusion and represents just one side of the business. The very complex interdependences of tasks that have to be done and decisions that have to be taken often remain hidden. Due to this 'catwalk' imagery, the economic significance of the industry is often underestimated or even dismissed. However, the fashion industry provides employment to many people and makes a significant contribution to the world's economies (Jackson and Shaw 2000).

6.2.1.2 The Fashion Consumer

Compared to the past, today's fashion consumers are segmented in significantly smaller social groups, which are influenced by different stimuli like music, sports, films, TV, and dance. This is what is called 'tribalism' in a marketing context. Due to the increasing variety of stimuli and an increasingly crowded and confused society, people try to find their place in society and want to be part of smaller social groups with clear values and symbols. Fashion became part of these symbols and helps the consumer to make a statement about their self-perception with the purpose to express and elevate theirselves physically and psychologically. Depending on the country, fashion and clothes play different roles. In some countries dressing up fashionably and properly is a norm for all classes in the society, whereas there are countries in which fashion does not play an important role and is very low on the social agenda (Jackson and Shaw 2000).

Consumer behavior implicates a consumer's selection, purchase, usage, and disposal of goods in order to satisfy his/her needs (Solomon and Rabolt 2004). A consumer's shopping goal can be driven by many different parameters. However, the core motivations are utilitarian and hedonic motivations underlying consumer behavior (Babin et al. 1994). Consumers are motivated by efficiency if they purchase products in the sense of utilitarianism (Childers et al. 2001). From a hedonistic perspective, the shopping experience itself is the main motivation to purchase a certain product (Holbrook and Hirschman 1982).

Workman (2009) distinguishes between two essential sorts of fashion consumers —the 'fashion change agents', including innovators, opinion leaders, and innovative communicators and 'fashion followers'. Fashion innovators are those consumers buying and wearing new fashion at first and encourage others. Fashion opinion leaders in turn are the ones convincing others to buy and wear new fashion. The innovative communicators combine both roles. Lastly, fashion followers follow the lead and wait till a fashion trend is finally accepted by the great masses (Workman 2009).

6.2.2 The Prosumer Concept

Alvin Toffler introduced the concept of the prosumer more than three decades ago in 1980. However, it took years till researchers and scientists began to understand the importance of this new idea and understanding of the fusion of production and consumption. Due to the fact that the process of presumption has changed and expanded recently, it becomes more and more traceable (Ritzer 2015).

Alvin Toffler was the first scholar mentioning the term prosumer. He defines the prosumer as a person, producing goods and services for his own consumption. Those people make their own clothes, cooking their own food or fixing their own cars. As all of these goods and services can be purchased in the marketplace, today customers mainly purchase them form others. That is, what is called the essence of being a customer—to buy goods and services from others. The essence of being a prosumer in turn is to produce most of the consumed products and services on one's own behalf.

The ladder got distinguished into the production for someone's own use and the production for the exchange with others. People, who produce for use, unite production and consumption in theirselves, whereas people producing for exchange separate production and consumption. Hence, these people produce in order use the earnings to buy other things they need.

Toffler considered that these two activities got separated during the Industrial Age. He differentiated between three major stages in human history (Kotler 1986).

6.2.2.1 The Rise of the Prosumer

The Prosumer Concept According to Toffler (1980)

The First Wave was dominated by agriculture. According to Toffler the majority of people were neither producers nor consumers in the usual sense at that time, as they are hunting and growing their own food, making their own clothes and being responsible for their own amusements. They were so-called prosumers. This group of people built the so-called Sector A—the societies' largest sector. Once a small number of people were specialized in some fields, like candlemaking and were trading with the obtained goods. These people—the consumers of the society—built Sector B. Hence, the First Wave was marked by self-production. Its society was kept together by kinship and friendship and driven by the will to survive.

The Second Wave was triggered by the Industrial Revolution in England. The industrial landscape was characterized by factories. More and more people got hired and spent their productive hours in factories instead of being back at home. They got used to work eight hour per day in order to spend their earned income in the marketplace for products their needed. It was the time when people started to produce for exchange, not for use. Therefore, the Sector A (prosumers) was decreasing, whereas Sector B (consumers) was increasing. However, there were still prosumers in the form of housewives, who cooked, cleaned, sewed, knitted, and shopped. Even if they did not get paid for their activities, they were still named prosumers. The Second Wave Societies were dominated by industrialization and marketization and marked by the establishment and extension of networks through which people got the opportunity to get goods and services they needed. The production of these goods was driven by efficiency and the consumption was marked by the sense of enjoyment. The societies' nexus were, apart from kinship and friendship, contracts and transactions, holding people together. People became very specialized producers during the industrialization and were almost unable to produce anything else besides their own specialization.

The Third Wave society was, according to Toffler (1980), dominated by demarketization and demassification. The society was marked by individual rather than mass consumption. Family and neighborhood were the things keeping the society together.

The key question was why would people move back to more prosumption activities? Toffler (1980) named several arguments: The workweek will continue to decline, as it was an 80–90 h week during the beginnings of the Industrial Revolution and became a 40 h week as of today. In the future it might be even fewer hours. Apart from that people will voluntarily work less than these 40 h per week in order to have more leisure time. The overall educational level will improve. People will be more highly educated. Their requirements on the jobs are getting higher. People do not want to be bored at work. The revolution in technology like computers and telecommunication will also lead them to (leisure) spend their time differently. Moreover, raising costs for goods and services will drive people to work more on their own and to develop own skills, especially if they are unemployed or underemployed. Due to the fact that work—especially in the field of technology—gets more mental, people need to find active recreation. Therefore, physical activities in private sometimes include self-production. Some people in society will recognize that they are able to produce better products compared to what is available in the marketplace. Especially those products will be self-contained produced, which are obviously declining in quality. The Dutch and the German are known for developing a high sense of quality and strong workmanship. They will undertake production and services, normally bought in the marketplace. Finally, people will more and more fulfill themselves by producing their own goods or providing their own services. They will attend courses on knitting, cooking, painting, etc. This prognosis follows the idea of Maslow's (1954) hierarchy of needs, which says that people will focus on self-actualization. This can be equated with the prosumption activity (Toffler 1980; Table 6.1).

Table 6.1 The main ideas of Toffler. Adapted from Kotler (1986)

	First Wave	Second Wave	Third Wave
Dominant institution	Agriculture	Industry	Home
Mix of prosumer and consumer	Many prosumer (large Sector A)	Few prosumer (small Sector A)	More prosumers (Sector A gets larger)
	Few Consumer (small Sector B)	Many consumer (large Sector B)	Fewer consumer (Sector B gets smaller)
Dominant process	Self-production	Industrialization	Deindustrialization
		Marketization	Demarketization
			Demassification
Norm	Survival	Efficiency (as producers)	Individualization
		Indulgence (as consumers)	
Social nexus	Kinship and friendship; tribe	Contracts and transaction; workplace	Family and friends; neighborhood

Further Development of the Prosumer Concept

Inspired by Toffler's work on prosumption, many scholars of different fields started to work on further theories underlying the changes named in the Third Wave. They used or created other concepts similar to Toffler's ideas. This in turn led to a wide range of concepts, having more or less similarities (Ritzer 2014):

> [...] do-it-yourself (DIY) (Watson and Shove 2008); craft consumption (Campbell 2005); Pro-Ams (Leadbetter and Miller 2004); co-creation (Prahalad and Ramaswamy 2004a, b); service-dominant logic (Vargo and Lusch 2004, 2008); commons-based peer production (Benkler 2006; Benkler and Nissenbaum 2006); collaborative capitalism involving both value co-creation and service-dominant logic (Cova et al. 2011); crowd- and open-sourcing (Howe 2009); putting customers to work (Ritzer 1993); Wikinomics based at least in part on the idea that businesses put consumers to work on the internet (Tapscott and Williams 2006); the complete collapse of consumption into production (Zwick and Knott 2009); Laughey's (2010) productive consumption; and the produser (Bird 2011; Bruns 2005, 2008).

However, Kotler (1986) was the first scholar studying possible consequences of the development of prosumption in the field of marketing research. Due to the fact that the Third Wave was characterized by deindustrialization, demassification, and demarketization, there ought to be consequences for marketing researchers and practices. As Toffler already named rising costs of labor, structural unemployment, the demand for higher quality goods and services, the development of new technologies enabling people to take part in designing customized goods, and a general increase in education and thus the desire for self-actualization, as triggers of more prosumption, Kotler hold the same idea. Apart from that, also counteracting forces like threatened stakeholders (e.g. producers of certain goods, trade unions or specialist professionals) have been identified. Their approach might be the slow down of the prosumer trend in order to protect their business.

Kotler considered that prosumers have to be assigned to certain prosumer groups and that these ones should be carefully studied. He also argued that the prosumer has to be seen as an emerging market segment. Nevertheless, he did not believe that the consumer will disappear eventually. On the contrary, Kotler explained that individuals would act as prosumer in one and as consumer in another situation.

Another further development of the concept of prosumption was introduced by Tapscott (1997). He stressed that digital technologies, the increasing use of the internet and the transition to the digital era were offering new opportunities that might change the conditions for prosumption. As Tapscott said, the gap between consumer and producer was blurring in the new economy. Mass customization was replacing mass production and retailers had to ensure to be able to satisfy individual consumer needs and requirements.

Recently, the prosumer concept has received swelling attention in the context of research focusing on digital technology. Bandulet and Morasch (2005) focused on prosumption behavior within e-commerce, whereas Stock (2007) mentioned prosumers in a Web 2.0 service context. Ribiere and Tuggle (2010) connected Web 2.0

with Enterprise 2.0 technologies and discussed the critical role customers can play in innovation processes.

Social media is one of the specifications within the context of digital media, where scholars have increasingly applied the prosumption concept. The roles of producers and consumers did not only blur, but also began to merge. In social media, the user is a supplier of content and supports or even provides the distribution of content and services through wikis, tagging, feedback, information-sharing, etc. Now the ideas of Toffler are becoming reality (Pascu et al. 2008).

Fashion Blogger as Prosumer

Lately, debates about the concept of prosumers and prosumption beg the question whether fashion blogger practices can be seen as forms of prosumption activities or not. Apart from that, it has to be questioned what role such activities can play in relation to emerging marketing practices. An obvious prosumption activity, identified in the context of fashion blogging, is the production and documentation of the fashion blogger's styles by combining different products and creating unique outfits. Those outfit creations are, according to Toffler (1980) and Kotler (1986), prosumption activities, as they are also posted publicly and enable the generation to cause and affect a diffusion of the fashion system. Hence, the role of prosumption activities in connection with both, production and consumption, becomes an issue that has to be studied in order to understand possible effects on the diffusion of the fashion system.

The contemporary directions and trends named in fashion theory provide a useful perspective for analyzing the phenomenon of fashion blogging and its dimensions and possible effects in order to understand how fashion is materialized when consumers get part of the digital context. One of these trends, the shift from mass/class fashion towards consumer fashion, underlines the role of the consumer's individual identity. It provides interesting ideas of the decentralization of the fashion system and the consumer's opportunities to create and develop his self-image.

6.2.2.2 A New Do-It-Yourself (DIY) Paradigm

Recently, a new Do-It-Yourself model, involving ordinary people, who invent, design, make, and sell goods, arose. The range of goods goes from jewelry over furniture to off-road vehicles. Fox (2014) named this new paradigm the Third Wave DIY, which is also based on Toffler's (1980) ideas. The Third Wave DIY is geared to the reading and writing functionalities of the internet and the therefore occurring digital-driven designs, enabling ordinary people to invent products themselves (Fox 2013). Nowadays, websites like Shapeways combine the read and write Web 2.0 functions with computer-aided (CAD) design. Fox (2014) has claimed that the Third Wave DIY is driver for prosumption, innovation, and entrepreneurship and that it can be carried out everywhere by anybody. In this context, prosumption is meant to be the individual producing for his own consumption, whereas innovation refers to new products and entrepreneurship aims at the invention of these new products to the market (Ritzer and Jurgenson 2010).

Web-Based Third Wave DIY

As mentioned before, Fox (2014) describes Third Wave DIY websites as tools that enable consumers to combine the read and write Web 2.0 functions with CAD tools and digitally driven manufacturing tools such as 3D printer. These websites encourage people at different locations to communicate with each other and to interact. OpenMaterials for example is a platform, where dividuals can share the latest DIY ideas and can source according to production materials (OpenMaterials 2016). Another website offering DIY ideas is Instructables (Instructables 2016). Users can upload their projects and are rated by other users of the community. Other platforms such as Tinkercad focus on the digital design of new types of products (Tinkercad—Create 3D digital designs with online CAD 2016). Apart from that there are websites enabling individuals to make and sell their own ideas. Finally, websites such as Betabrand, Crowdcube, and Kickstarter support individuals and offer a platform for crowdfunding (Betabrand—Crowdfunded Clothing 2016; Kickstarter 2016; Crowdcube 2016).

Betabrand is an online clothing community, based in San Francisco. They design, manufacture, and release new products nonstop. Brand new ideas are presented on their site every day. Fans co-design and crowdfund them into existence in a matter of weeks. Betabrand is home to Web hits like Dress Pant Yoga Pants, Bike-to-Workwear, Disco Hoodies, and many more. "Our customers are our models. Our customers are our designers. Our customers are helping us build a brand unlike any other" (Betabrand—Crowdfunded Clothing 2016; Fig. 6.1).

Fig. 6.1 Betabrand—the crowdfunding platform. Adapted from Betabrand—Crowdfunded Clothing (2016)

Workshop-Based Third Wave DIY

Apart from the web-based Third Wave DIY, workshop-based Third Wave DIY is also of growing importance. Handheld tools and digital-driven manufacturing equipment at fixed locations are involved in those workshops. Fixed locations include Fab Labs, Hackerspaces, Local Forges, and Techshops. Fab Labs for example are fixed facilities and mobile facilities, including computers; scanner; router; laser cutter; mini-mill; vinyl cutter; dust extraction system; vacuum cleaner; and a wide variety of electronics parts. They serve the purpose to enable DIY design, manufacture, and installation of goods. Although workshop-based Third Wave DIY is realized at fixed physical locations, the Web 2.0 functionalities of the Internet, also including blogs, forums, wikis, etc., build the basis for the workshops and enable individuals to digitally pre-network and follow-up.

Third Wave DIY Goods

The produced DIY goods range from small and simple to large and complicated. Small goods can be produced by manufacturing three-dimensional (3D) printer or laser cutter offered by Tinkercad and Shapeways for example. Items like jewelry and toy figures are preferably manufactured by these tools. Trade-offs between aesthetic creativity and production efficiency are reduced due to the usage of additive manufacturing (AM), such as 3D printers. While industrial machinery that are product-specific and efficient or is general-purpose and less efficient, AM machines are general-purpose and efficient.

Moreover, AM powders can be tailored, rather than mass produced. Hence, the production performance can be maximized, and the mechanical performance of what is manufactured can be improved. The manufacture of complicated aesthetic goods becomes finally much simpler. The production of jewelry for example needs only a low consumption of AM powders and enables an easy delivery through postal services. This, in turn, leads to lower production and retail prices of the goods. The range of available materials and production processes is not very extensive. Hence, the difficulty of material forecasts is reduced. However, it also reduces the potential for innovation (Fox 2014).

3D printers are becoming more and more mainstream and available for personal use. 3D printing will most likely influence fashion through homemade gadgets-like items that will pop up in the streets and on street style blogs (Fig. 6.2).

Third Wave DIY Prosumption

Subsistence DIY enables individuals to produce for their own consumption whatever they can imagine and are able to produce based on the materials and hand-crafted tools that are available. Due to the production process itself, but also due to the infrequent buying of external sources, the production efficiency is

Fig. 6.2 3D printing fashion items. Adapted from 3D Fashion (2016)

compared to industrialized production low (Waters 2006). Companies, using forms of industrial DIY within their business, often offer kits of mass-produced components in combination with standardized installation instructions in return for monetary payment (Woodcock 2006).

Millions of IKEA customers make use of the DIY transportation and installation, in order to save money. In economic means it is more efficient for them and they prefer to undertake DIY as they would pay the company for what they can do themselves. Apart from financial advantages, DIY can strengthen the feeling and pride of ownership (Williams 2004). Hence, the decision whether to make or buy a certain product or service for the individual also depends on his/her personal finances. Within industrial DIY, the ratio between production efficiency and prosumer imagination and creativity is very clear. IKEA for example provides very detailed standardized instructions to a very large clientele in order to prevent any issues. Hence, consumers' creativity is very limited. However, IKEA customers accept the concept of final assembly and installation by the prosumer himself.

On the contrary, a prosumer's ability of being creative is much higher, if he buys paints, timer boards, and metal fixings. It is then up to the prosumer to decide how the cupboard looks like. In this case, production efficiency is much lower, as specialized equipment and knowledge are only covered by the preparation of materials.

Nevertheless, due to Web 2.0 and more and more emerging blogs, forums, and wikis, individuals and communities are increasingly networked and have access to specialized equipment in order to increase efficiency. Networking individuals and communities are provided with digitally driven manufacturing services and have access to read and write functionalities. Hence, the individual creativity is strengthened on the one side while efficiency gets increased at the other hand. For example, Arduino micro-electronics software can be downloaded for free, and expertise in its application is continually adjusted and enhanced through its networked users. Thus, Fox (2014) states that "ThirdWave DIY prosumption goes beyond the buy or make decision of industrial prosumption to decisions about to what extent to contribute without buying anything or making anything".

Due to advances in data compression technologies, more digital data can be transferred via the Internet to mobile devices (Bhojani and Dwivedi 2012). This is essential for transferring CAD data over the Internet. Moreover, it is crucial to push the use of visual images instead of texts, in order to make Web 2.0 functionalities of blogs or forums, for instance, accessible and understandable for everyone (Ananian et al. 2012). Further, many new and easily understandable CAD tools are being developed (Song and Guimbretière 2009). With regard to the production itself, access to needed materials and equipment is essential. This can be achieved via mobile factories. Mobile factories are used for the seasonal processing and packaging of meat and fish, for instance, already. By the usage of additive manufacturing material wastage is radically reduced, as the material is added layer per layer only where needed (Horn and Harrysson 2012).

However, each approach also has its challenges. In order to be useful, powders need to be kept clean and dry and have to be protected from extremes of temperatures and humidity. Moreover, AM can only be used for the manufacture of specific goods. Another challenge is the need trained personnel, having the right skills in order to properly use manufacturing equipment inside mobile factories (Fox 2014).

6.2.2.3 Co-creation in Fashion Retailing

The term 'co-creation' also refers to situations in which consumers collaborate with companies or with other consumers to produce goods or services. Ramirez (1999) claimed that "customers create value; or more exactly, co-create and even co-invent it". These activities are not only co-creation but co-production. The customer is engaged as active participants in a company's work (Auh et al. 2007). Therefore, Humphreys and Grayson (2008) argued that the more companies encourage product users to participate in the production process, the smaller the gap between consumer and producer will become (Humphreys and Grayson 2008).

Today, more and more marketers face the issue that the gap between of what companies think consumers value and what consumer values actually look like is getting bigger. This mismatch appears due to the traditional sense of marketing that aims at making business by treating consumers as segments of whatever they get offered. Within the traditional idea the market is set as center of a company-centric model in which value extraction and exchange, and respectively the sales counter and the purchase act, are equal. According to a strategy like this, the product value and value creation are defined based on principles of cost-effectiveness. The interaction with the consumer is seen as a necessary evil. Thus, companies should create clever marketing concepts by which products seem as though they were utterly indispensable items.

Co-creation refers to a company's way of trying to engage with their customer by integrating their key consumer values into retail concepts and marketing strategies. In order to establish consumer loyalty a company has to address consumers on eyelevel and has to find out what motivations have led to final purchase decisions. Companies have to establish and maintain (inter-)active relationships with their consumers. In order to create long-term bonds marketers need to get involved in a dialog with consumers (Maltzahn 2015). The fashion industry in large is facing the issue of not engaging directly with their consumers. This is somehow surprising as barely no other consumer market is struggling so hard when it comes to consumers' retention. The industry is continuously faced with unsteady consumption habits of so-called mayfly shoppers.

Fashion consumers purchase products for aesthetic values and brand name reasons. However, this does not turn them automatically into loyal customers. Even if they were fully satisfied, they may not return as buying customers. If consumers often visit certain brands or retail stores on a regular basis, the decision to return is probably not based on their strong emotional involvement. Most fashion consumers follow the trend and purchase what is "hot", ignoring what is not respectively. Due to unpredictable influencing factors, this whole process is from season to season totally haphazard. However, this guiding principle behind the fashion industry in large can be questioned as cannot only be seen as an issue, as especially the fashion industry is living from fast changing trends and the industry's strongest asset is creativity. Thus, the "progressively changing face of fashion" and consumers' brand loyalty are not obligatory mismatching. Relevant questions that need to be answered by retailers are 'What is it that keeps me away from engaging with my consumers?' and 'How can I effectively make use of my consumers input?' One reason might be the unique role of the designer. He is the one staying ahead of his game. Apart from that, and perhaps even more important, there is the fast pace of fashion itself. Brands invent (sub-) trends nearly on a monthly basis. This poses a stumbling block, as the periods in between collections are simply too short to gather, translate, and implement consumer input into the organizational performance. In order to meet consumer needs and requirements best, fashion brands should focus on the creation and matching of unique experiences. They have to reflect consumers' own desired image of themselves and have to cater them accordingly, as consumers always want to be recognized as individuals. Moreover, fashion brands have to take opportunities offered by the Web 2.0. The internet does not only represent challenges, but can be also used as an effective tool for the interaction with consumers.

Nike, as one of the world's leading sportswear brand, is also one of the first retailers representing an interesting and effective case of consumer involvement by letting consumers be part of the design process. Consumers get the chance to customize different shoe models on a given template with multiple design options. Consumers and brand co-design/co-create new design options. When they introduced their latest running/lifestyle shoe Vomero II, Nike provided an in-store service at both own flagships and various franchise branches of footwear retailer

Footlocker around the globe. Consumers got the chance to choose from a range of designs and colors in order to customize their personal design of the shoe.

The in-store computer system specifications enabled consumers to register all relevant information into the system that was linked to the company's database. Based on these transferred information the shoe was fabricated and sent to the shop, the order was entered. On top of that, consumers also got the change to take part in a worldwide competition. They were encouraged to participate with their shoe designs in order to become one of three winning models. The winning models got prized with a small production series available at selected retail stores. Apart from the experience of having an individual choice when purchasing a Nike product and therefore experience a feeling of uniqueness, customers also got the chance to be involved in the creative process of product and brand. Hence, customers were integrated in the company's network. A resulting advantage for Nike was the enlarged data of consumer preferences. Nike could use the gathered information for future design of their own by filtering specific aesthetics prevailed in different regions or countries, for instance. Hence, consumer got involved in the creative process twice. First, they created their very own customized shoe model. Second, their preferences gathered during that promotion were consolidated and filtered in order to create new styles for the masses. Obviously, Nike's approach of letting consumers co-design individual products cannot be afford by a lot of brands or retailers as the whole processes become very cost-intensive (Fig. 6.3).

However, the dynamics of the co-creation process itself remain relatively stable. This means that co-creation eventually represents an option for every retailer, depending on the degree of integration of the approach. Companies can profit immensely from engaging in the co-creation process. They can enlarge their knowledge of their consumer base using the internet as source for direct appeal and interaction. Due to Web 2.0 technologies, consumers/users can participate in competitions like style contests. Winning outfits are prized and validated, so that consumers have experiences of both individual and collective emotional value.

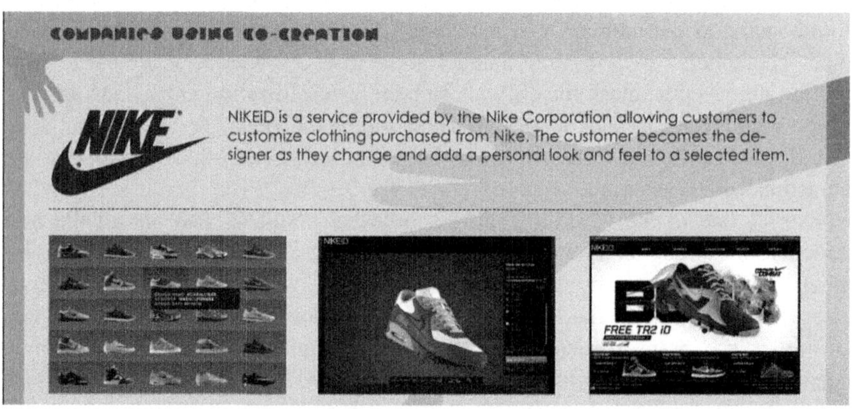

Fig. 6.3 NikeID—the online co-creation. Adapted from Co-creation (2016)

Apart from that, Web 2.0 enables consumers to create outfit ideas and their very own color and style combinations that can be tracked by brands and retailers. Hence, trend spotting and forecast can get easier for companies if they use these information in the right way.

Finally, the general approach and aim should not only be the integration of consumers into existing business models, but also the creation of experiences of emotional value. In order to turn consumers into loyal and returning purchasers, experiences of emotional value need to go beyond the purchase act as a point of exchange of financial sources and goods or services. If fashion brands want to create co-dependent relationships between themselves and their customers, they cannot only focus on products and product innovations, but also have to provide added values, that is, what personalizes brand and product experiences for consumers and what effectively turns them into contributors in the value creation chain.

6.2.3 The Decline of the Prosumer and Rise of Smart Prosuming Machines

It is mainly the internet providing space for the new world of prosumption. It made prosumption more common and the norm. Furthermore, it made new forms of prosumption possible. The website Wattpad, for instance, provides users the opportunity to directly insert public commentaries and suggestions into the serialized stories (Wattpad 2016).

However, even in its current and therefore newer forms, prosumption is very common in the material world. Automated teller machines (ATMs), for instance, have transformed consumers into prosumers. Individuals using ATMs for certain banking services prefer to interact with the machine that produce what they need, instead of relying on a paid employee at the counter.

Therefore, even the newest forms of prosumption are undergoing dramatic changes. One of the most important changes in this context is the beginning of the replacement of the prosumer by smart machines. The prosumer is more and more assisted by supportive tools and will be ultimately replaced by those smart machines someday. The prosumption process is increasingly automated, so that the prosumer himself lose more and more of his importance. A demise of the human prosumer is also possible, as for example, diabetics will be provided with "a one-device-does-all" that performs the pricking, measuring, and dosing automatically. Ironically, the decline of human prosumers has started just as companies are beginning to know and understand them. Smart machines are affecting prosumption, and may someday come to dominate the whole process. They may push human prosumers to the secondary status, while dominating the concept. Someday the prosumer may even be eliminated completely.

However, prosumption has long been automated to some extent and has been involved with smart machines in parts. For instance, after the human prosumer has set a smart machine, such as an ATM (that was invented in the early 1960s) in motion, the machine proceeds anything else automatically by itself in the background. Apart from that, there are many other types of prosumption, where the prosumer is eventually only the agent, who set an automated prosumption process in motion.

Due to an increasing number of instances in which the process occurs unknown by the involved prosumer, the discussion of the new world of prosumption is of growing importance (Ritzer 2015), that is, supported by companies that register people's/prosumers' keystrokes (Zwick and Knott 2009). Users produce keystrokes by searching for products or creating social media profiles, for instance. However, once entered, the keystroke is registered and archived. The user/prosumer only provided the impulse for millions of data bases to absorb the information. Thus, a series of automated processes was set in motion by the unknowing prosumer.

Another form of instances in which the agent providing the impulse neither knows about the process, nor takes any clear action in order to set it in motion, appeared through the invention of the smartphone. Smartphones disclose anonymized data of users' physical location. The user is only carrying the phone without providing any information by himself. However, the sires of recent social changes give even greater power to smart machines. Those smart machines have certain implications for the prosumer. Of course, these machines serve the purpose of supporting the prosumer and automatizing the process of prosumption. This goes beyond the examples of automated processes initiated intentionally or unintentionally by the prosumer. In fact, there is a variety of technological changes that lead to a process of prosumption initiated and controlled by non-human technologies. Therefore, the focus of the prosumption process and the study of it turns from the human prosumer to smart machines. One subset of those smart machines is the ones related to prosumption, the so-called smart prosuming machines. Existing smart prosuming machines operate on their own already, but still need to be set in motion by an agent. Apart from them, there are also smart prosuming machines working on their own without any human impact. As scholars as well as companies just begun to get aware of and understand the importance of human prosumer, this new development shows that they are more and more controlled or even replaced by smart prosuming machines (Ritzer 2015). The discussion of smart prosuming machines can be linked with the distinctions made between human and non-human technologies in the McDonaldization of Society (Ritzer 2011). People control human technologies, as they control the hammer while hammering. The hammer does what the user wants him to do. The user's motive for hammering is to get the nail in the wall, which represents the production part of the action. Apart from that, the hammering process also includes a consumption action, as the user has to purchase nails and the hammer itself. However, it is not the use of human technologies, but the use of non-human technologies that is of growing importance. Smart machines clearly belong to this category, as the major part of control is by the machine itself. Automated deep fryer in fast food restaurants, for instance, is also set in motion by a worker, but proceeding the rest mainly by itself. The machine lowers the fries into the oil, decides when the fries are done and lifts them out of the oil again. It is comparable to the ATM prosuming process.

There are a number of examples of these new smart machines, like those wearable technologies with sensors, monitoring various dimensions of a person's health. Furthermore, there are automobiles that park and will soon drive themselves. Technologies are moving from hybrid modes that involve human and non-human inputs to technologies that are working totally independent and autonomous without the need of human involvement (Ritzer 2015).

The discussion of smart (prosuming) machines goes even further, claiming that this development has to be seen in the context of ambient production. Rey (2011) discussed the ambient production as a byproduct that simply occurs as a result of one's presence. Hence, previously discussed smart machines involve ambient production as opposed to active production (Ritzer 2015).

However, to come back to the early stage of smart prosuming machines, another leading sportswear brand shows that the integration of those machines has begun within the fashion industry as well.

The German sportswear brand Adidas aims to open an automated shoe factory in Germany within this year. It would be the first fully automated shoe factory, which is part of an effort to bring manufacturing back closer to its consumers in more affluent countries. The sporting goods manufacturer signed an agreement to obtain technology from German engineering group Manz. This technology will enable the company to design and make custom-tailored shoe components in a new type of automated plant that is called Speedfactory. The fact that Adidas has been working with the German government, academics, and robotics firms underlines the project's importance. Especially, the company itself hopes that the new technology will trigger a significant shift in the footwear industry as the move to a production in Asia was led by its competitor Nike decades ago. The company wants to speed up delivery times to fashion-conscious customers and reduce shipping costs. The key to moving footwear manufacturing closer to Western markets are technologies that minimize the need for workers to assembly certain components of a shoe.

As part of that initiative, Adidas also invented a 3D printed running shoe sole that can be tailored to a person's foot. The 3D concept is part of the 'Futurecraft series'. This forward-looking initiative places open source collaboration and craftsmanship in the center of the design process in order to drive innovation across all elements of production. Adidas creates a unique combination of material and process that provides the ultimate personalized experience for all athletes. The production breakthrough will take the running shoe standard to the next level, offering individualized support, to enable athletes to perform at their best. The company's ambition is that customers walk into an Adidas store, running briefly on a treadmill and instantly getting a 3D-printed running. The linkage of existing data sourcing and footscan technologies opens unique opportunities for immediate in-store fittings.

"Futurecraft is our sandbox. It is how we challenge ourselves every day to explore the boundaries of our craft. Driving material and process innovation, bringing the familiar into the future. Marrying the qualities of handcrafting and prototyping with the limitless potential of new manufacturing technologies. Futurecraft is stripped back—fast, raw, and real—it is our approach to design". *Paul Gaudio, Creative Director, adidas.*

Both approaches serve the purpose to engage closely with customers and to turn them into co-creating prosumers, who take actively part in the design process. However, referring to the concept of smart prosuming machines, the consumer is exploited as agent, who set the machine/process in motion.

6.3 Conclusion

When fashion is seen as a system, professional actors such as designers and fashion journalists have been argued to maintain a key position. In the context of a concept like this, the role of the consumer has been integrated by arguing that fashion is a social process that creates value when can be achieved when requirements between actors are matching.

Among other dimensions of the concept of prosumption, it has been illustrated how consumers, by engaging in the practice of blogging, add new aspects to the understanding of fashion as a social process. Due to new technologies consumers were enabled to create (new) distribution channels for communicating fashion information and fashion advice, that is, to a great extent, competing with traditional fashion journalism. Hence, they not only take part in the fashion business but also become increasingly influential in terms of how ideas of fashionability are spread within the fashion industry. With regards to the concept of prosumers presented by Toffler in 1980 already, fashion blogger transformed from consumers to prosumers. The main prosumption activities consist of independently producing unique combinations of apparel, shoes, and accessories that are not only consumed by the bloggers themselves, but also documented for a widespread readership—so-called followers.

However, due to new technologies, the public posting of blog posts has enabled the spread of this form of documentation and has thus enabled it to become a part of the fashion system. Due to the fact that bloggers highly compete with the origin fashion journalism, it underlines that prosumption can assume large proportions and lead to great changes within the industry. However, by engaging in the different social practices that have emerged within the sphere of the fashion blogging and that surround blogging, fashion bloggers produce fashion advice for the consumers. This in turn is a process, taking place in the borderland between the fashion consumer and the fashion producer.

Fashion bloggers have a high degree of stylistic diversity. By presenting their ideas to a large readership and incorporating the tastes and concerns of the different social groups, they connect with their audiences. Hence, users/followers become passively integrated into the process of producing blog content. The fashion system has been transformed in such a manner that the concept of consumer fashion was pushed forwards in the direction of prosumer fashion. In contrast to consumer fashion, the shift towards prosumer fashion has been enabled and facilitated, by the adoption of technologies enabling the consumer to engage in the creation of distribution channels of the fashion industry. By engaging in the practice of blogging, consumers have been given the opportunity to participate in prosumption activities already. An increasing number of bloggers have also succeeded in generating large readerships. In doing so, they become, as already mentioned above, competitors of traditional fashion journalism. Besides the similarities to consumer fashion, the adoption of new technology has enabled actors who started out as consumers to further entering the production process of the fashion business. But even more important is that they have become incorporated into a value network. Hence, they have attracted the attention of commercial actors and traditionally influential actors within the fashion industry. The new ways of applying technologies have enabled them to become influential. In this context, the prosumers' engagement in the practice of blogging becomes a distribution channel. Due to this, new marketing strategies have emerged, which are not only influencing the consumer, but also and predominantly the traditional influential actors within the fashion industry. In order to understand the rise of prosumer fashion and use this new trend efficiently, the groups of actors that have played important roles in the process need to be included in new business strategies. Blog platforms and advertising companies were the first ones, understanding how the access to the practice of blogging could be facilitated and due to this, how commercial values could be generated. The invention of technical platforms enabling blogging is one of the main reasons why an organizational field centered on the practice of blogging appeared. Hence, the organizational field where these actors operate is highly characterized by its technical environment. Due to their key processes consisting of the controlling of technical systems, this is especially the case in relation to blog platforms. Hence, also advertising companies repositioned themselves in order to focus their activities on the practice of blogging. Here, the key of efficient repositioning is to understand how technical scripts can be constructed in order to integrate advertising into the technical system of the blog platforms.

However, the social practices that have emerged within the blogosphere have been the result the interactions of the users themselves, namely the fashion bloggers. Here, one important aspect needs to be highlighted. While the technical systems of the blogosphere have been create by entrepreneurs within the IT sector, the practices of the fashion blogosphere itself have been created, developed, and pushed to a high degree by the users of the technology. Hence, the main value-creating process of this field can be found through these social practices. In other words, by engaging in prosumption activities, the users of the technology enable the rise of an organizational field. Thus, without those prosumption activities, the field would not have been able to be created.

In the context of the fashion blogosphere, understanding of how the organizational field has emerged, and how traditionally influential actors within the fashion system have approached it in order to take part in the creation of a value network for the purposes of brand and market communication, has enabled an explanation of how value is created and appropriated.

By outlining in what ways the market has changed, the focus on the Internet is a major challenging force. Due to Web 2.0, consumers are enabled to almost immediate and direct interact with other users across the globe. Consumers have been turned into empowered individuals who increasingly strive for having their say in how brands and products should look like and be dealt with in the market. The process of co-creation underlies the idea to strike consumers into a dialog in order to get first hand access to their individual preferences and requirements which should lead to risk reduction and more transparency. How successful the integration of consumers into existing company structures can look like is illustrated by examples like Adidas and Nike. Co-creation has to be adapted to the general demands and dynamics of the market. Compared to other industries, fashion retail is market by fast-paced trends. Hence, there is little space to directly integrate consumer demands into existing structures because of too short cooling times between identification and realization. Therefore, companies have to create emotional value which extends the point of exchange away from one-directional purchase experiences towards individual experiences with brands and products, rather than directly integrate consumer input into fashion clothing. By addressing consumers as individually as possible, a different route towards consumer satisfaction has been tried to create, where experiences with fashion brands mean identification with lifestyle concepts. However, again companies such as Adidas and Nike try to integrate the consumer in the business process actively. In this context the Internet is used as a potent channel for consumer interaction. It has been noted that online environments are not a threat to the business, but can be seen as creative hubs and key sources of consumer interaction.

However, ironically the decline of human prosumers has started just as scholars and companies are beginning to know and understand them. Two claims exist in the context of Third Wave DIY: on the one hand it is revolutionary, and on the other hand could it be carried out by anybody anywhere. The claim that it is revolutionary has been investigated through the analysis of subsistence DIY (First Wave) and industrial DIY (Second Wave), which follows the ideas of Toffler of the Three Waves of prosumption. However, Third Wave DIY is not only revolutionary for prosumption, but also for innovation, and for entrepreneurship. Apart from that, the claim that Third Wave DIY could be carried out by anybody at any location is evaluated by the potential of Third Wave DIY in the context of local populations' lack of functional literacy, computer skills, and access to industrial manufacturing infrastructure. This is necessary, in order to determine how useful Third Wave DIY technologies are and whether they are appropriate technologies for regions where Third Wave DIY has its origins. Although Third Wave DIY may be increasingly possible through technical advances but it may never be fully realizable without high financial investments. Even if in many parts of the world, prosumption of basic goods may be more important compared to innovation of new sophisticated goods, the reality in fashion retail is different. Consumers are increasingly looking for new experiences. Consumers want to become actively involved and want to be turned into prosumer. This can be enabled by combining Third Wave DIY technologies with mobile production facilities, for instance. More recently, we entered the era of prosumer capitalism in which prosumers become increasingly influential. This is clear in many modern developments in the material world, but it is even clearer in the digital world where, those who consume products on eBay produce the orders for them. It is in prosumer capitalism that there is now a boom in smart prosuming machines that are by replacing the latest concept of (human) prosumers.

Complex interactions have taken place between different groups of actors within the fashion industry. Different social practices have been developed. Companies noted that there is a special need to address individual groups of actors in order to provide new and attractive solutions that lead to an improvement of the business. The concept of prosumption has not yet come to a final "solution". As the industries, the concept is adopted to, are progressive, rather than being static, the development of the prosumer concept will continue and companies will find further ways to catch the consumers'/prosumers' attention and to involve them even more in design and production processes. Apart from involving the human prosumer, the paper has also illustrated the occurrence of smart prosuming machines. The industry has to find a way to efficiently combine both tendencies of the concept, in order to improve the business in a long-term run.

References

3D Fashion D (2016) Retrieved from 8 Jan 2016. http://cubify.com/store/fashion

Ananian CS, Ball CJ, Stone M (2012) Growing up with Nell: a narrative interface for literacy. In Proceedings of the 11th international conference on interaction design and children (S. 228–231). ACM, New York. doi:10.1145/2307096.2307132

Auh S, Bell SJ, McLeod CS, Shih E (2007) Co-production and customer loyalty in financial services. J Retail 83(3):359–370. doi:10.1016/j.jretai.2007.03.001

Babin BJ, Darden WR, Griffin M (1994) Work and/or fun: measuring he-donic and utilitarian shopping value. J Consum Res 20(4):644–656

Bandulet M, Morasch K (2005) Would you like to be a prosumer? Information revelation, personalization and price discrimination in electronic markets. Int J Econ Bus 12(2):251–271. doi:10.1080/13571510500128038

Benkler Y (2006) The wealth of networks: how social production transforms markets and freedom. New Haven, Yale University Press

Benkler Y, Nissenbaum H (2006) Commons-based peer production and virtue. J Political Philos 14:394–419

Betabrand—Crowdfunded Clothing (2016) Retrieved from 8 Jan 2016. https://www.betabrand.com/

Bhojani RD, Dwivedi VV (2012) A novel approach towards video com-pression for mobile internet using transform domain technique. Int J Comput Appl 58(10):29–33. doi:10.5120/9320-3552

Bird SE (2011) Are we all produsers now? Cult Stud 20:502–516

Bruns A (2005) Gatewatching: collaborative online news production. New York, Peter Lang

Bruns A (2008) Blogs, wikipedia, second life and beyond: from production to produsage. New York, Peter Lang

Campbell C (2005) The craft consumer: culture, craft and consumption in a postmodern society. J Consum Cult 5:23–42

Childers TL, Carr CL, Peck J, Carson S (2001) Hedonic and utilitarian moti-vations for online retail shopping behavior. J Retail 77(4):511
Co-Creation (2016) Retrieved from 9 Jan 2016. http://prosumer.simdo.nl/wp-content/uploads/2015/11/Co-Creation.jpg
Cova B, Dalli D, Zwick D (2011) Critical perspectives on consumers' role as producers: broadening the debate on value co-creation in marketing processes. Mark Theory 11:231–241
Crowdcube (2016). Retrieved 8 January 2016, from https://www.crowdcube.com
Fox S (2013) Paradigm shift: do-it-yourself (DIY) invention and production of physical goods for use or sale. J Manuf Technol Manag 24(2):218–234. doi:10.1108/17410381311292313
Fox S (2014) Third wave do-it-yourself (DIY): potential for prosumption, innovation, and entrepreneurship by local populations in regions without industrial manufacturing infrastructure. Technol Soc 39:18–30. doi:10.1016/j.techsoc.2014.07.001
Holbrook MB, Hirschman EC (1982) The experiential aspects of consumption: consumer fantasies, feelings, and fun. J Consum Res 9(2):132–140
Horn TJ, Harrysson OLA (2012) Overview of current additive manufacturing technologies and selected applications. Sci Prog 95(3):255–282. doi:10.3184/003685012X13420984463047
Howe J (2009) Crowdsourcing: why the power of the crowd is driving the future of business. New York, Crown Business
Humphreys A, Grayson K (2008) The intersecting roles of consumer and producer: a critical perspective on coproduction, cocreation and prosumption. Sociol Compass 2(3):963–980. doi:10.1111/j.1751-9020.2008.00112.x
Instructables (2016). Retrieved 16 September 2016, from http://www.instructables.com/
Jackson T, Shaw D (2000) Mastering fashion buying and merchandising management. Palgrave Macmillan, London
Kickstarter (2016) Retrieved from 8 Jan 2016. https://www.kickstarter.com/
Kotler P (1986) The prosumer movement : a new challenge for marketers. In: Advances in consumer research. http://www.acrwebsite.org/search/view-conference-proceedings.aspx?Id=6542
Leadbetter C, Miller P (2004) The Pro-Am revolution. London: Demos
Maltzahn CV (2015) Co-creating individuals: a roadmap to value creation in fashion retailing online investing, equity crowdfunding, business finance: Crowdcube. (2016). Retrieved from 8 Jan 2016. https://www.crowdcube.com/
OpenMaterials (2016). http://openmaterials.org/
Pascu C, Osimo D, Turlea G, Ulbrich M, Punie Y, Burgelman JC (2008) Social computing: implications for the EU innovation landscape. Foresight 10(1):37–52. http://doi.org/10.1108/14636680810856017
Prahalad CK, Ramaswamy V (2004a) The future of competition: co-creating unique value with customers. Cambridge, MA, Harvard Business School Press
Prahalad CK, Ramaswamy V (2004b) Co-creation experiences: the next practice in value creation. J Interact Mark 18:5–14
Ramirez R (1999) Value co-production: intellectual origins and implications for practice and research. Strateg Manag J 20(1):49–65
Rey P (2011) Incidental productivity: value and social media—cyborgology. Retrived from http://thesocietypages.org/cyborgology/2011/11/17/incidental-productivity-value-and-social-media/
Ribiere VM, Tuggle FD (2010) Fostering innovation with KM 2.0
Ritzer G (1993) The McDonaldization of society. Thousand Oaks, CA, Pine Forge Press
Ritzer G (2011) The McDonaldization of society 6. Pine Forge Press, Thousand Oaks
Ritzer G (2014) Prosumption: evolution, revolution, or eternal return of the same? J Consum Cult 14(1):3–24. doi:10.1177/1469540513509641
Ritzer G (2015) Automating prosumption: the decline of the prosumer and the rise of the prosuming machines. J Consum Cult 15(3):407–424. doi:10.1177/1469540514553717
Ritzer G, Jurgenson N (2010) Production, consumption, prosumption the nature of capitalism in the age of the digital 'prosumer'. J Consum Cult 10(1):13–36. doi:10.1177/1469540509354673

Solomon MR, Rabolt NJ (2004) Consumer behavior in fashion. Prentice Hall, Upper Saddle River
Song H, Guimbretière F (2009) The ModelCraft framework: capturing free-hand annotations and edits to facilitate the 3D model design process using a digital pen. In: Proceedings of TEI'10 (S. 121–128). ACM
Stock WG (2007) Folksonomies and science communication: a mash-up of professional science databases and Web 2.0 services. Inf Serv Use 27(3):97–103
Tapscott D (1997) The digital economy: promise and peril in the age of networked intelligence, New edn. Mcgraw-Hill Professional, New York
Tapscott D, Williams AD (2006) Wikinomics: how mass collaboration changes everything. New York, Portfolio
Tinkercad—Create 3D digital designs with online CAD (2016) Retrieved 8 Jan 2016, von https://www.tinkercad.com/
Toffler A (1980) The third wave. Morrow, New York
Vargo SL, Lusch RF (2004) Evolving to a new logic for marketing. J Mark 68:1–17
Vargo SL, Lusch RF (2008) Service-dominant logic: continuing the evolution. J Acad Mark Sci 31:1–10
Waters T (2006) The persistence of subsistence agriculture: life beneath the level of the marketplace. Lexington Books, Lanham
Watson M, Shove E (2008) Product, competence, project and practice: DIY and the dynamics of craft consumption. J Consum Cult 8(1):69–89
Wattpad (2016) Wattpad—Geschichten, die du lieben wirst. Retrieved 9 Jan 2016, von https://www.wattpad.com/
Woodcock J (2006) Home truths on DIY then and now. Post Magazine, Yorkshire
Williams CC (2004) A lifestyle choice? Evaluating the motives of do-it-yourself (DIY) consumers. Int J Retail Distrib Manag 32(5):270–278. doi:10.1108/09590550410534613
Workman JE (2009) Fashion consumer groups, gender, and need for touch. Cloth Text Res J. doi:10.1177/0887302X09356323
Zwick D, Knott JD (2009) Manufacturing customers: the database as new means of production.

Chapter 7
The Second Hand Market for Fashion Products

Jochen Strähle and Linda Maria Klatt

Abstract The second hand concept indicates a growing trend in clothing recently, leading to growing numbers of second hand shops and developments of new second hand retail forms. This paper concentrates on the current second hand market for fashion products and presents the different motives toward second hand consumption as well as alternative consumption channels for second hand products. The findings of the paper are founded on literature research of academic articles and case studies. Results show that there is a high potential for the second hand market due to the increasing interest of consumers in buying second hand products. The paper concentrates on the second hand market for fashion product in the western society. This means that there was no research on second hand products for disadvantaged people in poor countries. Furthermore, the paper focuses the formal second hand retail channels to see what is already on the market.

Keywords Second hand market · Consumption behaviour · Sustainability · Distribution channels · Second hand fashion products

7.1 Introduction

7.1.1 Background

While the focus in the recent years was most on fast fashion, there is currently a change in consumers' consumption behaviour. The sustainability is a growing topic for consumers as well as for companies. The awareness of responsibility gives the consumers a reason for thinking about their consumption behaviour, which becomes an unsustainable and wasteful one in the last few years. The main trigger

---

J. Strähle (✉) · L.M. Klatt
School of Textiles and Design, Reutlingen University, Reutlingen, Germany
e-mail: jochen.straehle@reutlingen-university.de

L.M. Klatt
e-mail: Lindamariaklatt@gmail.com

J. Strähle (ed.), *Green Fashion Retail*, Springer Series in Fashion Business,
DOI 10.1007/978-981-10-2440-5_7

for the current over-consumption is an effect of the fast fashion trend (Dobscha 1998). Fast fashion induces consumers to buy "more clothes more frequently" what results into increasing disposal of garments (Joy et al. 2012). As a consequence companies need new strategies to fulfil the expecting corporate social responsibility. Most of the companies already improve their production and tried to make it more sustainable (Csanák 2014). Others set the focus on their working conditions and create a more sustainable working relationship (Vilimaviciute 2015). While the mentioned efforts focus on the upstream aspect of the value chain this paper analyses the downstream value chain-related issues. H&M is one example for a company that put effort into the named issue. H&M made a garment collection program to repurpose the unwanted garment. They request on their website the people to bring their unwanted clothes to the local shops; it does not matter which colour, brand, size or age. Depending on the condition of the goods H&M presents three possible procedures to repurpose the garments:

- Rewear—clothing that can be worn again will be sold as second hand clothes
- Reuse—old clothes and textiles will be turned into other products, such as cleaning cloths
- Recycle—everything else is turned into textile fibres, or other use such as insulation (H&M 2016)

Besides H&M there is another fashion brand that developed a sustainable program to reuse their products. Filippa K, "a high-quality, leading Swedish fashion brand" (Kant Hvass 2015, p. 14), opened up their own second hand shop where customer can bring back their worn and no longer wanted clothes, shoes and accessories for resell (Kant Hvass 2015). The business model of Filippa K is a new innovation towards sustainability. With a view of this business model idea the purpose of the paper is to offer a better understanding of the current demand on the second hand market for fashion products and to determine the different consumer motivations for buying second hand products. What are the consumer expectations of second hand shops and which kind of second hand channels can be found at the second hand market for fashion products?

7.1.2 *The Current Situation on the Second Hand Market*

In the last few years, there has been a change in purchase intention of second hand products. The second hand market was an important consumption source in Europe and North America until mass production and growing prosperity in the nineteenth century made buying new products possible for the rank and file. After that the second hand clothing became an export-oriented good until the last years (Jenss 2016). Whereas in the passed years the second hand market in the West was only an insignificant form of commerce which took place at flea markets, second hand markets and antique markets (Guiot and Roux 2010), it achieved special interest in the meantime: "[…] used clothing has drawn a bigger spotlight than ever" (Jenss 2016, p. 234).

The change in consumer behaviour comes from sources like for example ecological, economical, ethical and hedonic ones (Han 2013). Williams and Paddock (2003) mentioned the decreasing purchasing power of the middle classes since 1980s as a motive for them buying more second hand products. Iverson (2010) supported this aspect by naming the recession as a factor for the change in patterns of consumption and therefore a growing in the popularity of second hand market. Besides the economical aspect that changed the consumer patterns of consumption, there is also a more emotional one. The term second hand is in French *l'occasion* which comes from the Latin word *occasio* and describes a "stroke of luck" (Roux and Guiot 2008). Furthermore, it describes a "transaction that is advantageous to the buyer, as well as the object of this transaction, whose price is attractive due to the depreciation associated with its use or previous ownership" (Roux and Guiot 2008). Due to the bargain hunting or the pleasure of making a find or discovery the consumers of second hand products perceive a positive shopping experience. It needs to be stated that the economical and the recreational motives are linked to each other (Guiot and Roux 2010).

7.1.3 Definition of Second Hand Products

Second hand products are often thought of as being synonymous with vintage products. However, it is important to understand the difference between the terms. Vintage products are defined as previously owned, but not necessarily used items. Furthermore, vintage products can also be defined as products that are produced between the 1920s and the early 1980s (Cornett 2010). The value of the vintage product depends on the age, era and condition of the goods (Cervellon et al. 2012). Compared to second hand products the similarity is that both items are previously owned. The difference is that second hand items are "any piece of clothing which has been used before, notwithstanding the age of the clothes" (Cervellon et al. 2012). The financial value of a second hand clothing item is always lower than the original new product (Turunen and Leipämaa-Leskinen 2015). Within this paper the focus is on second hand products. For a better understanding of the following paper the term second hand market is to be defined according to Han (2013), as the market including all consumer durables given away, sold or swapped with or without intermediary or third party, after disposal by a household (Han 2013).

7.2 Consumer of Second Hand Products

7.2.1 Change in Patterns of Consumption

The consumer of today is strongly influenced by the fast fashion concept, which popped up few years ago. With frequently changing collections and the cheap prices of items retail brands like and H&M and Zara influence consumers heavily and generate a permanent drive to buy new clothes. In fast fashion the consumers' role is of fundamental meaning. The fast fashion model presents consumers with frequently renewed offers so they can make purchases when they require (Gabrielli et al. 2013). Furthermore, there is an emergence and rapid dissemination of the fast fashion phenomenon. It is more frequently recognized to socio-cultural changes in the daily behaviour of consumer. The consumer of today is constantly well informed about the latest fashion trends and feels need to adapt to the reality around him in an reasonable and dynamic manner (Cachon and Swinney 2011), which means that consumers nowadays are more demanding and fashion conscious. This phenomenon forces fashion retailer to provide the right product at the right time in the market, that is, the demand of fast fashion has been increased too (The future of fast fashion 2005).

Wang et al. (2010) found out that the most influential factor, which makes consumers choose fast fashion, is that the products are cheap and also trendiness is a key criterion. Another important factor is that even when the products are cheap it does not feel cheap. Especially, Zara stores are large, swish and centrally located. Large single-brand stores are positioned in the best locations in city centres. The distribution format of fast fashion is a persuasive aspect of fast fashion and can be more important than the product itself. By means of display solutions they attract the consumers. Many fast fashion consumers have impulsive buying tendencies. To avoid gilt they prefer buying inexpensive goods (Watson and Yan 2013). Clothes are in general hedonic products which are items purchased and consumed for their ability to provide positive feelings or pleasure rather than their utilitarian value. The hedonic shopping experience in fast fashion is derived from the excitement felt from the clothing and the store atmosphere (Watson and Yan 2013).

In fast fashion stores new products arrive and are replaced by the next collection quickly, so most consumers do not have the chance to see a product twice, which means that every time when they visit the store, they are attached by new clothes. Retailers limit product offerings to ensure short renewal cycles and make constant room for new merchandise. They create urgency in consumers' mind and limit consumers' freedom to delay decisions. Consumers do not have time to form any clear expect about the product and to avoid any regret they will buy it immediately. Thereby fast fashion retailers do not need to amortize their products during the season or resign it completely. In general, consumers look forward that the product will eventually be reduced in price. But in fast fashion they will not delay the purchase because of the danger that the product will be out of stock. Because of the quick turn of products in the store, the fresh new fashion stimulates the consumer's nerves every time again when they visit the store and arouse their shopping desires. This can raise the motivation of the personal relevance of a product in a particular context. This short renewal cycle does not only increase expectations for new products but also encourages frequent store visits (Byun and Sternquist 2011, p. 192). Furthermore, fast fashion consumers speak about time being short and the limited availability as a benefit. They benefit from the rarity principle, in which the company reduces the likelihood of others copying individual's created style by the shortness of supply. So the uniqueness is preserved. This rarity principle creates for customers a sense of urgency in the purchasing process and enjoyment in the post-purchase consumption process (wearing of the clothing) (Miller 2013, pp. 168–170). People who are buying fast fashion find generally pleasure in shopping and in creating different and new looks with their pieces. These consumers are mostly hedonic and emotional and enjoy the creativity and uniqueness of fast fashion.

Moreover, consumers of fast fashion do not expect purchases to be long lasting, and they expect their clothing to fall apart. People accept lower quality because they purchase unusual garments that they do not intend to wear very often. Consumer's attention is moving away from the intrinsic quality of a product to the experience and emotional charge of it (Gabrielli et al. 2013). Therefore, they are not dissatisfied when an item becomes damaged. However, Watson and Yan (2013) found out that fast fashion consumers are satisfied with the purchase of the clothing but not with the consumption. They are dissatisfied with the damageability even when it is fulfilling their expectations. Furthermore, because the clothing damages faster as a result of material and construction quality fast fashion consumers divest more. Fast fashion consumers divest their clothes based on damage or boredom (Watson and Yan 2013). They are bored because the very trendy clothes go out of style more quickly. Thereby fast fashion consumers frequently throw their clothing away. Morgan and Birtwistle (2009) found out that fast fashion encourages a throwaway culture, and furthermore it comes clear that fast fashion consumers are different in regards to resources.

According to a study conducted by IPSOS in 2014 the interest in second hand consumption is increasing (Fitzgerald 2015). More and more consumers like the idea of buying second hand items. The survey also shows that this change happened in the last 5 years and is still progressive (Fitzgerald 2015). The consumers on the British market show the highest interest in second hand items followed by Germany, USA and France (Fitzgerald 2015). Preuss (2015) justifies the change in patterns of consumption on the one hand with the consumer growing awareness of quality and on the other hand with the increasing demand for "smart buys". 55 % of the surveyed people named the good price-performance ratio as a reason for buying second hand products (Preuss 2015). For some reasons the consumer of today expects more for his spending money than ever before (Preuss 2015). This could be a first step away from the throwaway culture that came up with the fast fashion trend. The following section is going deeper into the motivations of second hand consumption and should show the main reasons for buying second hand items.

7.2.2 Consumer Buying Motivations for Second Hand Consumption

7.2.2.1 Economical Motivations

"Low prices have been a major reason for consumers to shop for second-hand clothing" (Xu et al. 2014, p. 670). The cause is that economical reasons, what is already explained in the beginning, were the trigger for the disposal beginning of second hand items. However, there are different aspects that are linked to economical motivations.

Fair Price

Consumers who buy in second hand shops on the basis of paying fair prices do not accept spending more money for products just because their new (Roux and Guiot 2008). When they buy second hand products they have the feeling of paying the right price (Roux and Guiot 2008). Compared to conventional stores where the prices of the products are higher then the actual production value (because of the brand value for example), the prices for second hand products are often base on their value of quality.

Frugality

Frugality is a characteristic of a special type of consumer behaviour. "Frugality is a one-dimensional consumer lifestyle trait characterized by the degree to which consumers are both restrained in acquiring and in resourcefully using economic goods and services to achieve longer-term goals" (Lastovicka et al. 1999, p. 88). People with a frugality lifestyle make smart choices, reuse their resources and spend their money carefully. The frugality aspect is therefore not only an economical motive but also an ecological and critical one (Roux and Guiot 2008).

7.2.2.2 Hedonic and Recreational Shopping Motivations

The entertainment in retailing has become an important role in the western society. Nowadays, the consumers perceived besides broad assortments, low pricing and extended store hours a pleasurable and entertaining shopping experience (Han 2013). Guiot and Roux (2010) found out that hedonic and recreational shopping motivations such as nostalgic feelings, need for uniqueness, treasure hunting and social interaction play an important role for consumers of second hand products. Browsing behaviours often cause the positive hedonic shopping motivation, which means to stroll through the second hand shops with no real intention of buying (Guiot and Roux 2010).

Nostalgic Pleasure

Nostalgic feelings arise mainly during shopping vintage items. Vintage products call back positive memories from the era they were manufactured. Especially consumers who lived during this era have greater intention to purchase vintage clothes (Cervellon et al. 2012). Nevertheless not only vintage products can cause nostalgic feelings. The nostalgic motivation consumers of second hand products have come from three different ways of thinking:

- I'm more attracted to old things than new ones. They recall memories, linked to my family and where they lived.
- Unlike new products, second-hand items have a history and that's what I obtain through them. I imagine it I reconstruct it through them.
- Overall, I don't like new products. It's rather sentimental, in fact. I like things that have lived, that have a soul, and you don't find that in new products (Guiot and Roux 2010).

Especially, the last point is a main factor for creating a hedonic shopping experience because it gives the used and old products something special and precious.

Need for Uniqueness

The desire for being individual and unique is a great topic in the contemporary western society. Consumers want to seek "differentness relative to others through the acquisition, utilization and disposition of consumer goods for the purpose of developing and enhancing one's self-image and social image" (Guiot and Roux 2010). The ambition to find unique items let the consumers shop in second hand stores. Compared to conventional stores the assortment includes only unique pieces. Besides the aspect of buying items nobody else owned, customers also try to find rare, unusual and historical items (Guiot and Roux 2010). It is important to mention that the need for uniqueness only exists in cultures "where individualism [is] valued" (Xu et al. 2014, p. 676). Xu et al. (2014) found out that the American consumers have a higher value of need for uniqueness than the Chinese consumers.

Treasure Hunting

Treasure hunting is the desire to find something rare, valuable and original while shopping in second hand shops. It is the hope of "unearthing something or other" (Guiot and Roux 2010). Consumers think of a treasure that is buried under a multitude of products they want to find it (Guiot and Roux 2010). Of course it is an arduous activity and often time-consuming but the chance of success gives the consumer the motivation to wandering and searching (Han 2013). Guiot and Roux (2010) define treasure hunting as "a mediator of the relationship between the traits related to uniqueness and nostalgia and the purchase intention of vintage clothing".

Social Interaction

The social interaction is as well a hedonic and recreational motivation for second hand consumers to visit second hand shops. The social interaction can be divided into two different types. The first type describes "[...] walking around the flea market, meeting people and looking at everyone", which is the pleasure of discovering (Guiot and Roux 2010). The second type is the "chatting with the sellers" and the "pleasure of joking and talking a bit about what they're selling" (Guiot and Roux 2010; Sherry Jr. 1990). Both of them are forms of amusement and more motivations for visiting second hand shops instead of buying second hand products.

7.2.2.3 Critical Motivations

The critical dimension is the answer of the throwaway culture and over-consumption due to the fast fashion model. More and more consumers have an aversion against the classic market system and the current consumption behaviour. Therefore, the upcoming and growing sustainability concept is a main trigger. The critical attitude is reflected in distancing and avoidance behaviours (Guiot and Roux 2010).

Distance from the System

The first aspect of the critical motivation of consumers is the avoidance of conventional channels as well as the desire of distance from the system. People feel like they can escape the consumption system by shopping second hand and distancing themselves from the current consumer society (Guiot and Roux 2010). Consumers buy second hand products to reduce consumption and stop the proliferation of products (McDonald et al. 2006).

Ethics and Ecology

Consumers with ethical and ecological motivations cannot understand why people throw away products that are still in good conditions (Guiot and Roux 2010). The actual way of consumption due to the continuous change of the assortments and the small prices of fashion brands like Zara and H&M results in masses of unused goods. Second hand buying consumers respond to the problem of proliferation by buying the products other consumers no longer wanted. It is a part of their waste reduction and environmental strategies (Bekin et al. 2007). All in all it is a way of taking responsibility for reusing functional products, reducing the depletion of natural resources and avoiding the unnecessary proliferation of products as Han

(2013) states. For young consumers who are not able to spend much money on expensive green fashion products, second hand shopping is a way to act sustainable (Xu et al. 2014).

7.2.3 Barriers of Second Hand Consumption

Besides consumers motivations of second hand consumption there are as well factors, which prevent consumers from buying used products. Not for every customer clothes are seen as products, which easily can change their owners. For some of them "clothing can only be worn by a sole owner, in the same way that food can only be eaten by one mouth" (Roux and Korchia 2006, p. 33). This way of thinking is linked to the idea that clothing is an extension of the body. These concerns over wearing used products include also the worries about contamination and germs (Xu et al. 2014). Another concern besides the hygienic aspect is fear of the unknown, which means people do not want to buy used products because by which they do not know the previous owner (Roux and Korchia 2006). Specifically thoughts about "disease, death, and social misfortune" (Roux and Korchia 2006, p. 33) are associated with second hand products. A third concern along with hygienic and fear of the unknown is the need for status. It is negative related to the purchase of second hand items because the people around could think that second hand buyers cannot afford new clothes (Roux and Korchia 2006).

7.3 Provider on the Second Hand Market

7.3.1 Second Hand Retail Channels

There are a variety of different providers on the second hand market, including "vintage shops/boutiques, consignment shops, thrift/charity shops, online auction sites, flea markets, antique fairs, swap meets, garage sales, car-boot sales, and classified ads" (Han 2013). However, studies on the different provider models are still lacking what made it difficult to find something about them. The many provider models are divided into informal and formal ones (Mhango and Niehm 2005). In this paper the focus is set on the formal retail channels in order to find out how they satisfy the current demand. On the formal site there are brick-and-mortal stores like vintage shops or boutiques, consignment shops, thrift stores organized by charity organizations as well as online shops and online auction sites. To get an overview of the different shop concepts they will be described in the following. The focus is set on the market positioning, supply chain and sourcing, customer relationships and interaction with the customers and the unique approach.

7.3.1.1 Vintage Stores

Vintage shops and vintage boutiques are specialist in selling original vintage clothes from the early 1920s until the 1970s. As described in Sect. 7.1.3 Definition of Second Hand Products, the vintage products are already owned before but not necessary used before. As a result of the high interest in vintage clothes, which came up in the last few years, vintage shops are flourishing. The rise in vintage fashion is based on the vintage-oriented collections of different designers, the aversion against fast fashion consumption, the consumer interest in ethical clothing as well as different celebrities wearing vintage, the media for example movies like "The Great Gatsby" or television series like "Mad Men" and last but not least the Internet, which made the access to vintage products possible for a wider audience. The sources of the products are "antique fairs, vintage fairs, swap meets, thrift stores, auctions, and rag houses" (Han 2013). Referring to the look of the vintage items and to the nostalgia, vintage shops often have an old, retro look. By the organization and presentation of the merchandise, vintage shops arranged them mostly according to their "time periods, styles or types, and range from inexpensive shirt to luxury fur coats" (Han 2013). The shops are located in side streets where students meet up and people drink coffee and visit galleries and where rental cost are not that high. However, due to the rise in vintage fashion there are vintage shops that are in the meanwhile located in higher frequented and more central shopping streets. The main target group age is between 20 and 40 (Cassidy and Bennett 2012; Han 2013).

7.3.1.2 Consignment Stores

The second form of formal second hand shops is the consignment shop. Consignment shops act like a mediator between seller and customer (Han 2013). In other words, the consignment shop presents the platform where sellers can bring their unwanted clothes and sell them throughout the shop. When the clothes are not sold during a determined period of time, they are returned to the owner. The owner and the consignment shop divides the sales' profit. On the one hand this cooperation enabled normal people to offer their clothes to consumers without any effort like advertising, display space or storage costs. And on the other hand the consignment shops get goods without paying money for them and without the risk of a full storage. It is important to mention that the goods have to be in a good condition that means without "stains, tears, mothball and smoke odors, animal hairs, missing buttons or broken zippers" and they have to be "current" and "seasonal" (Han 2013). These conditions are important for more security in sale. Due to the non-specialization consignment shops have a wide product range. But there are of course consignment shops with a more specialized assortment, the kind of specialization often influences the location of the store. Han (2013) listed young mothers, seniors, entire families and college students as potential customers for consignment shops.

7.3.1.3 Thrift Stores

Thrift stores represent the third form of second hand retail channels. There are three different types of thrift stores: non-profit, charity-affiliated for profit and for profit thrift shops. The type is depending on their operating system. Non-profit thrift stores are operated by "major national charities [...], religious groups, charities or cultural benefactors" (Han 2013). The goods came from donators and the sales' profit go directly or indirectly to charities. The only requirement towards the clothes is the usability of them. Charities affiliated for profit thrift shops are buying their goods from the charity organizations and sell them for profit. The third form is the for-profit thrift shop, which buy goods from individuals or wholesalers and sell them for profit. The claims toward quality, wear and tear, age and style of the products refer to the form of thrift shop. The main target groups for thrift shops are seniors, entire families and people who need social services (Han 2013).

7.3.1.4 Online Second Hand Shops

On the online marketplace there are two different types of business models. It can distinguish between swap meets and sale marts. Both of them unable a resale of used clothes in most instances between private persons. The online platforms play a role as a mediator between seller and customer. In other words, the online platforms deliver the infrastructure for the resale or change of used goods. This includes for example support in the search-, contact- and the settlement phase.

Swap Meets

Swap meets unable customers to purchase second hand products for free. The function of the platform is to mediate supply and demand between the private persons. The Internet made it possible to exchange second hand products supra-regional and quiet easy (Scholl et al. 2013). Besides the biggest swap meet "Kleiderkreisel", there are others like for example "Swapstyle" and "Mädchenflohmarkt" (Kleiderkreisel 2014).

Sale Marts

Supra-regional online sale and resale platforms enable private persons local independent purchasing of second hand products. The difference to the swap meets is the payment of the goods instead of the exchange. The mediation service is mostly financed by sale provision or registration costs. The online auction site eBay is the most popular one with a variety of providers (Scholl et al. 2013).

7.3.2 *Opportunities of Second Hand Retail Channels*

Based on the corporate responsibility in value chains, fashion brands work on different ideas to improve their sustainability. Most of the efforts relate to the upstream aspect in value chains and less on the downstream opportunities like for example reuse, remanufacturing and end-of-life of garments. Nevertheless, Kant Hvass (2014) presents in his study retail sustainability concepts of fashion brands that based on the downstream idea. The concept is described as a "resell/reuse platform[s] for prolonging the life of garments and thereby capturing the resell value" (Kant Hvass 2015, p. 14). This means that fashion brands invite their customers to donate their used and no longer wanted garments from the brand in the stores, which are resold in own branded second hand shops. Current fashion brands, which already work and developed this idea, are described below.

7.3.2.1 Filippa K

Filippa K is a Swedish premium fashion brand with timeless and high-quality products. The company was founded in 1993 by Filippa Knutsson and Patrik Kihlborg, which had the ambition to build up a sustainable company with a company philosophy relating to style, simplicity and quality. Filippa K made collaboration with an expert of second hand retailing and founded Filippa K second hand. The store has the same characteristics like a consignment store. Filippa K customers can sell their second hand clothes in the Filippa K second hand shop and if the product found another owner they get 50 % of the sales' profit. If this is not the case, the clothes go back to the owner or are donated to a charity. Besides the second hand products, Filippa K offers collection samples (Kant Hvass 2015).

7.3.2.2 Boomerang

Another Swedish premium fashion brand invites their consumers to donate their used clothes in the store. There the garments are resold or upcycled for Boomerang home collection products (Kant Hvass 2014).

7.3.2.3 Patagonia

Patagonia is a brand for high-quality outdoor clothes from the USA. Patagonia offers their customers an in-store product take-back opportunity. Instead of resell them in an own second hand store Patagonia made a resell platform for their customers in collaboration with eBay, where they can buy second hand Patagonia products (Kant Hvass 2014).

The examples show that second hand retail activities enable fashion brands to offer their customers new opportunities to interact with the brand. On the one side customers have the possibility to bring back their used clothes and on the other side customers can buy clothes from the brand, which are cheaper than the new ones. This means for the company first the opportunity to interact with their existing customers and second to enter a new customer segment. Due to the fact that most of the retail brands have no knowledge about the needs of second hand customers or about the second hand market, they need to enter a partnership with an expert from the new segment. Just like Filippa K and Patagonia, both of them collaborate with someone from the second hand market and benefit from the know-how as well as from the resources of them.

7.4 Discussion

Currently, there is a huge makeover in consumer way of thinking towards consumption behaviour and pre-owned fashion. So if there is a change in thinking could it be possible that second hand is the new fast fashion? Can consumers refrain from the small prices and continuous offer of new products? And how can the fashion brands react on this trend? The survey conducted by IPSOS showed that over 60 % of British women surveyed already bought second hand fashion products. So there is a change and if this change will go on it is important to understand the second hand fashion market to satisfy the consumers' demand (Fitzgerald 2015). Furthermore, it is important to know the motivation of second hand consumption. Guiot and Roux (2010) indicate that there are three main motivation aspects which lead to buy second hand products: Economical, hedonic and recreational and critical motivations. These motivations refer first to the characteristics of second hand products and second to the channels that sell them. Due to the fact that the three motivation dimensions "are extensively interwoven" (Guiot and Roux 2010), all of them play an important role for the investigation of consumer behaviour. Greater research into the theme contained the following second hand consumption motivations. Economical-driven motivations are on the one side the expectation of fair prices. This does not necessarily mean cheap prices, but prices that are reasonable. On the other side there is the lifestyle trait frugality that has a positive effect on buying second hand goods (Lastovicka et al. 1999). This special consumer behaviour has positive impact on other motivations lead to second hand consumption. Frugality means that the consumer has the ambition to make smart choices, spend their money carefully and reuse their resources and thus lead to sustainable acting (Roux and Guiot 2008). Compared to the aspects of fast fashion it can be said that the first motivation "need to pay fair prices" also increase the intention to purchase fast fashion products because the price-performance ratio of them is quite good. Nevertheless, it can be said that the lifestyle trait of frugality consumers is not an antecedent to fast fashion purchases. Although the products are cheap, the frugality of consumers expects a sustainable use of resources and the short lifetime of fast fashion products does not fulfil that. The second motivation is the hedonic and recreational one, which includes nostalgic pleasure, the need for uniqueness, treasure hunting and social contact. All of these motivation factors are important for the ambition to buy second hand products. But can the consumer find these factors also in fast fashion products? Facing the nostalgic pleasure, the need for uniqueness as well as the social contact, the answer is negative. Fast fashion products are mass products and therefore sell to a wide audience. The service personal in fast fashion stores is not specialized of customer service, and therefore the social contact and the interaction with people during the shopping are not given. Fast fashion products can as well satisfy treasure hunting with the aim to find something special. The last motivation dimension, which has to be compared to fast fashion motivations, is the critical aspect. Due to the fact that the critical motivations are the response of the throw away culture, it can be said that this dimension enhances the idea that second hand could be the new fast fashion. The logical result of the comparison shows that the interest in second hand purchasing can be a good alternative to fast fashion consumption.

When the patterns of consumption are going to change, retail brands should find new strategies to response the needs. The second hand market offers second hand consumers already different types of retail channels and forms of consumption. What consumers can find actually on the formal second hand market are vintage stores, consignment stores, thrift stores and online second hand shops (Han 2013). Besides the option to buy second hand products, consumer can swap their used products with other consumers. New are the branded second hand shops of retail brands like for example Filippa K and Patagonia (Kant Hvass 2014). This retail concept makes it possible for retail brands to control the second hand market of their products in a way. Another positive affect of the concept is the interacting with the consumers. Retail brands can thereby increase customers' loyalty and tap into new markets. Essential requirements for branded second hand shops are on the one hand products with good quality and long lifetimes. On the other hand retail brands should find a partner that knows the market.

7.5 Conclusion

The second hand market for fashion products gains more importance from the consumers and providers' site. Due to the increasing important issue sustainability and the aversion against fast fashion and over-consumption the interest in second hand products increases. In the last few years, the fashion industry strongly focused the fast fashion market. Changes in the way of thinking of resources and waste had already an impact on consumers' behaviour towards consumption. The rising interest in second hand products has economical, hedonic and recreational, and critical motives. Especially the critical motives, which includes the need of dissociate from the system and ethical and ecological motivations, are main triggers for a change in patterns of consumption. Second hand retail channels gain in importance due to this. Furthermore is increasing interest in second hand products a new opportunity for premium retail brands. Some of them already perceived the chance and work on concepts to enter the second hand market.

References

Bekin C, Carrigan M, Szmigin I (2007) Beyond recycling: 'commons-friendly' waste reduction at new consumption communities. J Cons Behav 6(5):271–286

Byun S-E, Sternquist B (2011) Fast fashion and in-store hoarding the drivers, moderator, and consequences. Cloth Text Res J 29(3):187–201. doi:10.1177/0887302X11411709

Cachon GP, Swinney R (2011) The value of fast fashion: quick response, enhanced design, and strategic consumer behavior. Manag Sci 57(4):778–795. doi:10.1287/mnsc.1100.1303

Cassidy TD, Bennett HR (2012) The rise of vintage fashion and the vintage consumer. Fash Pract J Des Creat Process Fash 4(2):239–262. doi:10.2752/175693812X13403765252424

Cervellon M, Carey L, Harms T (2012) Something old, something used: de-terminants of women's purchase of vintage fashion vs second-hand fashion. Int J Retail Distrib Manag 40(12):956–974. doi:10.1108/09590551211274946

Cornett JE (2010) What is considered vintage clothing? http://www.ehow.com/about_6593593_considered-vintage-clothing_.html

Csanák E (2014) Eco-friendly concepts and ethical movements in the fashion industry. Presented at the International Textile, Clothing and Design Conference—Magic World of Textiles, Dubrownik, Croatia

Dobscha S (1998) The lived experience of consumer rebellion against marketing. Adv Consum Res 25(1):91–97. http://eds.a.ebscohost.com/eds/detail/detail?vid=3&sid=a72dd4a2-dba7-4ac6-8b5e-261c55408af6%40sessionmgr4004&hid=4211&bdata=Jmxhbmc9ZGUmc2l0ZT1lZHMtbGl2ZQ%3d%3d#AN=83386409&db=bth

Fitzgerald B (2015) Second-hand fashion is growing in popularity, study reveals—Le Souk. http://www.lesouk.co/articles/material-inspiration/second-hand-fashion-is-growing-in-popularity-study-reveals

Gabrielli V, Baghi I, Codeluppi V (2013) Consumption practices of fast fashion products: a consumer-based approach. J Fash Mark Manag 17(2):206–224. doi:10.1108/JFMM-10-2011-0076

Guiot D, Roux D (2010) A second-hand shoppers' motivation scale: antecedents, consequences, and implications for retailers. J Retail 86(4):355–371. doi:10.1016/j.jretai.2010.08.002

Han J (2013) Understanding second-hand retailing: a resource based perspective of best practices leading to business success—viewcontent.cgi. Iowa State University. http://lib.dr.iastate.edu/cgi/viewcontent.cgi?article=4643&context=etd

H&M (2016) Garment collecting. http://about.hm.com/en/About/sustainability/commitments/reduce-waste/garment-collecting.html

Iverson E (2010) Millennial perspective: vintage fashion and the twice-around economy. http://www.newgeography.com/content/001343-millennial-perspective-vintage-fashion-the-twice-around-economy

Jenss H (2016) Secondhand clothing. In: Berg encyclopedia. pp 232–237. http://www.bergfashionlibrary.com/staticfiles/Encyclopedia/Secondhand-Clothing-Global-Fashion.pdf

Joy A, Sherry JF, Venkatesh A, Wang J, Chan R (2012) Fast fashion, sustainability, and the ethical appeal of luxury brands. Fash Theory J Dress Body Cult 16(3):273–296. doi:10.2752/175174112X13340749707123

Kant Hvass K (2014) Post-retail responsibility of garments—a fashion industry perspective. J Fash Mark Manag Int J 18(4):413–430. doi:10.1108/JFMM-01-2013-0005

Kant Hvass K (2015) Business model innovation through Second Hand Retailing. J Corp Citizsh. doi:10.9774/GLEAF.5001.2015.ma.00005

Kleiderkreisel (2014) Kleiderkreisel und Co.: Die besten Mode-Tauschbörsen im Internet—Star-News. http://www.focus.de/kultur/vermischtes/kleiderkreisel-kleiderkreisel-und-co-die-besten-mode-tauschboersen-im-internet_id_4160567.html

Lastovicka JL, Bettencourt LA, Hughner RS, Kuntze RJ (1999) Lifestyle of the tight and frugal: theory and measurement. J Consum Res 26(1):85–98

McDonald S, Oates CJ, Young CW, Hwang K (2006) Toward sustainable consumption: researching voluntary simplifiers. Psychol Mark 23(6):515–534

Mhango MW, Niehm LS (2005) The second-hand clothing distribution channel: opportunities for retail entrepreneurs in Malawi. J Fash Mark Manag Int J 9(3):342–356. doi:10.1108/13612020510610462

Miller K (2013) Hedonic customer responses to fast fashion and replicas. J Fash Mark Manag 17 (2):160–174. doi:10.1108/JFMM-10-2011-0072

Morgan LR, Birtwistle G (2009) An investigation of young fashion consumers' disposal habits. Int J Consum Stud 33(2):190–198. doi:10.1111/j.1470-6431.2009.00756.x

Preuss S (2015) Könnte Second Hand zur neuen Fast Fashion werden? https://fashionunited.de/nachrichten/mode/ko-nnte-second-hand-zur-neuen-fast-fashion-werden/2015051818131

Roux D, Guiot D (2008) Measuring second-hand shopping motives, anteced-ents and consequences. http://content.ebscohost.com/ContentServer.asp?T=P&P=AN&K=48553278&S=R&D=bth&EbscoContent=dGJyMNLe80SeprE4wtvhOLCmr02eprdSsau4SbCWxWXS&ContentCustomer=dGJyMPGqtUyvrrdJuePfgeyx43zx

Roux D, Korchia M (2006) Am i what i wear? An exploratory study of symbolic meanings associated with secondhand clothing. Adv Consum Res 33(1):29–35. http://eds.a.ebscohost.com/eds/pdfviewer/pdfviewer?sid=1542cd11-ffeb-4dee-91b4-566991d66909%40sessionmgr4004&vid=2&hid=4213

Scholl G, Gossen M, Grubbe M, Brumbauer T (2013) Vertiefungsanalyse alternative Nutzungskonzepte. http://www.ressourcenpolitik.de/wp-content/uploads/2013/04/PoLRess_ZB_AP2-Vertiefungsanalyse_alternativ-eNutzungskonzepte.pdf

Sherry JF Jr (1990) A sociocultural analysis of a Midwestern American flea market. J Consum Res 17(1):13–30

The future of fast fashion (2005) The economist. http://www.economist.com/node/4086117

Turunen LLM, Leipämaa-Leskinen H (2015) Pre-loved luxury: identifying the meanings of second-hand luxury possessions. J Prod Brand Manag 24(1):57–65. doi:10.1108/JPBM-05-2014-0603

Vilimaviciute I (2015) Two years on, how are global fashion supply chains changing in the wake of rana plaza? | sustainable brands. http://www.sustainablebrands.com/news_and_views/supply_chain/ieva_vilimaviciute/two_years_how_are_global_fashion_supply_chains_changi

Wang H, Gargano C, Lukac S, Jackson A, Beals C, Smiley P, Reicin A (2010) An enhanced bunionectomy model as a potential tool for early decision-making in the development of new analgesics. Adv Ther 27(12):963–980. doi:10.1007/s12325-010-0084-8

Watson MZ, Yan R-N (2013) An exploratory study of the decision processes of fast versus slow fashion consumers. J Fash Mark Manag 17(2):141–159. doi:10.1108/JFMM-02-2011-0045

Williams CC, Paddock C (2003) The meanings of informal and second-hand retail channels: some evidence from Leicester. Int Rev Retail Distrib Consum Res 13(3):317–336. doi:10.1080/0959396032000101372

Xu Y, Chen Y, Burman R, Zhao H (2014) Second-hand clothing consumption: a cross-cultural comparison between American and Chinese young consumers: second-hand clothing: a cross-cultural comparison. Int J Consum Stud 38(6):670–677. doi:10.1111/ijcs.12139

Chapter 8
Collaborative Consumption 2.0: An Alternative to Fast Fashion Consumption

Jochen Strähle and Carina Erhardt

Abstract The purpose of this paper is to evaluate consisting consumption patterns caused by fast fashion with a new appearing form of consumption and retaining potentials as an alternative as well as sustainable form of fast fashion consumption. This research is set up on a theoretical background of scientific literature including governmental as well as press releases in order to evaluate the status quo of consumption and answering the research question. A new consumption pattern as well as an appearing economy of sharing can be stated including potential aspects of raising businesses and sustainable alternative forms of fast fashion. The framework of the research is limited to the textile and fashion industry in industrialized countries focusing on consumption in the twenty first century.

Keywords Consumer behaviour · Fast fashion · Sharing economy · Collaborative consumption · Product service system

8.1 Environmental Aim for New Consumption Patterns

The inappropriate consumption patterns of these days lead to a massive exploitation of resources and appearing environmental problems such as waste. Despite the knowledge about the finiteness and threatened limitations of global resources, an increasing use can be observed. According to statistics another planet is needed by 2030 in order to fulfil the upcoming needs. In western industrialized countries where consumption is a natural part of everyday life and overconsumption is widespread,

---

The original version of this chapter was revised: The spelling of the second author's name has been corrected. The erratum to this chapter is available at 10.1007/978-981-10-2440-5_15

J. Strähle (✉) · C. Erhardt
School of Textiles and Design, Reutlingen University, Reutlingen, Germany
e-mail: jochen.straehle@reutlingen-university.de

C. Erhardt
e-mail: carinaerhardt@gmx.de

J. Strähle (ed.), *Green Fashion Retail*, Springer Series in Fashion Business,
DOI 10.1007/978-981-10-2440-5_8

resources are overexploited enormously. In order not to jeopardize the needs of further generations, strategies of a more sustainable consumption are craved.

> One of the major challenges currently facing academia, business, society and the political sector is the issue of resource efficiency and conservation. In this connection, the necessity to achieve an economic, effective use of natural resources—in order to continue to generate wealth in the future and to sustain human life in general. (Leismann et al. 2013)

This research will bring the challenge into the section of the textile and fashion industry, where unsustainable patterns of consumption and production embellishing the daily agenda of fast fashion. The focus is set on a sufficiency strategy of sustainable consumption. Due to an appearing phenomenon of a sharing economy and the possibilities of communication and information technologies, new business models are occurring on the fashion market such as luxury renting, fashion flat rates or fashion libraries revolutionizing the old forms of collaborative consumption and challenging the patterns of fast fashion consumption.

This research will evaluate the revolutionized collaborative consumption (2.0) as an alternative to the existing form of fast fashion consumption including sustainable aspects (social, economical, ecological) and focusing on the following questions:

- Why do we consume?
- How do we consume?

8.2 Status Quo of Consumer Behaviour

8.2.1 Consumer Behaviour: Between Consumption, Consumerism and a Consumer Society

The use of goods and services in order to satisfy present needs is defined as **consumption**. Before a service or good is used or consumed, the need of the consumer is awakened and the purchase is planned (e.g. gathering information) and executed. Afterwards the good is disposed. These phases before (prepurchase), during (purchase) and after (postpurchase) the purchase are building the stages of the **consumption process** (Solomon 2011) (Fig. 8.1).

Consumer behaviour refers to the behaviour of the consumer during the selection, purchase, use or dispose of economical goods or services to satisfy needs (Solomon 2011; Gabler Wirtschaftslexikon 2015). Compared to further developments in this area, focusing on the purchase interaction between consumer and producer, also defined as buyer behaviour, marketers have recognized that consumer behaviour is an ongoing process involving surrounding steps (e.g. prepurchase) (Solomon 2011).

The reasons of consumption these days are crossing basic needs and can be divided after (Lange 2004) into the three sections of rational, conspicious and compensatory consumer behaviour (Lange 2004). **Rational consumer behaviour** is about the satisfaction of needs suggested by Maslow. These consist of **basic**

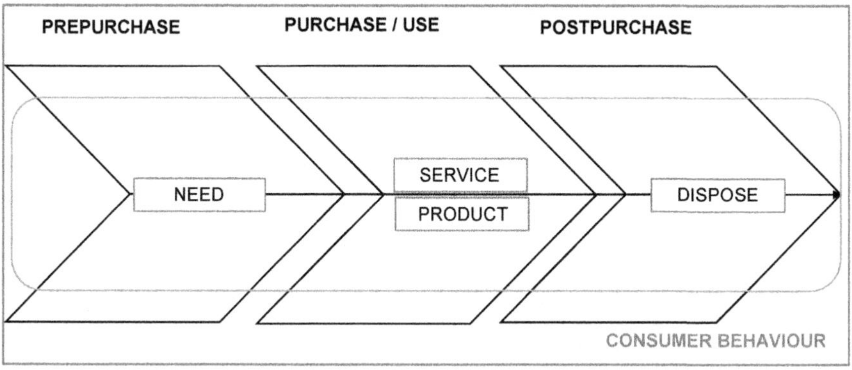

Fig. 8.1 Consumption process. Adapted from Solomon (2011)

needs: physiological (e.g. thirst) and safety needs (e.g. security), **psychological needs:** social needs (e.g. social acceptance) and esteem needs (e.g. self-esteem) as well as **self-fulfilment needs** (e.g. self-actualization) (Evans et al. 2010; Lange 2004). **Compensatory consumption behaviour** is focusing on the use and consumption of goods as compensation for deficiencies. With the purchase of goods, occupational and personal stress can be reduced or disappointments can be comforted away. Feelings of frustration, discomfort and defeat are compensated. Here the focus is set on the act of purchase, as this provides a short period of satisfaction and distraction from negative emotions (Lange 2004). **Conspicious consumer behaviour** concentrates primarily on self-realization. Here the purchase and consumption of goods and services should strengthen the social recognition (e.g. achieving a positive impression on friends and family) (Lange 2004).

Consumption not only servers for the satisfaction of basic needs (hunger, thirst, etc.), but also of self-realization and therefore goods should be functional as well as underline personality. This function of consumption as self-realization and the associated lifestyle is called **consumerism** (Tully 2012; Schulz 2013).

In **developed countries** consumption is a natural part of everyday life. Most of these people belong to a so-called **consumer society**. This society is characterized by a high power of purchase, material wealth and low-cost goods due to mass production. The interests and needs of the consumer are defined by mass media as well as advertisements (Kleinschmidt 2008; Schulz 2013).

8.2.2 Limitations of Consumption: Finiteness of Resources

Global environmental problems e.g. climate change, increasing amount of rubbish, loss of biodiversity, water and soil pollution, deforestation, etc. as well as social problems, e.g. human right abuses are caused due to unsustainable patterns of consumption and production (Boulanger 2010; Schulz 2013). WWF (2015) stated

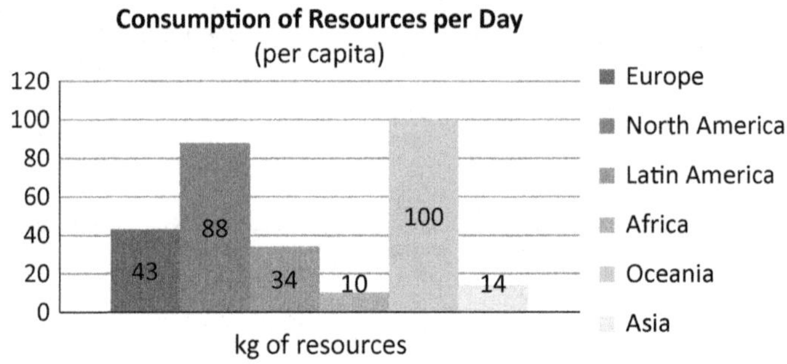

Fig. 8.2 Consumption of the world's natural resources. Own illustration adapted from Friends of the Earth Europe (2009)

in their Living Planet Report that another planet is needed by 2030 in order to fulfil the upcoming needs. Despite the awareness of a **finiteness of resources**, a global increasingly use of resources can be observed (Schulz 2013). In Western societies, overconsumption is widespread and leads to an exhaustion of precious natural resources (Manchiraju 2013; Schaefer and Crane 2005). Furthermore, these rich countries, compared to the poorest countries, consume per capita ten times more natural resources (Fig. 8.2).

Regarding the different phases of the product life cycle, about two-third of the environmental pollution are caused during the consumption, where as barely one-third is caused by production and almost nothing by disposal (Quack and Rüdenauer 2007; Scholl et al. 2013). Due to these facts "it is therefore imperative that consumers in industrialized countries adopt more sustainable consumption patterns" (Boulanger 2010).

8.2.3 Resource Saving Consumption: Alternative Strategies

To reduce the present amount of natural resource usage, resource efficiency in production as well as technological innovations is necessary but insufficient. Therefore, social innovations along with additional and equivalent-valued **strategies of sustainable consumption** are necessary (Leismann et al. 2013).

Focusing on the environment and future generations, sustainability along with sustainable consumption has gained substantial attention by scientists, governments and companies in the last few years (Manchiraju 2013). Ofstad (1994) defined "the use of goods and services that respond basic needs and bring a better quality of life, while minimizing the use of natural resources, toxic materials, and emissions of waste and pollutants over the life cycle, so as not to jeopardize the needs of future generations" as **sustainable consumption** (Boulanger 2010; Ofstad 1994). Schulz (2013) stated that the consumer is set on the end of the supply chain and decides with his individual consumer behaviour on how and with which consequences

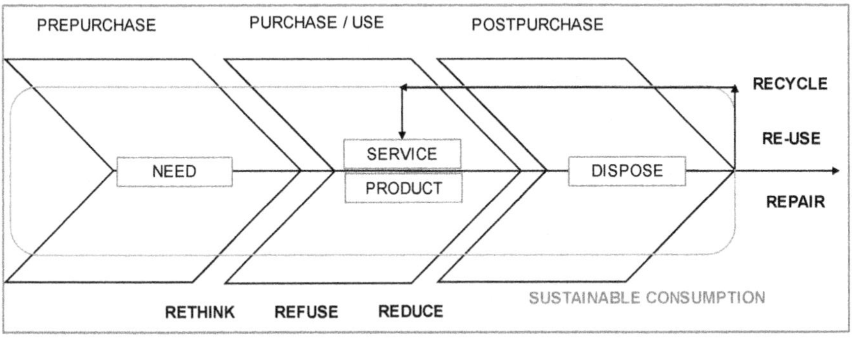

Fig. 8.3 Sustainable consumption along the process of consumption. Adapted from Solomon (2011) and Schulz (2013)

goods are produced and how many resources are used for his individual consumption lifestyle (Schulz 2013). The idea of sustainable consumption in this aspect is to expand the mentioned thought with the conviction that each consumer has the force to change things due to his purchase or non-purchase decision by involving the relevant principles of rethink, refuse, reduce, reuse, repair and recycle (Schulz 2013) (Fig. 8.3).

These principles are also defined as the 6R rules of sustainable consumption involving the following aspects of behaviour (Alatervo 2013; Schulz 2013):

- **Rethink**: Considering and questioning consumption habits
- **Refuse**: Making the choice to not generate waste (e.g. rent, share, swap)
- **Reduce**: Making decisions in order to decrease the amount of produced waste (e.g. economy size packages)
- **Reuse/Repair**: Expanding the life of product
- **Recycle**: Reclaiming the raw material

Three strategies of sustainable consumption have been discussed in order to prevent the crossing of ecological limits by reducing the use of resources and energy caused by decent consumption patterns (Misereor 2011; Schulz 2013).

The **eco-efficiency strategy**, stated by the World Business Council for Sustainable Development (1992), is aimed to raise the productivity and consumption of goods and services, by a lower input of resources and lower creation of waste and pollution (Boulanger 2010).

The **consistency strategy** is focusing on ecologically compatible technologies (e.g. cradle to cradle approach) (Boulanger 2010; Misereor 2011; Schulz 2013).

The **sufficiency strategy** puts an end to the excessive use of resources by changing lifestyle and economic activity (**reduction of consumption** and production) (Boulanger 2010; Misereor 2011; Schulz 2013).

From the perspective of consumption the eco-efficiency strategy leads to a rebound effect, where consumer uses the saved resources for other investments (Boulanger 2010; Misereor 2011). In the context of this research the strategy of sufficiency (consumption reduction) is focused and evaluated in the following sections.

8.3 Fast Fashion Consumption

8.3.1 The Concept of Fast Fashion: Transformation of the Fashion Industry

In the last few years, a new concept appeared in the apparel market next to the traditional models of high fashion and prêt-à-porter and revolutionized the traditional 6-month cycles of ready-to-wear fashion (Gabrielli et al. 2013). This concept which has been acknowledged in the fashion press as well as in academic literature as **fast fashion** aims to reduce the lead times of the supply chain in order to satisfy the needs of the customers by providing a constantly renewing assortment of apparel (Barnes and Lea-Greenwood 2006; Gabrielli et al. 2013). Furthermore, the concept is defined as a production principle where mass produced fashion is thrown into a saturated market, cheaper and within shorter rhythms (Thöne 2015).

Important components are stated as Quick Response and Enhanced Design which are combined in the fast fashion system. **Quick Response (QR)** describes a technique, which leads to a shorter production and distribution time in order to match supply with uncertain demand. Through the combination of local production, sophisticated information systems, for simplifying inventory monitoring and replenishment, and accelerated methods of distribution, short lead times are established and supply and demand are effectively matched. The **Enhanced** product **Design** technique improves the match of supply and "uncertain" demand by developing highly fashionable products that contain the latest consumer trends. Here design lead times are reduced and consumer and industry tastes are monitored for unexpected fads (Cachon and Swinney 2011).

For modern fashion retailers such as Zara, H&M and New Look, fast fashion has been established as a key business strategy for success (Barnes and Lea-Greenwood 2006).

8.3.2 Consumer-Driven Approach: Raising Demand of the Customer

Back in the mid 1980s the textile and fashion industry was based on **standardized styles** without a frequent change of the assortment due to factory design restrictions. Low-cost mass production was the success of fashion retailers these days (Bhardwaj and Fairhurst 2010). In order to compete in the market, forecasting of consumer demand and fashion trends were used long before the actual beginning of consumption. Basic apparel was preferred by the customer due to a lower sensitivity towards style and fashion. Fashion exhibitions, fashion shows and trade fairs consisting of basic patterns of Spring/Summer and Autumn/Winter were visited in order to develop a seasonal range in one whole year (Bhardwaj and Fairhurst 2010).

At the same time **fashion-conscious customers** appeared, leaded by an increased import of fashion-oriented apparel. The sales during the forecasted seasons failed and ended up with an increase in mark-downs in the apparel market. This change strengthened the importance of fashion-oriented apparel and retailers started with the beginning of the 1990s to focus on updated products in order to expand their product range. Instead of only cost-efficient manufacturing, **refreshing products** were provided due to a quicker response towards appearing fashion trends (Barnes and Lea-Greenwood 2006; Bhardwaj and Fairhurst 2010).

Moving into the twenty first century **changes in consumer lifestyle** and the **demand for new fashion items** set established supply chain formats under pressure. These days, fashion companies focusing on price, are now shifting towards fast response to altering fashion trends and consumer demand as competitive advantage. **Changes in lifestyle** as well as the **need for individuality** result in constantly changing products to respond the appearing consumer needs (Barnes and Lea-Greenwood 2006). These shifts are caused by **socio-cultural changes** creating a faster speed of living and the orientation of mass society towards constant change and progress. One main influence of the rapidly changing consumer needs is seen in mass communication. Here the consumer gains access to appearing trends and styles affected nowadays, in contrary to former 12 months before used mood boards or trend forecasting agencies, by culture (incidents on the street, in clubs, hotspots, etc.) and fashion "flash points" (e.g. fashion week). Music, film, television and other channels of media are shaping consumer needs additionally (Barnes and Lea-Greenwood 2006). To sum it up, the consumer today is seeking for similar fashion items seen for example on catwalks as well as on celebrities and having the ability to purchase them immediately (Alexander 2004). Customer needs are thus changing more frequently and wardrobes are revised appropriately (Barnes and Lea-Greenwood 2006). By introducing fast fashion, styles from the runway can be adopted and interpretational designs can be introduced to the market in a minimum of 3 to 5 weeks (Barnes and Lea-Greenwood 2006). Thus fast fashion can be seen as "a **competitive response** to changes in consumer demand" (Barnes and Lea-Greenwood 2006).

8.3.3 Consumption Behaviour: Disposable Clothing and the Throwaway Mentality

Images of the fashion shows in New York, London, Milan or Paris go around the world in a minimum of time. Weeks later the must-haves can be seen in the windows of fast fashion retailers (Carolin Wahnbaeck 2015). Furthermore, the new lines are introduced every 2 to 3 weeks at very low prices, in order to raise sales through impulsive purchases of consumers (Bianchi and Birtwistle 2012). Owing to the rapidly changing assortment and the idea of "Here Today, Gone Tomorrow", consumers are visiting stores more often and are expecting constantly changing assortments (Bhardwaj and Fairhurst 2010). The provided cheap prices, additionally caused by price wars of the retailers, result in a uncontrolled purchase behaviour, increasing purchase rate and the fact that the consume gets used to cheap goods (Bianchi and Birtwistle 2012; Carolin Wahnbaeck 2015).

A statement of a Hong Kong consultant gets to the heart of the whole problematic: "[…]Affordable prices mean that consumers are buying more clothes more frequently. But it also means they are truly disposable. You may keep an item after ten washes, but the item may lose its lustre by then, or it may have gone out of fashion" (Joy et al. 2012). Fast fashion companies stated ten washes, as a benchmark for providing "original" value, due to poor-qualitative utilized resources and production techniques (Joy et al. 2012). Regarding the different generations, for consumers of **generation Y**, a higher amount of low-quality of fast fashion items is preferred, in contrast to **baby boomers** preferring a smaller amount of high-qualitative items (Bhardwaj and Fairhurst 2010). The **willingness to lose quality and making a poor choice** by purchasing fast fashion products on one hand is accepted by consumers due to the fact that they can save money, build up their own individual style (self-realization) and the ease of occasionally wearing different items with the effort of the variety of rapidly changing, affordable, fashionable and luxury-related items on the other hand (Gabrielli et al. 2013). Thus the occurred behaviour of consumers influenced by fast fashion can be characterized as follows:

- Frequent purchase of clothing
- Frequent change of clothing
- Significant amount of unworn clothing in their wardrobe
- Rapid disposal of unworn clothing

Restricted considerations about environmental impacts of consumption behaviour (Jørgensen and Jensen 2012).

Changing consumer preferences and poor-qualitative utilized resources shorten the time of usage of fast fashion items, between 1992 and 2002 by half, and cause **a rise in clothing disposal** (Bianchi and Birtwistle 2012; Joung 2014; Vaitheeswaran 2007). Consumers choose different ways of disposal "such as donations, drop-off, resale, swap, pass-on, reuse and throwaway" according to linked efforts like gaining money or helping others (Joung 2014). Greenpeace stated that the annual average of purchased items is set by 60. Whereby half of this amount is worn, but the other half is recycled (best case) or thrown away (worst case). Regarding the **throwaway mentality** in Europe, 5,8 tonnes of apparel are thrown away annually and 75 % are burned or end up on the garbage heap (Carolin Wahnbaeck 2015; Garcia 2013). Garcia (2013) states this problematical behaviour of the consumer as follows: "For the eager consumer there is no great attachment to stuff. It's only a matter of using it and throwing it away, like chewing gum" (Garcia 2013). Fast fashion and its disposable clothing is therefore a supporting model leading to a reduction of personal ownership (Allwood 2006; Garcia 2013).

8.4 Collaborative Consumption

8.4.1 Change in Consumption: Driving Forces Towards Non-ownership

In the twentieth century, the western society was shaped by the industrial economy churning out consumer goods and proclaiming for an ownership society (Walsh 2011). "Someday we will look back on the twentieth century and wonder why we owned so much stuff" (Walsh 2011). This statement illustrates the ongoing transformation in our society nowadays **moving away from hyperconsumption** (Botsman and Rogers 2010; Walsh 2011).

The consumer of today is fed up with the normal purchase of items and seeks for a **non-owing usage of goods** in order to gain greater freedom and capacity to make flexible as well as quick decisions. Therefore, the goods these days are increasingly expected to be available right at the appearance of the need, even when the price cannot be afforded by the consumer (Garcia 2013). Additionally, appearing **environmental concerns**, cost consciousness, **importance of community**, from the consumer perspective as well as the **appearance of social technologies**, are moving the consumer away from the old forms of consumerism (Botsman and Rogers 2010) (Fig. 8.4).

Besides an increasing amount of purchased goods, a new consumption movement can be stated testing alternative forms of consumption such as sharing, swapping or borrowing and changing the fast moving goods into circulation goods (Jungblut 2013). Nowadays, the consumer has the ability to rent private living spaces of other consumers (e.g. Airbnb), uses the car from his neighbour (e.g. take my car) or swaps clothes with others (e.g. Kleiderkreisel) (Jungblut 2013). The mentioned examples show that this upcoming idea of consumption gains ground in the society rather than being a short trend. Furthermore, it can be stated that this idea of consumption is a growing movement involving millions of people around the world (Jungblut 2013). Nielsen (2014) stated that there is a high participation willingness worldwide differentiating between 52 % in America, 54 % in Europe up to 94 % in China in these collaborative forms of consumption. Consumers are willing to share as well as rent different types of goods such as cars, power tools or clothing with others (Fig. 8.5).

Fig. 8.4 Alternative forms of consumption. Adapted from Botsman and Rogers (2011)

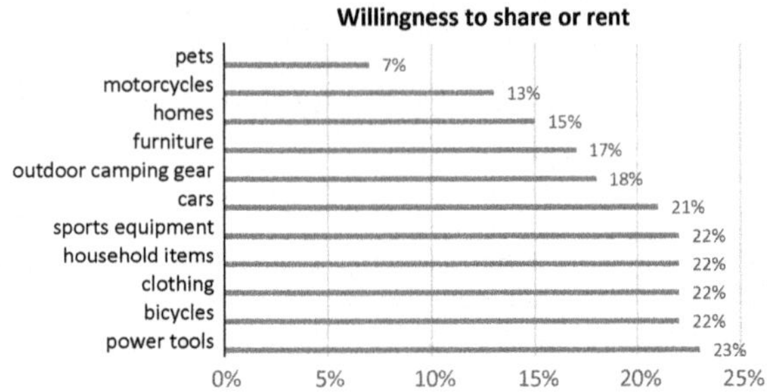

Fig. 8.5 Sharing and renting willingness of consumers. Own illustration adopted from Nielsen (2014)

8.4.2 *Classifications and Approaches: Collaboration as Core Idea*

In academic literature as well as in the press, different terms for this movement are used synonymously just like sharing economy, collaborative economy or collaborative consumption (Bagó 2015; Botsman 2013; Piscicelli et al. 2015). The ongoing text will classify the different terms and their meanings. In context of this research the focus is set on the term of collaborative consumption.

For Botsman (2013), the creator of the term of collaborative consumption in *What's Mine Is Yours: How Collaborative Consumption is Changing the Way We Live*, the meanings of these terms are different but include overlapping common core ideas (Table 8.1).

Figure 8.6 perfectly classifies the differences of the terms. The **collaborative economy** builds the framework of the whole issue. This economy, basing on the competition of individual as well as community networks and institutions, can be divided into the parts of production (e.g. collaborative designing), education (e.g. person-to-person learning), finance (e.g. crowd-driven investing) and consumption (e.g. redistributing). The sharing economy is set in the consumption part of the collaborative economy and is more often mentioned in connection with peer-to-peer markets and business-to-consumer models (Botsman 2013).

Table 8.1 Definitions of sharing economy, collaborative economy and collaborative consumption [Own illustration adopted from Botsman (2013)]

Collaborative economy	Sharing economy	Collaborative consumption
"An **economy** built on distributed networks of connected individuals and communities versus centralized institutions, transforming how we can produce, consume, finance, and learn" (Botsman 2013)	"An **economic model** based on sharing underutilized assets from spaces to skills to stuff for monetary or non-monetary benefits. It is currently largely talked about in relation to P2P marketplaces but equal opportunity lies in the B2C models" (Botsman 2013)	"An **economic model** based on sharing, swapping, trading, or renting products and services, enabling access over ownership. It is reinventing not just what we consume but how we consume" (Botsman 2013)

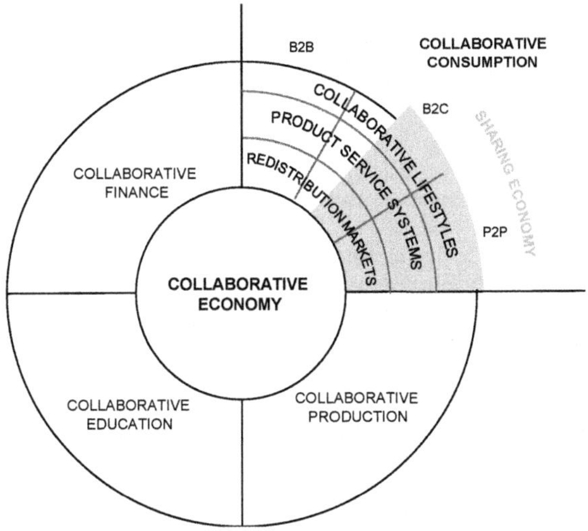

Fig. 8.6 Classification of sharing economy, collaborative economy and collaborative consumption. Own illustration adopted from Botsman (2013)

Collaborative consumption (CC) builds the consumption part in the collaborative economy focusing on product and service sharing, bartering, gifting, swapping, trading, lending, borrowing or renting. Furthermore, these consumption forms can be divided into the models of redistribution markets, product service systems or collaborative lifestyles, stated in Sect. 8.4.3 (Botsman 2013; Botsman and Rogers 2011).

The idea of collaborative consumption is not a new phenomenon. Back in the 1970s, "use rather than own" was the slogan of an ecological movement. Furthermore, shared flats, libraries, laundrettes and returnable bottles are examples for common forms of usage existing for decades. From a historical approach this occurring revival of collaborative consumption, appearing from a time and society where not almost everything was achievable for cheap prices, is redefined through information and communication technology and the resulting community marketplaces (Botsman and Rogers 2010; Jungblut 2013). It also can be seen as conscious approach to consumption where resources are consumed less and used in a more efficient way (Garcia 2013).

In 2015, Botsman redefined the stated definitions according to a perfectly adaption and integration of containing behaviour, business models, economical principles as well as businesses correlated to these terms (Botsman 2015) (Table 8.2).

8.4.3 *New Consumption Models: Variety of Supply*

Botsman and Rogers (2011) categorized the alternative formats of consumption (renting, swapping, sharing, etc.) into the three systems: collaborative lifestyle,

Table 8.2 Improvements of sharing economy, collaborative economy and collaborative consumption definitions

Collaborative economy	Sharing economy	Collaborative consumption
"An **economic system** of decentralized networks and marketplaces that unlocks the value of underused assets by matching needs and haves in ways that bypass traditional middlemen" (Botsman 2015)	"An **economic system** based on sharing underused assets or services, for free or for a fee, directly from individuals" (Botsman 2015)	"The reinvention of traditional market behaviours —renting, lending, swapping, sharing, bartering, gifting— through technology, taking place in ways and on scale not possible before the internet" (Botsman 2015)

Own illustration adapted from Botsman (2015)

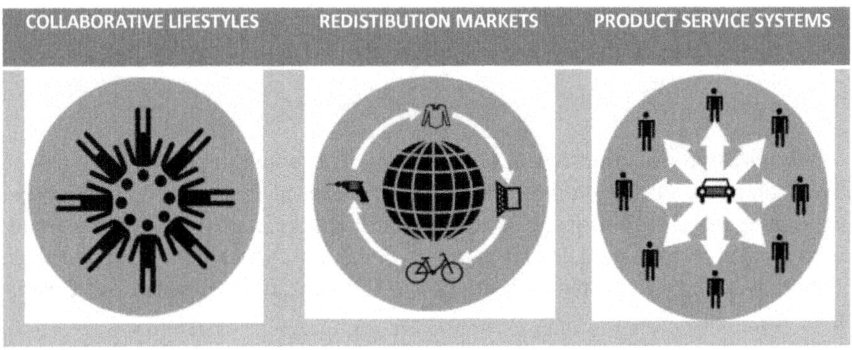

Fig. 8.7 Collaborative consumption models. Botsman and Rogers (2011)

redistribution markets and product service systems (Botsman and Rogers 2011) (Fig. 8.7).

Collaborative lifestyles deal with exchanges of less tangible goods. People with similar interests or needs can share and exchange for example time, space, skills or money (Botsman 2010; Bagó 2015; European Sharing Economy Coalition 2015).

Redistirbution markets provide customers with required goods (used or pre-owned) from where they are not needed. The items can be redistributed for free, swapped or sold for cash. It also can be seen as one of the 6Rs mentioned in Sect. 8.2.3 extending the life cycle of the product and thus reducing waste. Good examples of this system are exchange markets and second hand (Botsman 2010; Bagó 2015; European Sharing Economy Coalition 2015).

Product service systems (PSS) or product-based systems enable customers to use a product without the need for owning it by paying for the benefit of use (Leismann et al. 2013; Botsman 2010; Bagó 2015; European Sharing Economy Coalition 2015). From the perspective of ownership, the different forms of collaborative consumption can be divided into **approaches based on ownership** (e.g. reuse of goods by donating, selling or swapping) and **substituting ownership** (e.g. lending, renting, sharing, leasing) (Scholl et al. 2013). Regarding the

involving target groups the variety of supply can also be categorized in consumer-to-consumer formats (peer-to-peer (P2P)), company-to-consumer formats (B2C) or business-to-business formats (B2B) (Jungblut 2013).

8.4.4 Potentials for the Fashion Industry: Arising Business Models

When the financial crisis found its peak in 2008, by a breakdown of the investment bank Lehman Brothers and continued in the European debt crisis leading to a recession, high-street clothing retailers were sensing the upcoming dark days in the fashion retail market (Heinrichs and Grunenberg 2012; McDonald 2009). At this time, when consumers tightened their belt due to financial concerns, innovative websites occurred on the market offering fashion for almost free by **requiring** just the **willingness to share** (e.g. Bigwardrobe) as well as renting websites requiring **small percentages of the retail price** (e.g. Rent the Runway). The necessarily thrift of the consumer leads to a boom of this arising businesses, basically setted up focusing on eco-consciousness which appeared in this case as a positive side effect (McDonald 2009; Mayer 2013; Cauthen 2014). The trend behind this new innovational business is not unusual. People always have "rented, swapped and re-sold thir clothes" (Wosskow 2014) but these days technologies are retransforming these habits by giving the consumer access to a variety of people and "items they might want" or the others might own (Wosskow 2014).

Today, the consumption trend is focusing on a sustainable and ecological consumption, first appearing in the food market and nowadays appearing in the fashion industry (Mayer 2013). In this context collaborative consumption can also be stated as **sustainable countertrend to fast fashion consumption**, where costumers still have the possibility to shop the latest trends, but furthermore gaining access to unaffordable luxury fashion, reducing unneeded one-time purchases for occasional events or the rasing amounts of clothes caused by overconsumption due to fast fashion (Pedersen and Netter 2015; Cauthen 2014; Mayer 2013). Therefore, innovative business models reducing environmental impacts (e.g. waste) are needed these days in the fashion industry. **Clothing product service systems** can provide many opportunities for this purpose (Pedersen and Netter 2015, p. 259). Furthermore, examples of these systems can be stated such as Rent the Runway **focusing on renting** of fashion, working well in the fashion industry, where users rented in the first half of 2014 around 300 millions of dollars (Wosskow 2014). Or Le Tote founded in 2012 and gaining a revenue grow of 600 % in 2014. Also the consumer fluctuation can be mentioned, where 94 % of users are retaining Le Tote over several months (Peters 2015a, b; Taylor and Contributor 2015) (Table 8.3).

Rent the Runway offers the consumers the access to a wardrobe full of high luxury fashion for occasional usage. This business model focusing on **renting luxury items** for a "rarely" amount can satisfy the need of consumers for luxury

Table 8.3 Renting systems in the fashion industry

Company	Rent the Runway	Le Tote	LENA
Business model	**Luxury Renting** Renting of clothes, for 4-or 8-day periods, 30–800 $ per item	**Fashion Flat rate** Renting flat rate based on 2 until 12 months, 45–55 $ per month	**Fashion Library** Renting clothes, for 5 days to unlimited period, 20–50 € per month, 50 € for 500 points
CC model	PSS, B2C, online shop and stores	PSS, B2C, online shop	PSS, B2C/P2P, store
Value proposition	Variety of luxury fashion, fashion consultancy, free second size, lowering price of second item, cleaning service	Variety of fashion items, free shipping, cleaning service, insurance service	Qualitative and eco-friendly products, possibility of buying and swapping
Assortment	Variety of high-price dresses, beauty, shape wear, accessories	Variety of fashionable items and accessories	High-quality collection of vintage wear, upcoming designer and eco-labels

Own illustration adapted from Cauthen (2014), Lena-library (2015), Le Tote (2015) and Rent the Runway (2015)

fashion without possessing a high amount of money and furthermore, reduce superfluous purchases by satisfying occasional needs due to upcoming events. Rent the Runway is these days focusing on an additional format of flat rate system (Rent the Runway 2015; Cauthen 2014). With the slogan "Always have something new to wear" Le Tote is attracting consumers to their business model of **renting monthly clothes as often as needed** also stated as "Netflix for clothing" (Taylor 2015). The consumer gains the ability of prioritizing items and obtaining at least one of it in the monthly delivery of three garments and two accessories. Furthermore, the consumer can get this package delivery as often as wanted by sending the previous received items back (Le Tote 2015; Cauthen 2014) (Fig. 8.8).

"Wear beautiful clothing, but not at the expense of people and the environment. Being able to select by style and cut, but saying no to bulging wardrobes and fast fashion. Enjoy the feeling of new but without having to possess everything" (Lena-library 2015). This statement defines the purpose of the fashion library called

Fig. 8.8 Point system of the fashion library Lena. Adapted from Lena-library (2015)

Lena, which is placed in Amsterdam. Here consumers have the opportunity to borrow or buy clothes from a compiled collection of the owners. The whole system is based on membership, where consumers are swapping clothes with a library card filled with points. Each item is characterized with points. Additionally, the consumer has the ability to swap his own clothes at the library or buy preferred items (Lena-library 2015; Peters 2015a, b).

These examples show the **potentials of renting systems** in the fashion industry, which can be stated as follows (Leismann et al. 2013):

- Lease for a variety of needs (occasional, fads, individuality)
- Reduction of cost per consumption of garment
- Maintenance of garment after consumption returnment
- (Allwood 2006)
- Extension of product life
- Avoidance of mispurchases

Furthermore, Jungblut (2013) stated another potential by mentioning a resource saving consumption against overproduction and the throwaway mentality caused by fast fashion. Thus a production reduction of around 100,000 tones (e.g. UK), "more than 7,000 terra joule in energy, 20 billion litres of water and 27 % reduction in land filled clothing waste" could be saved due to renting systems (De Brito 2007).

8.4.5 Limitations in the Fashion Business: Humanity Barrier

In the apparel market the reasons of consumption are crossing basic needs such as the protection from a concrete environment, temperature or cleanliness. These days fashion is consumed in order to improve personality, attract the social environment, comply to a gender, group or hierarchy or to clarify status (Allwood 2006). In Western Europe the reasons of consumption are stated as follows (Allwood 2006):

- Following changes in fashion
- Being attracted by cheap prices
- Aiming to dress for special occasions
- Being attracted by a brand or label
- Having regular buying habits
- Being in need of replacing old worn-out clothes

By focusing on status as reason of consumption, product service systems can limit the part of clarifying status through possession in specific situations by replacing ownership with non-ownership (De Brito 2007; Botsman and Rogers 2011). From an **economical** perspective, cheap prices only can be offered by renting systems according to the market prices of their supplier causing a dependency and leading to fluctuating prices. Additionally, operational costs as well as cleaning and washing costs have to be included causing a higher price than existing offerings of cheap fast fashion retailers. Regarding these limitations, it can be said that the impact is not that high and leasing prices still can be affordable for price conscious consumers (De Brito 2007). The **sustainable** purpose of fashion product service systems mentioned in Sect. 8.4.4 is limited by the framework conditions of each business model. For example, the transportation route from consumer-to-consumer or business-to-consumer differs between the distance of them and thus can lead to a higher usage of resources and energy. Furthermore, the sustainable savings of lending can be limited to the kind of product used. For example, inefficient products result during the consumption time and the extended life cycle time in comparison to new items to a worse environmental balance mentioned in Sect. 8.2.2. In the textile and fashion industry, regarding the use phase of a cotton shirt 80 % of energy can be related to laundry care (Leismann et al. 2013). In the case of renting systems, the product is washed more often by the consumer as well as the company, implying at least two more washes causing negative environmental impacts. Regarding other options of dry cleaning, the impacts can be at least reduced (De Brito 2007). A rebound effect can appear additionally, where consumers invest their savings due to the affordable prices into other products causing further resource exploitation. In the end the consumer has to make the decision of accepting and using these alternative systems, according to the point that they are more attracted by financial benefits rather than the **ecologica**l side effects (Jungblut 2013).

8.5 Discussion of the Fast Fashion Alternative

Previous sections presented the state-of-the-art of academical, political as well as press-related views on the research topic, building a scientific basement to the issue. The ongoing discussion will clarify the research question of the paper: Collaborative Consumption 2.0: An Alternative to Fast Fashion Consumption from social, economical as well as ecological aspects focusing on **Why do we consume?** and **How do we consume?**

8.5.1 Why Do We Consume?

In western industrialized countries, the usage of goods and services is a natural part of everyday life. Consumption, these days, is based upon basic needs (e.g. protection from temperature) in order to compensate defencies (e.g. personal stress) and for self-realization. Furthermore, the society is characterized by a high power of purchase and material wealth. The consumer is used to low-cost goods and influenced by mass media. According to changes in consumer lifestyle and technology, the consumer gains access to ongoing fashion events and thus are seeking for fashion items seen on catwalks and being able to purchase them immediately. The need for individuality, as well as shifting needs, due to appearing trends, can be satisfied by renting systems. Here the consumer has the access to a variety of items and can also rent for occasional reasons. Also the need for "wearing different items, different days" can be satisfied with rental systems such as fashion flat rates. In the context of quick responding to changes in consumer demand the fast fashion system can be stated competitive, whereas renting businesses are dependent on their supplier and their response to appearing fads. Regarding the price consciousness of the customer, the fast fashion system builds again a competitive position. However, the consumer gains via renting systems access to preferred luxury goods seen on the catwalks, he normally cannot afford, for an affordable price of renting. Occurring concerns of the consumer towards the environment, the fast fashion system with its disposable fashion as well as production techniques lacks in ecological competitiveness. Collaborative consumption can be seen more sufficient towards ecological aspects. According to the system of renting, the life cycle of product is extended and less goods are used.

8.5.2 How We Consume?

Consumer demand and purchase poverty as well as the seeking for economical industrial wealth leaded to a society of hyperconsumption. In this time, fast fashion retailers gained their success with a strategy revolutionizing traditional fashion systems of 6-month cycles by reducing lead times along the supply chain in order to quickly respond to altering fashion trends and consumer demand. Rapidly changing assortments and low prices appeared on the market and lead to a frequently purchase as well as raising amount of clothes on the side of the consumer. Low qualitative fashion items and production techniques are connected to the strategy where accepted by the consumer creating a lack of qualitative consciousness besides the adaption towards low prices. Impulse purchases were placed at the purchase phase but due to poor-qualitative items the consumption process was shortened and the postpurchase phase recorded a high amount of unworn and unusable disposal of clothes. Thus, the consumption patterns of western industries can be characterized these days as inappropriate as well as unsustainable. Resources are exploited enormously. But from a social aspect, there is an upcoming economy supported by communication and information technologies. This consumption shift is fed up with the conventional form of purchasing items and seeks for a non-owing usage of goods gaining greater flexibility, importance of community as well as included an environmental concern shifting towards a sustainable consumption. Appearing formats, of this change in consumption patterns, like sharing, swapping, renting or borrowing, lead to a sustainable countertrend to fast fashion. Regarding to a lack of social as well as ecological aspects, the fast fashion system can be revolutionized in the approach of how we consume these days and in the near future.

Via luxury renting, fashion flat rates or fashion libraries the consumer still gains access to the latest trends for affordable prices. But furthermore the unneeded one-time purchases can be reduced due to the appearing opportunity of renting clothes for one-time occasions. Thus, regarding ecological aspects, the amount of disposal can be reduced, by extending the lifecycle of fashion items and reducing deployed resources according to a decreasing need of produced items. Saved amounts of resources during the consumption process are stated to be shifted in aspects of transportation, cleaning as well as a rebound effect in the investment of consumers to other resource usage which is mentioned weakening the poverty as an sustainable consumption pattern but not stealing the thunder as an alternative to fast fashion consumption.

8.6 Implications for the Fashion Business and Future Research

The previous discussion demonstrates the competitiveness of collaborative consumption, in the environment of industrialized western countries as well as in the section of textile and fashion industry, as a sustainable alternative of fast fashion consumption. Moreover, it can satisfy the needs occurring in the economy as well as maintaining a business of renting resolving upcoming economical problems of waste and resource exploitation. Nevertheless, the system of fast fashion is undisputed in the section of consumer response. But the environmental as well as social aspects demonstrate a need for a shift to sustainability, due to the finiteness of resources and adaption to appearing patterns of a sharing environment removing from a society of ownership to a society of non-ownership.

> One of the major challenges currently facing academia, business, society and the political sector is the issue of resource efficiency and conservation. In this connection, the necessity to achieve an economic, effective use of natural resources—in order to continue to generate wealth in the future and to sustain human life in general. (Leismann et al. 2013).

Regarding the stated quote in Sect. 8.1, the challenge can be solved in the fashion and textile industry with businesses based on an approach of value usage rather than ownership just like luxury renting, fashion flat rates or fashion libraries. Furthermore, the concern of sustainable consumption patterns these days can be solved by this as an sustainable alternative of fast fashion consumption reducing waste and resource usage (ecological), connecting people with unneeded items with people with required needs of the items (social) and additional maintaining high profitability (economical). Regarding the stated 6R rules of sustainable consumption in Sect. 8.2.3, renting has to be stated next to reuse and recycle as own "R" strengthening its mention behind refuse and extending the loop of the consumption process.

Regarding further research, collaborative consumptions such as product service systems or distribution markets have to be put on the research agenda of green fashion retail.

References

Alatervo SJ (2013, February 2) The 6 Rs: making a sustainable impact. Retrieved January 4, 2016, from http://prosperouswaydown.com/6rs-making-sustainable-impact/

Alexander H (2004, June 9) Store wars: fast fashion. BBC. Retrieved from http://news.bbc.co.uk/2/hi/business/3086669.stm

Allwood JM (2006) Well dressed? The present and future sustainability of clothing and textiles in the United Kingdom. University of Cambridge Inst. for Manufacturing

Bagó AC (2015) Introduction to collaborative consumption. Retrieved from http://www.consumocolaborativo.com/download/

Barnes L, Lea-Greenwood G (2006) Fast fashioning the supply chain: shaping the research agenda. J Fash Mark Manag Int J 10(3):259–271. doi:10.1108/13612020610679259

Bhardwaj V, Fairhurst A (2010) Fast fashion: response to changes in the fashion industry. Int Rev Retail Distrib Consum Res 20(1):165–173. doi:10.1080/09593960903498300

Bianchi C, Birtwistle G (2012) Consumer clothing disposal behaviour: a comparative study: Consumer clothing disposal behaviour. Int J Consum Stud 36(3):335–341. doi:10.1111/j.1470-6431.2011.01011.x

Botsman R (2010) The case for collaborative consumption. Sydney. Retrieved from http://www.ted.com/talks/rachel_botsman_the_case_for_collaborative_consumption

Botsman R (2013) The sharing economy lacks a shared definition. Retrieved December 10, 2015, from http://www.fastcoexist.com/3022028/the-sharing-economy-lacks-a-shared-definition

Botsman R (2015) Defining the sharing economy: what is collaborative consumption–and what isn't? Retrieved December 10, 2015, from http://www.fastcoexist.com/3046119/defining-the-sharing-economy-what-is-collaborative-consumption-and-what-isnt

Botsman R, Rogers R (2010, September 14) Goodbye hyper-consumption, hello collaborative consumption. Retrieved December 22, 2015, from http://www.huffingtonpost.com/rachel-botsman/goodbye-hyperconsumption-_b_716107.html

Botsman R, Rogers R (2011) What's mine is yours: the rise of collaborative consumption. In: Botsman R (ed.). Retrieved from https://www.scribd.com/book/163617594/What-s-Mine-Is-Yours-The-Rise-of-Collaborative-Consumption

Boulanger P-M (2010) Three strategies for sustainable consumption. S.A.P.I.EN.S. Surveys and Perspectives Integrating Environment and Society, (3.2). Retrieved from https://sapiens.revues.org/1022

Cachon GP, Swinney R (2011) The value of fast fashion: quick response, enhanced design, and strategic consumer behavior. Manage Sci 57(4):778–795. doi:10.1287/mnsc.1100.1303

Cauthen CE (2014) 5 Websites to rent clothes from right now. Retrieved from http://thestylishstandout.com/2014/04/16/where-to-rent-clothes-from/

De Brito MP (2007) Towards sustainable supply chains: a methodology. In: The continous challenge for innovation and integration, SIMPOI/POMS proceedings, Rio de Janeiro, Aug. Gertulio Vargas Foundation. Retrieved from http://repository.tudelft.nl/assets/uuid:b0be1d7a-23d6-4bb9-b4b9-883d733a7d2b/203812.pdf

European Sharing Economy Coalition (2015) Retrieved from http://www.euro-freelancers.eu/european-sharing-economy-coalition/

Evans M, Jamal A, Foxall G (2010) Consumer behaviour (2 edn., reprint). Wiley, Chichester

Friends of the Earth Europe (2009) Overconsumption—our use of the world's natural resources. Sustainable Europe Research Institute (SERI) and GLOBAL 2000 (Friends of the Earth Austria). Retrieved from https://www.foeeurope.org/publications/2009/Overconsumption_Sep09.pdf

Gabler Wirtschaftslexikon (2015) Käufer-und Konsumentenverhalten. In Gabler Wirtschaftslexikon. Springer Gabler Verlag. Retrieved from http://wirtschaftslexikon.gabler.de/Definition/kaeufer-und-konsumentenverhalten.html

Gabrielli V, Baghi I, Codeluppi V (2013) Consumption practices of fast fashion products: a consumer-based approach. J Fash Mark Manag Int J 17(2):206–224. doi:10.1108/JFMM-10-2011-0076

Garcia H (2013) Consumption 2.0. Futurist 47(1):6–8

Heinrichs H, Grunenberg H (2012) Sharing economy: Auf dem Weg in eine neue Konsumkultur? CSM, Centre for Sustainability Management. Retrieved from http://www.staedtetag.de/imperia/md/content/dst/internet/fachinformationen/2013/fdw_2014_hp_vortrag_heinrichs.pdf

Jørgensen MS, Jensen CL (2012) The shaping of environmental impacts from Danish production and consumption of clothing. Ecol Econ 83:164–173. doi:10.1016/j.ecolecon.2012.04.002

Joung H-M (2014) Fast-fashion consumers' post-purchase behaviours. Int J Retail Distrib Manag 42(8):688–697. doi:10.1108/IJRDM-03-2013-0055

Joy A, Sherry JF, Venkatesh A, Wang J, Chan R (2012) Fast fashion, sustainability, and the ethical appeal of luxury brands. Fash Theory J Dress Body Cult 16(3):273–296. doi:10.2752/175174112X13340749707123

Jungblut I (2013) Sharing is caring—liegt die Zukunft im kollektiven Konsum? Retrieved December 16, 2015, from https://reset.org/knowledge/sharing-caring-%E2%80%93-liegt-die-zukunft-im-kollektiven-konsum

Kleinschmidt C (2008) Konsumgesellschaft. Vandenhoeck & Ruprecht, Göttingen

Lange E (2004) Jugendkonsum im 21. Jahrhundert. Wiesbaden: VS Verlag für Sozialwissenschaften. Retrieved from http://link.springer.com/10.1007/978-3-322-80952-0

Leismann K, Schmitt M, Rohn H, Baedeker C (2013) Collaborative consumption: towards a resource-saving consumption cultureving consumption culture. Resources 2(3):184–203. doi:10.3390/resources2030184

Lena-library (2015) The fashion library. Retrieved December 8, 2015, from http://www.lena-library.com/english/

Le Tote (2015) How it works. Retrieved August 12, 2015, from https://www.letote.com/how-it-works

Manchiraju S (2013) Predicting behavioral intentions toward sustainable fashion consumption: a comparison of attitude-behavior and value behavior consistency models. In: Kaufmann H-R, Panni MFAK (eds) Handbook of research on consumerism in business and marketing: concepts and practices (pp. 225–243). IGI Global. Retrieved from http://services.igi-global.com/resolvedoi/resolve.aspx?doi=10.4018/978-1-4666-5880-6

Mayer J (2013, August 6) Getragene Kleidung muss nicht im Müll landen. Welt Online. Retrieved from http://www.welt.de/regionales/koeln/article118737287/Getragene-Kleidung-muss-nicht-im-Muell-landen.html

McDonald A (2009) FRUGAL FASHION Beg, borrow and style. New Statesman, 19

Misereor (2011) In den Grenzen von morgen—Für ein neues Verhältnis von Entwicklung, Wirtschaftswachstum und Umwelt (No. 9-2011) (p. 28). MISEREOR. Retrieved from https://www.misereor.de/informieren/publikationen/?tx_infothek_infothek%5Baction%5D=list&tx_infothek_infothek%5Bcontroller%5D=Product&cHash=d42428dbf7aaebcc2b1314b85d373837

Nielsen (2014) Global survey: sharing. Retrieved December 23, 2015, from http://www.nielsen.com/lb/en/press-room/2014/global-consumers-embrace-the-share-economy.html

Ofstad S (ed) (1994) Symposium: sustainable consumption. Ministry of the Environment, Oslo

Oxford Dictionary of Business & Management (2009) A dictionary of business and management: [clear, concise, and up-to-date defintions for 7.000 terms]. In: Law J (ed.) 5 edn. Oxford University Press, Oxford

Pedersen ERG, Netter S (2015) Collaborative consumption: business model opportunities and barriers for fashion libraries. J Fash Mark Manag 19(3):258–273. doi:10.1108/JFMM-05-2013-0073

Peters A (2015) At this fashion library, you check out clothes instead of buying them. Retrieved December 16, 2015, from http://www.fastcoexist.com/3045366/change-generation/at-this-fashion-library-you-check-out-clothes-instead-of-buying-them

Peters R-H (2015) Fashion-flatrate: der unendliche kleiderschrank. Retrieved December 21, 2015, from http://www.stern.de/wirtschaft/news/mode-und-fashion-in-der-flatrate-sharing-economie-im-kleiderschrank-2190636.html

Piscicelli L, Cooper T, Fisher T (2015) The role of values in collaborative consumption: insights from a product-service system for lending and borrowing in the UK. J Clean Prod 97:21–29. doi:10.1016/j.jclepro.2014.07.032

Quack D, Rüdenauer I (2007). Stoffstromanalyse relevanter Produktgruppen. Freiburg: Öko-Institut e.V. Retrieved from http://www.ecotopten.de/sites/default/files/ecotopten_endbericht_stoffstrom_2007.pdf

Rent the Runway (2015) How renting works. Retrieved August 12, 2015, from https://www.renttherunway.com/how_renting_works#step-3

Schaefer A, Crane A (2005) Addressing sustainability and consumption. J Macromarket 25(1):76–92. http://doi.org/10.1177/0276146705274987

Scholl G, Gossen M, Grubbe M, Brumbauer T (2013) Vertiefungsanalyse 1: Alternative Nutzungskonzepte - Sharing, Leasing und Wiederverwendung. Institute für Ökologische Wirtschaftsforschung. Retrieved from http://www.ioew.de/publikation/alternative_nutzungskonzepte_sharing_leasing_und_wiederverwendung/

Schulz C (2013) Verantwortungsvoller Konsum: Wir können auch anders! Hamburg: Agrar Koordination. Retrieved from http://www.konsum-welt.de/fileadmin/dateiupload/KonsUmwelt/Bildungsmappe_I_Verantwortungsvoller_Konsum.pdf

Solomon MR (2011) Consumer behavior: buying, having, and being (9. ed., global ed). Pearson, Boston

Taylor C, Contributor (2015) Clothing rental startup le tote lands $8.8 M In Series A Funding. Retrieved from http://social.techcrunch.com/2015/02/05/clothing-rental-startup-le-tote-lands-8-8m-in-series-a-funding/

Thöne E (2015) Mode-Ausstellung “Fast Fashion”: Shoppen und Jagen. *Spiegel Online*. Retrieved from http://www.spiegel.de/kultur/gesellschaft/mkg-mode-konsum-ausstellung-fast-fashion-in-hamburg-a-1024526.html

Tully CJ (2012) Nachhaltiger Konsum, 51–56

Vaitheeswaran V (2007) Revving up. The Economist. Retrieved from http://www.economist.com/node/9928259

Walsh B (2011) 10 Ideas That Will Change the World. *Time*. Retrieved from http://content.time.com/time/specials/packages/article/0,28804,2059521_2059717_2059710,00.html

Wahnbaeck C (2015) Fast Fashion versus grüne Mode. Retrieved December 26, 2015, from https://www.greenpeace.de/themen/endlager-umwelt/fast-fashion-versus-gruene-mode

Wosskow D (2014) Unlocking the sharing economy: independent review - Publications - GOV.UK. Government UK Department for Business, Innovation & Skills. Retrieved from https://www.gov.uk/government/publications/unlocking-the-sharing-economy-independent-review

WWF (2015) Sharing economy. Retrieved December 26, 2015, from http://www.wwf.de/aktiv-werden/tipps-fuer-den-alltag/vernuenftig-einkaufen/sharing-economy/

Chapter 9
Cross-Industry Learnings: What Fashion Retail Can Learn from the Used-Car Industry

Jochen Strähle and Maria Höhn

Abstract The purpose of this research is to explore current boundaries of the fashion industry's second hand market and which solutions and approaches can be adopted from the used-car industry. The paper is based on the study of existing literature which deals with sustainability in combination with second hand markets in general and adaptable feature of the used-car industry. Adaptable features are found using the business model canvas. The key finding of this study indicates that the fashion industry faces immense social and environmental challenges which can be partly solved by the development of the second hand market. Used-car industry can be seen as role model for fashion retail. In this study only aspects of used-car distribution are highlighted; therefore, characteristics of the recycling of used cars are not examined.

Keywords Sustainability · Fashion · Used-car industry · Second hand · Cross-industry learnings · Reverse logistics · Extended loop

9.1 Introduction

The companies of the fashion industry are facing more and more social and environmental challenges. Therefore, an increasing number is experimenting with new products, processes, and business models. Enterprises are exploring alternatives to conventional methods and new technologies, for instance, the reduction of water consumption in the finishing process, cradle-to-cradle principles, various take-back systems or the transformation of used materials into new products. By initiatives like this, fashion industry is breaking with its current strait system and stops fashion products end up on the landfill. These innovations are not only a benefit for society

J. Strähle (✉) · M. Höhn
School of Textiles and Design, Reutlingen University, Reutlingen, Germany
e-mail: jochen.straehle@reutlingen-university.de

M. Höhn
e-mail: maria.hoehn@gmx.net

J. Strähle (ed.), *Green Fashion Retail*, Springer Series in Fashion Business,
DOI 10.1007/978-981-10-2440-5_9

and environment, but can also be an advantage for the economic situation of a fashion company, its profit, growth and competitiveness (Pedersen and Andersen 2013). However, up to now there is no perfect solution to meet the expectations of consumers, non-profit organizations (NGOs) and companies. Fashion industry's second hand market contains a lot of potential though. In the past the second hand market was limited to some small second hand shops, consolidated containers and clothing stores. Nowadays, it is becoming more and more a profitable business model. Nevertheless, there are no statistics, numbers or sales figures of this branch since the business sector of the second hand market is underdeveloped. But how is it possible to take even more advantage of these opportunities? Also other industries are facing the challenge of being sustainable and are trying to close the loop of their supply chain. Therefore, a more elegant solution already exists in the used-car industry, though they are not recognized yet as a possibility. The automobile sector has created a very profitable market for their second hand products. The used-car market in Germany is more than twice the volume of the new-car market and growing (Ebel and Hofer 2014). Especially, the branded used-car market can be a role model for companies in the fashion industry. In order to make an effective use of the knowledge of the automobile industry, it has to be adapted, modified, and customized toward fashion. In reflection of this the focus of this article is on fashion industry's second hand market, its challenges, and opportunities, in which way cross-industry learnings of the used-car industry can lead to seize the chances.

9.2 Background

The concept of sustainability is not new, and although almost every company is facing the connected challenges, it is difficult to define. In the Brundtland report of the United Nations, sustainable development was defined as "[...] development, that meets the needs of the present without compromising the ability of future generations to meet their own needs" (Imperatives 1987). It expresses that economic development should be possible without harming the environment and depleting natural resources. Nowadays for understanding the concept of sustainability, it is necessary to see development from three perspectives: the social impact for the people, the environmental impact for the planet, and the economic impact for the profit. Those three impacts are also called the triple bottom line. It is an advancement from seeing the economic impact of economic development only (Bloemhof et al. 2009). Companies of all industries are trying to reach a higher level of sustainability. Although it is not clear to define what makes economic development sustainable, the waiver of virgin, non-renewable resources are evident.

In the last few years, changing of old supply chain patterns was mostly in the focus of sustainable development strategies. By recycling or reusing old products, businesses are trying to close their loop of materials and getting more sustainable. Closed-loop supply chains are an integrated management of all processes in the supply chain, including procurement, production, and distribution as well as the processes for reintroducing returned products, parts, and materials into the supply chain. Closing the supply chain is about reusing products, parts, materials, and energy (Bloemhof et al. 2009). The reuse of products in second hand markets does not fulfill all aspects of a closed-loop supply chain, since most of the garments are used longer, but do not come back to the loop of the supply chain after its distribution to the customer. Thus it is an extended, but not closed loop.

9.2.1 *Extended Loop and Reverse Logistics*

For Beamon (1999) "the fully integrated, extended supply chain contains all of the elements of the traditional supply chain, but extends the one way chain to construct a semi-closed loop, which includes product and packaging recycling, re-use, and remanufacturing operations." This means an additional level of complexity for the supply chain and also new requirements and considerations. In Fig. 9.1 the increasing complexity of an extended supply chain (dashed lines) compared to the traditional supply chain (solid lines) is evident. The two basic operational and strategic issues of an extended supply chain are mostly uncertainty arising from the replacement and recovery process like quality and quantity of returned products. Moreover, it is the reverse distribution process itself which means the collection and transportation of used products (Beamon 1999).

Due to the rise of complexity in the extended supply chain, reverse logistic management is gaining a big importance. It is a way of controlling the flow of raw

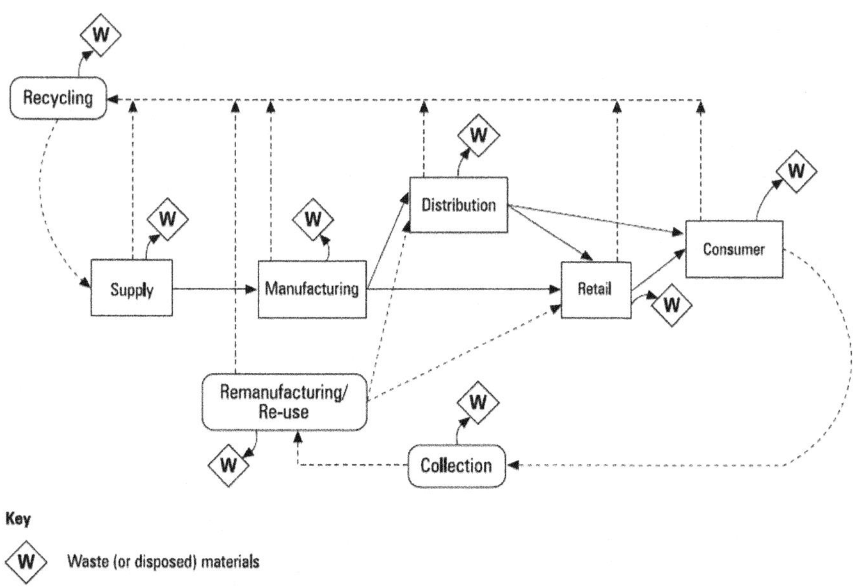

Fig. 9.1 The extended supply chain. Adapted from Beamon (1999)

materials, inventory, finished goods, and also information in an efficient and cost-effective way from the consumer to the point of origin. It includes the handling of returned goods, restock, product disposals, and all related processes (Venkatesh 2010). For many companies the return process is still not an integral element of the management in their supply chain, although it is an integral aspect of the product life cycle. Part of this process is also the disposal or the reselling of the product by the customer. As reverse logistic can be seen as part of the extended supply chain, also challenges in the quality management, routing network, and cost transparency may arise. In the reverse process the returned products are from a wide variety, and the distribution network is not clear and costs are not transparent since the processes are not standardized (Venkatesh 2010). Furthermore, Venkatesh (2010) cites that reverse logistic is a process which needs all information from the purchase to the return of the product to make the most out of it. There are two divisions of the recovery. They are direct recovery, which involves reuse and resale, and process recovery, which includes also the cleaning or assembling of the returned products. By innovations in handling of the returned goods an increase of effectiveness of the supply chain and a combined increase of the profit takes place. By reverse management the gain of input materials, cost reduction and value-added recovery processes are direct advantages for the company. Moreover, market protection, a greener image, and customer loyalty are indirect benefits, which lead to customer satisfaction, loyalty, and goodwill.

The key drivers for an extended supply chain and the integration of a reverse logistic management are environmental and green concerns, economics, and corporate citizenship. Nowadays, people are more focused on green and environmental issues and are amendable to green branding. This is also related to waste disposal handling, as used products may not necessarily be waste but can be reused or recycled. Therefore, less waste will be generated, which protects the environment (Chan et al. 2012). Moreover, reverse logistic improves economic factors, since it makes processes more efficient and cost effective. By repairing, remanufacturing, or recycling valuable product or materials, companies can gain more profit and opportunities. Chan et al. (2012) clarified that extending the loop and reverse logistic lead to an improvement of brand reputation, positive marketing for the company's products, and a competitive advantage. Today customers expect more and more a Corporate Social Responsibility. Now it is imperative to analyze the whole life cycle effects of a product and is no longer acceptable to consider only the local and immediate effects. Due to public pressure, customer's expectations and changes in the state of the environment, there has been a shift in manufacturing business (Beamon 1999).

9.2.2 The Second Hand Market and its Consumer

Beamon (1999) defines reuse as "the process of collecting used materials, products, or components from the field, and distributing or selling them as used. Although the ultimate value of the product is also reduced from its original value, no additional processing is required." In the fashion industry the term second hand is mostly used for the reuse of products. It means, regardless to the age of the clothes, any garment which has been used before (Cervellon et al. 2012). Like the consumers interest in sustainability, there has been a steady growth in the second hand clothing market. A long time it was linked to students or families with low incomes, which are unable to afford higher priced apparel (Beard 2008). However, during the last years the clothing second hand market has experienced a turnaround.

9.2.2.1 Buying Motives in the Second Hand Market

Although the society's high need for status is influencing the purchase intention of second hand clothing negatively, there is a high growth in the second hand market of garments. The general buying motives for second hand consumers are price sensitivity, frugality, environmental awareness, and hedonic and economic motives. Guiot and Roux (2010) found out that price consciousness has a positive predictor of second hand shopping behavior. It is obvious that second hand products have a lower price than the new ones in most purchasing contexts. However, in cases of vintage clothing or old-timer cars, second hand products can be more cost effective (Cervellon et al. 2012). Another by (Cervellon et al. 2012) identified motive, frugality, is strongly connected to price consciousness. It can be seen as a complete lifestyle that is characterized by consumers, which try to make smart choices, reuse their resources, and spend their money carefully. They do not purchase unnecessary things and save their money and time for more valuable acquisitions in the future. In general, they are less materialistic.

The reuse and recycling are considered by consumers as effective ways to reduce waste and the negative impact on the environment (Bianchi and Birtwistle 2010). Consumers are more and more concerned about the impact of products and waste on their health, environment, and the total of the society. Therefore, the motives behind donating and consuming second hand products issue from the idea that this is a way of extending the lifespan of a product and thereby limiting waste (Guiot and Roux 2010). The purchase of second hand pieces can also occur from the rebellion against the throw-away behaviors of the society and the belief that the purchase of second hand pieces slows down the production and selling of unnecessary goods (Cervellon et al. 2012). Moreover, purchasing second hand products can lead to a socially conscious self and can be regarded as assign of opposition to consumerism (Roux and Guiot 2008). For the consumer second hand, ethic and sustainability are strongly related, because recycling a product, by passing it on for resale in a second hand market, means gaining a new, stylish item, and at the same time reducing the consumption of a new clothing item and saving the environment. However, often second hand clothing items need to be dry-cleaned, necessitating the use of chemicals, which may negate the positive environmental element of a second hand purchase. Moreover, it is the aftercare aspect of clothing that has the most demonstrable negative impact on the environment as a whole (Bianchi and Birtwistle 2010).

After all Roux and Guiot (2008) describe hedonic and economic motives as the main drivers for second hand consumption. They cite the recreational aspects as the heart of the second hand shopping experience, as they include the social contact with salespeople, the entertaining aspect of the shopping activity itself, and the serendipity from the unexpected finding of exceptionally products. The purchase and shopping motivation is often related to bargain hunting or to find the lowest price for a special piece. Shoppers derive from this money saving activity recreational benefits. They derive pleasure from finding the product at a much lower price than in general (Cervellon et al. 2012). However, there are also negative feelings related to second hand products (Fig. 9.2), like the contamination of clothes and the penetration of the privacy (Roux and Guiot 2008).

9.2.2.2 The Challenge of Adverse Selection and Risk

One of the main challenging aspects for the purchase of second hand products and for the whole market is the concept of adverse selection. Akerlof (1970) explains in his model a form of market failure, which results from asymmetrical information between vendor and purchaser before the conclusion of contract. Consumers often do not have all necessary information of the product to feel secure in the purchasing process and therefore cannot distinguish the suppliers. On the other hand, most suppliers know the exact quality of their products. This lack of information is getting bigger in second hand markets due to the increasing number of suppliers. Therefore, the purchaser assumes an average quality of all offered products. Suppliers of good, high-quality products will claim a higher price than suppliers of bad products with low quality. However, suppliers of low-quality products will state, due to the lack of information, the same quality as good suppliers to a lower price. Because the price will be higher than the average price, which is expected from the consumer, they will not buy the high-quality product and the suppliers

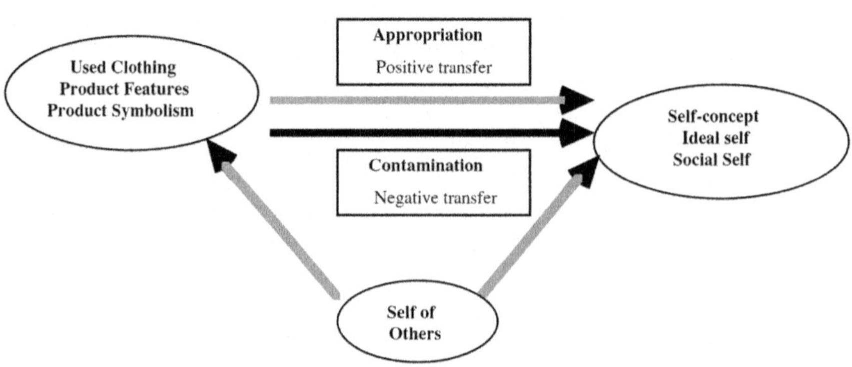

Fig. 9.2 Negative and positive transfer between used clothing and consumer self-concept. Adapted from Roux and Korchia (2006)

need to leave the market. As a result only low-quality products will be offered in the market. This form of market failure is a consequence from the asymmetrical information. If purchasers would be totally informed about the quality of the product, it would achieve a price depending on its quality.

Strongly related to the concept of adverse selection is the aspect of risk in consumer's purchase behavior. An individual will adopt specific cognitive and behavioral strategies in order to reduce the risk that their purchase will have negative consequences. For the consumer a purchase includes a risk in case it has a negative impact for performance, psychology or social, financial, and physical aspects of the consumer (Gabbott 1991). To lower the risk the individual will develop a risk reduction strategy (RRS). The first strategy is the information search in the presence of perceived risk by gaining pre-purchase information about the product and the supplier. Furthermore, Gabbott (1991) states the use of salient product or situational cues to infer other product characteristics, which may be unavailable as the second RRS. Such characteristics could be quality or performance attributes. In second hand markets the attributes of products are linked with the history of the individual product, which adds an extra dimension. Therefore, products in second hand markets cannot be considered similar with any degree of certainty, as they are used in different areas, by different people, in different circumstances. The older and more owner the product has got the higher the risk. Traditional product cues like brand and price are not reliable anymore, because of the insecure term of product usage history.

9.3 The Actual State: Fashion Industry Extending the Loop

Fashion industry today has a serious image problem as it is seen mostly as irresponsible, unsustainable, and unethical. Clothing causes 5–10 % of environmental impacts across the European Union. This means it is the most significant category after food and drink, housing, and transport. Each year more than 1.1 million tons of clothes are consumed and disposed. 48 % is reused, 14 % recycled, 7 % incinerated, and 31 % ends up in landfill (Buttle et al. 2013). In Pedersen's and Anderson's (2013) study about sustainable fashion one of the participants even pictured it to a ticking time bomb with a person standing next to it, unsure what to do. Fashion industry seems to have reached a level that is unsustainable for a long time and unsure of what is the best thing to do. Media, non-governmental organizations, and social networks are increasingly active in spreading news about social and environmental mismanagement among fashion brands and influencing the consumer in its thinking about sustainability. Therefore, an increasing number of companies are experimenting with new products and processes to cope with the social and environmental challenges. Various take-back systems, recycling arrangements, and cradle-to-cradle initiatives are introduced to break with the current linear system through which fashion products end up on the landfill. One example is Filippa K facilitating sales of its second hand products in their own store. Initiatives like this help change consumer behavior and can be a valuable component in a sustainable strategy (Pedersen and Andersen 2013).

9.3.1 Sustainability in the Fashion Industry

"Sustainability challenges span across the entire life-cycle of a piece of garment" (Pedersen and Andersen 2015). The extensive uses of resources, products' short life cycles, the high volume of fashion consumption, and the complex supply chains are currently the basis of the fashion industries business model, which leaves behind several social and environmental impacts (Hvass 2015). For fashion companies it is often difficult to make enlightened decisions about sustainability, as they are perceived of having little knowledge of what goes on in the supply chain of their products. Fletcher (2014) clarifies the higher complexity and transparence of fashion industry's supply chain compared to the food industry for example. An individual garment can consist of several fabrics like interlinings, zip, buttons, and finishing techniques, which all may be produced in different places. However, as there are upstream challenges, also the problems arising from the throw-away consumer culture, cause challenges in the downstream of products. The amount of garments being produced and sold is one of the main causes of unsustainability in the industry. The consumption of 80 billion new garments a year exceeds human needs and planetary boundaries (Pedersen and Andersen 2013). From this over-consumption stems the key environmental impacts of fashion, especially fast fashion. Furthermore, these clothes are made in developing eastern countries, necessitating long distances of travel to reach the distribution market (Shambu 2015). The idea and decision to buy a garment, purchase, usage, maintenance, and disposal of clothes are stages of the fashion consumption. Consumers think that the main sustainability issues do not occur in those stages, but in the stage of manufacturing. It is part of fashion companies' responsibility to tackle sustainability issues within their sphere of influence. However, laundering, which is part of the use stage, has the greatest impact on society and environment, and also disposal is a key issue of sustainability (Fletcher 2014). Nowadays, garments contain mostly of synthetic fibers, such as polyester and viscose, which are difficult to recycle and take a very long time to decompose in landfills (Shambu 2015). Therefore, the change in dominant consumer values, attitudes, and behavior is very important for the progress and success of sustainable fashion. There is a big knowing–saying–doing gap in consumer's behavior, which needs to be bridged. The positive views of socially and environmental-friendly products are rarely transformed into concrete buying and consumption behavior. Due to the complexity and combined non-transparency of supply chains, consumers are unaware and thus unwilling to pay more for alleged sustainable fashion (Pedersen and Andersen 2015). On the contrary, consumers are getting more and more accustomed to the low prices of garments and accept the poor quality. Therefore, there is a fall in the value of a fashion product, due to the oversupply of cheap, poor-quality garments.

Against this background reaching a higher, profitable level of sustainability is becoming very difficult. Since most customers base their purchase decisions on price rather than on sustainability, only few companies will see a business case for sustainability, since there are no other reward or sanction systems for it.

Against the background of challenges, like the impact of resource constraint, higher raw material costs, and therefore a higher need for sustainability, the industry needs new innovative business models, which may be better suited to this economy (Buttle et al. 2013). An increasing number of new closed-loop business models, like various take-back systems, are getting more and more important. However, it is necessary to know more about how consumers perceive the sustainability innovations to evaluate their potentials for a systemic change in the fashion industry. People have become much more aware of exchanging, borrowing, and reusing clothing as well as upcycling, customizing, and repairing what they have (Pedersen and Andersen 2013), also if it is still a small number of customers.

9.3.2 *Fashion Industry's Second Hand Market and its Innovative Business Models*

The second hand market for clothing is recently dominated by non-profit organizations, textile recycling firms (Hansen 2004), privately owned consignment and vintage stores, online reselling platforms, and clothing libraries. People mostly give their used clothes to friends or family, donate it for charity, or sell it online (Bianchi and Birtwistle 2010). However, in the last few years the number of used-merchandise retail establishments has grown at about ten times the rate of other stores (Beard 2008), which shows its huge potential.

Although the issues related to garment consumption and its disposal are emerging the whole fashion industry, the downstream value chain-related issues (reuse, remanufacturing, end-of-life solutions) have received less attention from the big companies until now. Supply chains that serve the second hand market have often been ignored or not seen as part of the industry (Svensson 2007). Moreover, after the final sale to consumers the fate of garments has not been relevant for fashion companies in the past, though the used energy to collect, sort, and resell second hand garments is between 10 and 20 times less than needed to make a new item (Hvass 2015). Further, Buttle et al. (2013) found out that according to current estimates, the reuse of one garment displaces 0.6 new garment purchases on average and delivers savings estimated at 13 tons of CO_2 equivalent per tons of clothing reused. Fletcher and Grose (2012) state that only by bringing the fashion industry into the discussion of textile waste and making producers accountable for future disposals, changes in the logic will occur. Moreover, it will lead to an extension of focus beyond the upstream manufacturing, to downstream actions, and resource flows. In the last few years, there are two new set up ways how companies address the downstream value chain now. The first strategy is instore take-back schemes for fiber recycling purposes, allowing consumers to drop off their used garments in exchange for a discount voucher. Even though this strategy seems to be a positive solution, vouchers can lead to a stimulation toward more consumption rather than to a more sustainable one. The second strategy is the developing of resell or reuse platforms to extend the life of the used garments. This strategy is mainly chosen by premium brands with products of good quality (Hvass 2015). The retailer will take back all clothes brought back by customers, but will only buyback garments of a sufficient quality for restyle or resale. Returned clothes which are not of good enough quality to be resold are sent to a commercial recycler (Buttle et al. 2013). Garments which are sold in a second hand store need to be long-lasting through design and quality and therefore need to have a style and materials that live for more than one season. These strategies have the core aim for the retailer to take responsibility for the full life cycle of the garments, and moreover it generates positive returns and profits. Pre-owned collection of ABC retail promotes the reuse of clothes by offering an incentive for customers in terms of a gift voucher for the collection to return their used garments to a store. Those garments are then sold in the store after sorting, cleaning, and re-styling. It shall encourage the consumers to change their purchasing patterns and generate commercial gain for the retailer.

Also, the term of collaborative consumption is gaining more and more attention and relevance in the second hand market of fashion products. New companies with new, innovative business models based on sharing, swapping, bartering, trading, or renting are becoming part of the industry. These old market behaviors are enabled by technology and peer communities on a scale, which was never possible before. It is disrupting outdated modes of business and reinventing not just what people consume, but how they consume (Buttle et al. 2013). Online communities for collaborative consumption, like Refashion, are popular today. It brings together consumers in search of sustainable living and those with a passion for designer clothing without having the money to buy expensive garments. This requires a certain level of social trust (Pedersen and Andersen 2013). Also the Canadian outdoor company Mountain Equipment allows people to sell and swap outdoor products on their website. Moreover, several companies have built a business on transforming used materials into new products. For instance, the brand Globe Hope creates new fashion products using existing materials like seat belts and Christopher Raeburn is using military fabrics to create garments that are functional and intelligent (Buttle et al. 2013). After all it is necessary to raise a stronger relationship to the customer, since they become the supplier of the second hand store. The supply of merchandise is a key success factor for a fashion company, and therefore second hand stores are highly dependent on their current customers to return their unwanted garments (Hvass 2015). Buttle et al. (2013) found out that the biggest barrier for consumers to buy second hand clothing was the choice of second hand clothes available. The consumers would buy more if the choice will be improved. Moreover, most of the participants of the study would consider using a retailer buyback system. Especially in the categories formal wear, designer clothing and clothes for going out this model would be interesting for consumers. These innovations are not only a benefit for society and environment, but can also be an advantage for the economic situation of a fashion company, its profit, growth, and competitiveness (Pedersen and Andersen 2013). However, up to now there is no perfect solution to meet the expectations of consumers, NGOs, and companies. Fashion industry's second hand market contains a lot of potential though.

9.4 Analytical Framework and Methodology

The study is grounded in an analytical framework of the business model canvas on a higher industry level in the context of the used-car market and aspects which can be beneficial for fashion industry's second hand market. Osterwalder and Pigneur (2010) have applied a pragmatic perspective to the concept that helps to assist in understanding how a firm or industry does business, for analyses comparison, performance assessment, management, communication, and to assist firms in their innovation. The concept contains a four-pillar frame work for clarifying the business model that is additionally broken down into nine building blocks (Fig. 9.3). The framework is also known as the Business Model Canvas. It constitutes the essential elements of a company's or in this case industry's value creation process (Osterwalder and Pigneur 2010).

This article is a conceptual discussion based on data gathered from literature review in the context of the fashion industry and its sustainable aspects in combination with an analysis of the used-car market and its beneficial, assumable aspects for the fashion industry. By analyzing the used-car industry in the framework of Osterwald's Business Model Canvas (2004) transferable aspects of the

Pillars	Building blocks	Description
Product	Value proposition	Overview of products and services and their inherent value a company offers to its customers
Customer interface	Target customer Distribution channel Relationships	Description of segments of customers a company wants to offer value to and means to build a strong relationship with them
Infrastructure management	Value configuration Key configuration Partnerships	Key activities, internal and external resources that are necessary to create value
Financial aspects	Cost structure Revenue model	The revenue model, the cost structure and the business model's financial sustainability

Fig. 9.3 Four-pillar template for business models. Adapted from Osterwalder (2004)

industry for fashion are figured out. It can be assumed that findings of existing research addressing the used-car market and its aspects can be partly transferred to the fashion industry and expose requirements for the initial process.

9.5 Findings and Discussion

This section highlights the main findings about the transferable aspects of the used-car industry to fashion. Through the analysis of case studies, research papers, reports, and standard literature a set of occurring themes and questions arose, which are linked to different business model elements. The automobile sector has created a very profitable market for their second hand products. The used-car market in Germany is more than twice the volume of the new-car market and growing (Ebel and Hofer 2014) and especially the branded used-car market can be a role model for companies in the fashion industry. Used-car distributors, which are linked directly with the car manufacturer, are in the following named as branded used-car dealers. Most car manufacturers, like Audi, BMW, or Mercedes, are distributing their used car on their own, differentiated by an additional brand name. They are less likely to wholesale them to independent used-car dealers, since once customers get used to the comfort and feel of luxury they often step up and buy or lease a brand new car (Alger 1995).

9.5.1 Product

The value proposition is the central issue of each business model. It consists of five stages, which are value creation, value purchase, value use, value renewal, and value transfer (Osterwalder et al. 2005). The buying experience created during a purchase or during the actual consumption of a product is the main value of a customer. But also value renewal and transfer stages can provide additional opportunities for the consumers (Hvass 2015).

The value propositions of used-car distributions are good quality products in a good physical condition related to their age and mileage ratio, since customers expect a long-lasting product (Peterson and Schneider 2014). Especially, branded used-car distributors are creating security for quality and therefore a value creation by giving warranties and guarantee services in their branded workshops. In case of a damage or low quality of the car the purchase does not have negative consequences for the buyer. The brand and the positive image are leading to a feeling of risk reduction for the costumer. The used-car distributor, in most cases, prepares, checks, and repairs the car after the buyback from the previous owner and creates a value compared to the private sale of used cars for the future customer. Also the introduction of Certified Pre-owned (CPO) programs or the MOT badge can create a bigger value for the customer (Kooreman and Haan 2006). In addition due to the mostly strong connection between branded used-car distributor and its previous customer or supplier of the used car, the asymmetry of information can be reduced. This is necessary since the customer expects at least a good experience in buying and owning a used car as they do with a new one (Cooke 2010). Moreover, one of the biggest values the used-car industry can offer, related to the new-car industry, is the price. Since a type of car is mostly produced over a longer time and is used by its owner in different ways, there are differences in age, mileage ratio, and in the history of the car. This causes a wide variety and flexibility in prices for the same type of car and also expands the assortment. In addition, branded used-car dealers are expanding their assortment and therefore the selection for the consumer, since their main business is mostly the sale of new cars from one or more brands.

On the other side there is he commercial gain for the consumer, due to the trade-in of used cars. The fall in value for cars is relatively low compared to consumer goods like food and fashion. Therefore, the residual value after the use of a car is very high and cars with good physical condition are mostly resold. Moreover, car parts of unusable cars can also be sold profitable. Used-car distributors are therefore creating value for the costumer by payment or by taking the car as a deposit for a new car or another used car.

Some aspects of the value proposition of the used-car distributors are already adapted by the second hand market for fashion. Also for garments a high quality is necessary to be valuable for the customer. However, due to the lower complexity the aspect of a good physical condition is easier to see for the consumer. Since mostly premium fashion brands can guarantee a high quality and a good physical condition for a longer time period, especially here the cross-industry learnings have a great importance. The aspect of giving warranties and the possibility of an after-sale service can also in the fashion industry be a great aspect to create more security and value for the costumer, not only in the second hand market, but also for the sale of new garments. High premium labels, like Louis Vuitton, already offering an after-sales service for their customers. Moreover, the presale treatments in the fashion industry should be guaranteed with regard to hygiene and personal space. Also the concept of the branded used-car dealers can be a great opportunity for the complete fashion industry. They are expanding their assortment in the matter of price flexibility and selection and by this gaining a value for the customer. So a store-in-store concept seems to have great benefits. Even more the commercial aspect for the customer is very important. The used-car industry is either pay out the value of the car or grants in case of the combined purchase of a new or other used car in most cases even a higher value or better financing possibilities (Zerres 2010). This can also in the fashion industry be an incentive for bringing back clothes and purchase others. Vouchers like mentioned above are not usual in the used-car industry. The concept of financing for customers can also be relevant especially for premium labels in the upper price segment.

9.5.2 Customer Interface

To analyze the interface with the customer, there are three main areas. These are the target customer and how to engage with it, the customer relationship and how to motivate them to return their used car to the distributor and buy another one and the distribution channel which means the analysis of the retail format.

In general, the used car is a product for the mass market. The target customer for used cars is very price sensitive and perceives his financial situation as rather bad than good (Zerres 2010). Branded used-car dealers expanding their customer segment, since used cars are mostly an additional business line besides their new-car operations. Customers who cannot afford a new car of a special model might afford a used one. If a customer feels connected with the supplier in a positive way the business relation can gain importance, independent from the economic aspects.

Therefore, the relationship management between used-car dealer and costumer is very important. With the purchase of a car in general much additional services, like repair, insurance, etc., are linked. Most car brands are offering these services in their branded distributions centers and guarantee therefore a continuing contact and strong relationship with their customers. These services are also offered for their branded used cars. Since customers must be involved and are responsible partners in the value creation process (Lüdeke-Freund 2009, cited from Hvass 2015) of second hand markets in general, due they are the suppliers for used cars, this strong relationship is a great benefit for the car companies. Also the grant of a warranty can have positive impacts for the relationship between customer and company, since it reduces the risk for the customer. Moreover, the purposeful instrument of sales financing directly at the branded used-car dealer means a raise of loyalty with the brand and can lead to the earlier purchase as actually wanted or that a new or higher priced car is bought (Stenner 2010).

Consumers of used cars are informing themselves mostly online, where they have a complete price transparency and a big choice. Traditional suppliers and branded used-car dealers are mostly linked with these online portals (Diehlmann and Häcker 2013). The time of purchase is often linked to the residual value of the owned new car, since if they are low the purchase will be more expensive (Ebel and Hofer 2014). The used-car industry has several sales promotion methods, like special financing models or advertising. However, this leads to a habituation effect, to more and more intensive sales promotions and finally to a lower valence of the product (Ebel and Hofer 2014). The actual purchase is often a combination with the after sales of the pre-owned car. The consumer is purchaser and supplier. The owner of a used car in good condition can expect a good price and have therefore mostly a higher willingness to pay. Moreover, caused by the trade-in of the car by the used-car dealer, the customer gains temporal and organizational benefits, which result from the elimination of the sale by its own (Zerres 2010). The residual value of the traded-in cars is created in relation to life cycle, technical condition, and the further development of the successor model (Stenner 2010). Cars which are no longer usable are sold in parts to scrap dealers or recycling enterprises.

In the second hand market for fashion, also non-economic factors like the need of visual stimulation, the avoidance of the classical market system or the real hunt for bargain can be aspects for the target customer. Especially, the ecological and ethical consciousness in the used-car industry seems to be very low compared to fashion's second hand market. In contrast, new, hybrid, or electronic cars are seen as less harmful for the environment than older used cars. However, the used-car industry is aware that especially young customers can become more profitable over the years, if there is a good customer relationship management. Additional after-sale services like in the car industry can also be offered in the fashion industry and in its second hand market. For example, a customer service for suits, formal wear, shoes, or hand bags include professional treatments, repair, or cleaning. Moreover, also the financing aspect can lead to a strong connection and could be offered especially for high-quality and premium garments. A strong relation to the customer has also great economic impacts. The recruitment of a new customer is five times more expensive as the preservation of a consisting one (Ebel and Hofer 2014). Furthermore, there can also be cross-industry learnings from the distribution process. Especially, the link between online platforms and offline suppliers can be very beneficial. Up to now these portals are mostly linked with private suppliers and not with second hand stores or even branded second hand stores. Also, the model for overpricing a returned garment in case the customer trades it in for a new garment can be a good model for fashion. While after a payoff the customer might spend the money for other goods, with the possibility of trade-in a new garment with a higher margin might be sold and also maybe additional products. Maybe vouchers can be seen as the fashion industries trade-in concepts. Moreover, especially branded and high-quality second hand stores can pay a higher price for used garments than regular stores, because of the higher reselling prices caused by their brand or image and can therefore have a return flow of value into their supply chain.

9.5.3 Infrastructure Management

For the development reuse or resell initiative from an infrastructure management perspective, the industry's current resources and activities, logistical practices and partnerships need to be analyzed. One of the key aspects to engage with the take-back of used products is the reverse supply chain management (Hvass 2015). However, a reverse supply chain is, as seen above, more complex as a traditional one. Most literature of the car industry's reverse logistic systems are dealing with the end of the life cycle of cars, with its scrapping and recycling. Used cars are seen as integrated part of the supply chain and the reuse as part of their life cycle, which is not analyzed here. However, also the take-back methods of used-car dealers are part of the reverse logistic management. For the used-car industry special software packages were invented, which contain complete vehicle management solutions for take-back and distribution of the car. It contains for example an interface to the valuation software, the database to set up the product in the used-car dealers program, the automatic creation of price tags, or the interface to online distribution platforms. Moreover, branded used-car dealers have the possibility to round their assortment by getting used cars from the manufacturer or other used-car dealers, which distribute the same brand. The car industry has a wide variety of partnerships or own programs, which support the branded used-car dealers. Automobile banks facilitate the purchase of second hand cars and also the financing offerings for customers. Furthermore, the dealers profit from sales financing programs, free trainings, or the support in advertisement by the automobile bank. Moreover, there is a network of partner workshops for service and repair, which creates a benefit for the customer. Also insurances are part of the after-sales product range and often realized with partners.

The support of software to assure a standardized system and to simplify the take-back and resale process is a very important part to develop in fashions second hand market. To establish second hand products in the fashion industry as normal and profitable as used cars, a solid background and partnerships are needed. If a big retail brand would establish second hand clothes as part of their assortment, the stores could also use database to round their assortment in relation to visual merchandising, sizes, or selection. Through the cooperation with banks and insurances, especially in relation with high end garments, the purchase would cause less risk for the customer. Even more high-priced garments could be available also to fashion-addicted people with low budget.

9.5.4 Financial Aspects

The business model pillar refers to revenue streams, costs, and other financial aspects. By the distribution of used cars the seller has an increased financial value by an increased customer base and increased loyalty due to a very good customer relationship management. Moreover, specially branded used-car dealers profit from the brand name and its image and are able to claim a higher price due to warranties and after-sale services. It must be outlined that the highest value and profit creation is made after the purchase for repair, accessories, and financing. Loyal customers are creating a huge immaterial value for the company (Ebel and Hofer 2014). The biggest costs are arising from the reverse logistic process like trade-in, potential repairs, preparation, and the distribution. Moreover, also price reductions related to trade-ins are causing costs (Zerres 2010). Furthermore, it is critical for car manufacturers and their distributors to control the resale of used cars so prices do not nose-dive. If used-car prices drop, new-car sales suffer (Alger 1995) due to the financing aspect.

Also fashion industry can gain a financial value through an increased customer base and increased loyalty. Moreover, it results a better supplier–purchaser relationship, since customers need to come to the store more often not only for purchase but also to sell their used items. However, since the second hand market in fashion industry is not linked as strongly with the market for new garments as in the used-car industry the control of resale does not seem to be very relevant, although new garment sales can be pushed to another level by including the aspect of trade-in.

References

Akerlof GA (1970) The market for "lemons": quality uncertainty and the market mechanism. Q J Econ 84(3):488–500

Alger A (1995) Luxury, slightly used. Forbes 156(7):238–239

Beamon BM (1999) Designing the green supply chain. Log Inf Manag 12(4):332–342. doi:10.1108/09576059910284159

Beard ND (2008) The branding of ethical fashion and the consumer: a luxury niche or mass-market reality? Fash Theory J Dress Body Cult 12(4):447–468. doi:10.2752/175174108X346931

Bianchi C, Birtwistle G (2010) Sell, give away, or donate: an exploratory study of fashion clothing disposal behaviour in two countries. Int Rev Retail Distrib Consum Res 20(3):353–368

Bloemhof J, van der Laan E, van Wassenhove L, Zuidwijk R (2009) A sustainable future for closed loop supply chains. https://rzuidwijk.files.wordpress.com/2012/11/26-chap-15-bloemhof-et-al-rz.pdf

Buttle M, Vyas D, Spinks C (2013) Evaluating the financial viability and resource implications for new business models in the clothing sector. WRAP. http://www.wrap.org.uk/sites/files/wrap/Clothing%20REBM%20Final%20Report%2005%2002%2013_0.pdf

Cervellon M, Carey L, Harms T (2012) Something old, something used: determinants of women's purchase of vintage fashion vs second-hand fashion. Int J Retail Distrib Manag 40(12):956–974. doi:10.1108/09590551211274946

Chan FTS, Chan HK, Jain V (2012) A framework of reverse logistics for the automobile industry. Int J Prod Res 50(5):1318–1331. doi:10.1080/00207543.2011.571929

Cooke P (2010) Used cars—a dealer's sales action plan for recession and recovery. BCA. http://www.buckingham.ac.uk/wp-content/uploads/2011/05/pnc-used-car-sales-action-plan.pdf

Diehlmann J, Häcker J (2013) Automotive management (2nd ed). Oldenbourg-Verl, München

Ebel B, Hofer MB (eds) (2014) Automotive management. Springer, Berlin. doi:10.1007/978-3-642-34068-0

Fletcher K (2014) Sustainable fashion and textiles: design journeys. Earthscan, London [ua]

Fletcher K, Grose L (2012) Fashion and sustainability: design for change. Laurence King Publishing, London

Gabbott M (1991) The role of product cues in assessing risk in second-hand markets. Eur J Mark 25(9):38–50. doi:10.1108/EUM0000000000623

Guiot D, Roux D (2010) A Second-hand Shoppers' Motivation Scale: antecedents, consequences, and implications for retailers. J Retail 86(4):355–371. doi:10.1016/j.jretai.2010.08.002

Hansen KT (2004) Helping or hindering? Controversies around the international second-hand clothing trade. Anthropol Today 20(4):3–9. doi:10.1111/j.0268-540X.2004.00280.x

Hvass KK (2015) Business model innovation through second hand retailing: a fashion industry case. J Corpor Citizsh 2015(57):11–32. doi:10.9774/GLEAF.4700.2015.ma.00004

Imperatives S (1987) Report of the World Commission on Environment and Development: our common future. United Nations. http://www.ask-force.org/web/Sustainability/Brundtland-Our-Common-Future-1987-2008.pdf

Kooreman P, Haan MA (2006) Price anomalies in the used car market. De Economist 154(1):41–62. doi:10.1007/s10645-006-6485-z

Lüdeke-Freund F (2009) Business model concepts in corporate sustainability contexts. From rhetoric to a generic template for business models for sustainability. Centre for Sustainability Management, Lüneburg

Osterwalder A (2004) The business model ontology: a proposition in a design science approach. http://www.uniempre.org.br/user-files/files/TheBusiness-Model-Ontology.pdf

Osterwalder A, Pigneur Y (2010) Business model generation: a handbook for visionaries, game changers, and challengers. Wiley. https://books.google.com/books?hl=en&lr=&id=UzuTAwAAQBAJ&oi=fnd&pg=PA9&dq=%22models+and+design+tomorrow%E2%80%99s%22+%22model+lens,+and+5+a+generic+process%22+%22business+model%22+%22and+talk+about+the+same+thing.+The+challenge%22+%22This+concept+has+been+applied+and%22+&ots=yWGOyhJ9Zy&sig=xzZ6d2XqhHZ5SFp0sRNaYaZZXps

Osterwalder A, Pigneur Y, Tucci CL (2005) Clarifying business models: origins, present, and future of the concept. Commun Assoc Inf Syst 16(1):1

Pedersen E, Andersen K (2013) The SocioLog.dx experience: a global expert study on sustainable fashion. Mistra Future Fashion. http://www.mistrafuturefashion.com/en/publications/Documents/CBS%202014-01-23%20Report%20Project%201.pdf

Pedersen E, Andersen K (2015) Sustainability innovators and anchor draggers: a global expert study on sustainable fashion. J Fash Mark Manag Int J 19(3):315–327. doi:10.1108/JFMM-08-2014-0059

Peterson JR, Schneider HS (2014) Adverse selection in the used-car market: evidence from purchase and repair patterns in the Consumer Expenditure Survey. RAND J Econ 45(1):140–154. doi:10.1111/1756-2171.12045

Roux D, Korchia M (2006) Am I what I wear? An exploratory study of symbolic meanings associated with secondhand clothing. Adv Consum Res 33:29–35

Roux D, Guiot D (2008) Measuring second-hand shopping motives, antecedents and consequences. Recherche et Applications En Marketing (English Edition) 23(4):63–91

Shambu G (2015) Using the fast fashion supply chain to teach sustainability. Bus Educ Innov J 7 (1). http://search.ebscohost.com/login.aspx?direct=true&profile=ehost&scope=site&authtype=crawler&jrnl=19450915&AN=108948448&h=psi2HWje8W1R618%2Ba3OSbr%2FKSuLjyAAIetojPuJnWit8R9ME0ayno%2BRukIxPNB0AuNuMogdRh3YmfFh7%2FBNm8g%3D%3D&crl=c

Stenner F (2010) Handbuch automobilbanken. Springer, Berlin, Heidelberg. Retrieved from http://link.springer.com/book/10.1007%2F978-3-642-01582-3

Svensson G (2007) Aspects of sustainable supply chain management (SSCM): conceptual framework and empirical example. Supply Chain Manag Int J 12(4):262–266. doi:10.1108/13598540710759781

Venkatesh VG (2010) Reverse logistics: an imperative area of research for fashion supply chain. IUP J Supply Chain Manag 7(1/2):77–89

Zerres C (2010) Preis-Promotions im Automobilverkauf. Gabler, Wiesbaden. doi:10.1007/978-3-8349-8642-9

Chapter 10
The Value Chain of a Branded Second Hand Store—Possible Activities to Be Integrated by a Conventional Fashion Brand

Jochen Strähle and Franziska Sophie Matthaei

Abstract The purpose of this paper is to study the recycling form of reusing second hand clothing from a conventional fashion brand's perspective. It should clarify which measures and activities a fashion company needs to integrate in its value chain in order to offer branded second hand merchandise in a self-operated store. The research paper relies on a desk-based research and aims to illustrate the topic by means of a descriptive approach, processing the existing literature. Key findings demonstrate that fashion brands need to integrate complete lifecycle strategies, sustainability communication, and reverse logistics structures, like take-back schemes, for offering second hand clothing. The main limitations evolve from the research design. Further, empirical evidences need to be conducted for a more fundamental understanding of the new business model.

Keywords Second hand fashion · Reverse logistics · Post-consumer textile waste · Product take-back · Value chain model

10.1 Introduction

Corporate Social Responsibility (CSR)—a strategic model that unites ethics and economic sustainability, that is oriented towards the environment, the people, the society as well as the present systems, resulting in a lower environmental and social impairment—is not unfamiliar to the fashion industry (Dickson et al. 2009). CSR strategies include, amongst others, the thought of prolonging the product lifecycle by implementing different recycling practices in order to preserve the created value and to diminish the textile waste.

J. Strähle (✉) · F.S. Matthaei
School of Textiles and Design, Reutlingen University, Reutlingen, Germany
e-mail: jochen.straehle@reutlingen-university.de

F.S. Matthaei
e-mail: franziskamatthaei@googlemail.com

J. Strähle (ed.), *Green Fashion Retail*, Springer Series in Fashion Business,
DOI 10.1007/978-981-10-2440-5_10

Since 2013 H&M collects used garments from its customers in order to recycle the materials for using these within the production of new clothing (Godelnik 2012). In cooperation with Goodwill the denim producer Levi Strauss & Co launched the initiative "A Care Tag for our Planet is Goodwill's." Here, product care tags were provided with messages that encouraged customers to donate their unwanted, worn clothes (Kant Hvass 2014). These are examples where specific, unsustainable problems are tackled within business models of conventional fashion brands that are not solely sustainable.

The Swedish premium brand Filippa K follows a rather holistic approach, which is rarely represented in the present fashion industry. The company is pioneering the closed-loop mindset as they shifted their linear production model to a circular scheme, which aims to prolong the product life to the highest extend. By following the "FOUR R's" that determine the Filippa K philosophy—REDUCE, REPAIR, REUSE, and RECYCLING—the company established an holistic system wherein "every product is built to be worn, adjusted, mended, loved and recycled" (Filippa 2014). "Long-lasting product of quality, style and simplicity, as well as enabling [their] products a second life" are the key strategies to ensure a long-term sustainable success (Filippa 2014). Their customers have the possibility to make use of the repair service for refurbishing damaged clothing. Used Filippa K garments are resold in a self-operated second hand store in Stockholm and currently a leasing service for apparel is tested, whereby the company promotes new, sustainable consumption patterns. As every garment has its end, Filippa K encourages their customers to bring back their worn-out Filippa K clothes, which are then recycled in "the most suitable way" (Filippa 2014). All these activities lead to the minimization of textile waste within the business model. This model is rather exceptional. As already mentioned, "[c]lothes are often discarded when much of their potential lifetime is left" (Farrant et al. 2010).

Since much of the potential useful life of a discarded garments remains, reoffering second hand fashion in order to extend the product lifetime, profiting from the remaining product value and at the same time decreasing the textile waste, has to be considered as a possible CSR strategy in the fashion industry. Changing the patterns of consumption through a business model, where fashion brands reoffer their own-branded products in self-operated retail stores, is relatively scarce. But which problems and trends evoke the appearance of second hand distribution formats? How does the second hand market look like? Which business structures are fundamental for performing in this segment? The answers to these questions will be derived from literature and research findings in order to present the superordinate topic and give an answer to the following question: The value chain of a branded second hand store—What are activities to be integrated in the value chain of a conventional fashion retailer in order to resell own-branded clothing in a self-operated second hand store?

10.2 Literature Review

10.2.1 Value Chain Model

The concept of the value chain was introduced by Michael Porter in course of his book "Competitive Advantage: Creating and Sustaining Superior Performance" in 1998 (Porter 1998). Organizations seek for creating a competitive advantage in order to generate profits and to differentiate from their competitors. Porter defines the value chain as a system for analyzing the sources of a firm's competitive advantage by describing and evaluating all functions, within a business model, that create or add value to a product or service that is delivered to the customer. These internal strategically relevant activities (p. 33) contribute to the overall strategic success of a company. Porter names them value activities (p. 38). He claims that the competitive advantage cannot be deduced from taking the company as a whole. In his model he therefore explains that many "discrete activities a firm performs in designing, producing, marketing, delivering, and supporting its product" have to be taken into consideration when assessing the competiveness. Predominantly, Porter discusses two strategies that create competitiveness: on one hand he describes the cost advantages and on the other hand he justifies the differentiation as a possible strategic approach (Porter 1998).

Each business model has its own value chain that reflects how inputs are turned into outputs. Hence, the value chains of various competing companies within an industry have to be considered individually and differ from one another. Individual execution and coordination of the activities within a value chain can create a competitive advantage against rivalry (Porter 1998).

Furthermore, Porter (1998) introduces the term value system, which is described as the interlinked individual value chains of different entities that exchange products and services, for example, the supplier's or buyer's value chain that are connected and/or depended on the value chain of the considered company. The value system influences a company's value chain by determining for example what the buyers need. According to Porter these trade relationships might lead to coalitions when individual value chains are coordinated or shared between buyers and suppliers. Generally, Porter asserts "value is the amount buyers are willing to pay for what a firm provides them" (p. 38). The definition of the term value is further extended that value is always a subjective experience when needs are met through the provision of products, services, or resources and the meaning of value always depends on the context. Moreover, they derive that it flows from the customer and "value occurs when [his] needs are satisfied through an exchange of products and/or services for some form of payment" (Feller et al. 2006). In the value chain concept by M. Porter "value" is expressed through the total revenue. The revenue derives from the value multiplied with the units sold. Profitability is realized when the profits exceed the total costs of a company. The margin represents the difference between the total costs and the total value. As a result the monetary value the customer is willing to pay must be higher than the costs for creating the product or delivering the service and must be the objective of every generic strategy (Porter 1998).

Value activities are the basic, technologically and strategically distinct elements, which are performed by a firm to create a valuable product or service for the customer. Moreover, these activities entail information in form of data. Depending on the way of execution, the performance and the operating efficiency, "economics" (p. 38), of each value chain activity, the cost structure, and the degree of differentiation determine the strength of a firm's competitive advantage. According to M. Porter, each value activity can be categorized into direct, indirect, or quality assurance activities (p. 44). Direct activities, as the naming implies, are directly involved in the value creation process, whereas the indirect activities continuously support the former. Furthermore, Porter distinguishes quality assurance from quality management. He exemplifies that quality assurance ensures the quality of each step in the value chain by monitoring, reporting, or testing for example and that it is nearly present in all departments of a firm. M. Porter distinguishes between primary and support activities that are divisible into sub-divisions depending on industry considered (see Fig. 10.1) (Porter 1998). Primary activities are client facing and are directly contributing to the margin, the physical product creation, the sale, and the maintenance of the products or services. Porter divides these activities into five categories: inbound logistics, operations, outbound logistics, marketing, and sales as well as service. The importance of each category within a business model depends on the industry. The inbound logistics include the supplier relationships as well as all activities that are needed to receive, store, and dispose the inputs. Outbound logistics complement these activities by collecting, storing, and distributing the outputs or finished goods to the buyers. Operations contain all steps that transform the procured raw materials into the finished goods or services. The Marketing and Sales departments provide the channels that enable and motivate the consumer to buy the product or service and to enhance sales, for example promotions, pricing, or advertisement. The final part of the primary value chain forms service, right after the customer has bought the product. This activity functions as the service "to enhance or maintain the value of the product" (p. 40), for instance a repair service, customer care institution, or the installation (Porter 1998).

Support activities are executed firm wide and they sustain the primary activities. M. Porter distinguishes between firm infrastructure, human resource management, technology development, and procurement. He clarifies that the infrastructure supports the entire value chain, whereas the other three support elements "can be associated with specific primary activities as well as support the entire chain" (p. 38). This is illustrated with the dotted lines in Fig. 10.1. Procurement includes all purchases that are made within the different steps in the value chain to enable the organization to execute the business and to provide the service or product. The technology development implies the know-how needed, the implemented procedures, and processes or the design as well as the technological equipment that is required. Human resource management includes all "activities involved in the recruiting, hiring, training, development, and compensation of all types of personnel" (p. 42). Additionally, every value chain includes an infrastructure that

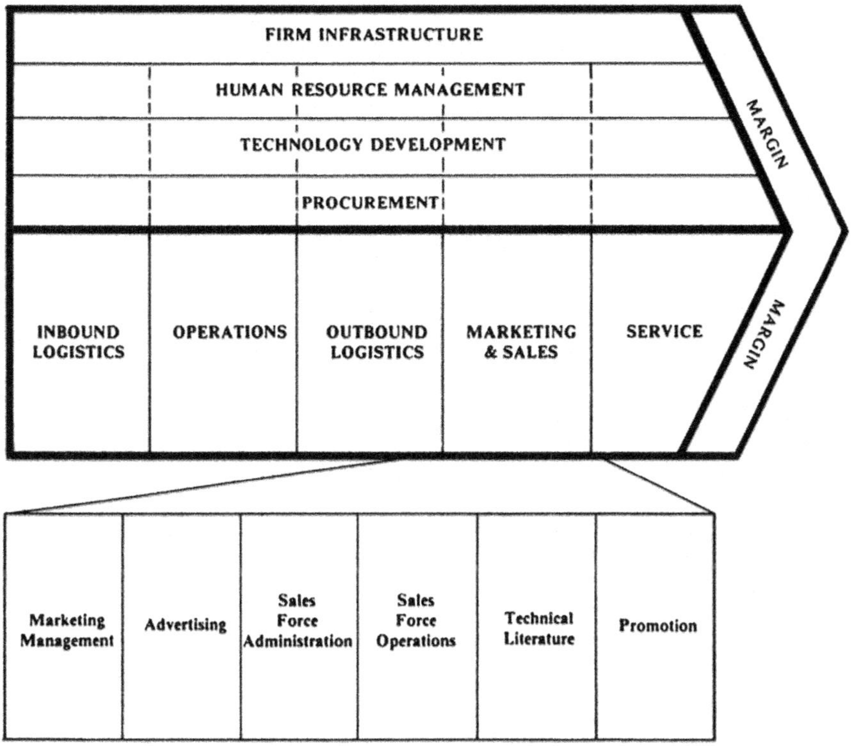

Fig. 10.1 The generic value chain and an exemplary subdivision. Adapted from Porter (1998)

might be built of general management structures, planning, finance and accounting, the legal department, or the quality management. These activities support the entire value chain (Porter 1998).

By applying the analysis tool by M. Porter, activities that create value to the customer can be clarified. It shows how the value is added an each category and activities can be evaluated for changes that would create more value and/or efficiency (Fig. 10.2).

In the course of this seminar paper the Value Chain Model by M. Porter serves as the frame for the depiction of the value chain of a branded second hand fashion store.

10.2.2 *Definition of the Term Brand*

In his work "Strategic Brand Management: building, measuring, and managing brand equity" Keller (2013) cites the definition of the term brand, given by the American Marketing Association (AMA), which says that a brand is a "name, term,

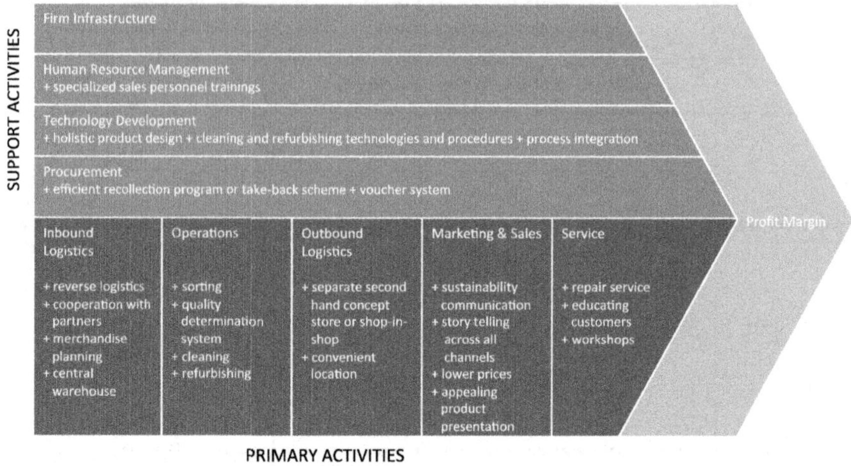

Fig. 10.2 Activities to be integrated in the value chain of a branded second hand store. Own visualization based on Chan et al. (2015), Choi et al. (2015), Connell and Kozar (2014), Kate Fletcher and Grose (2012), Kant Hvass (2014), Pal (2015), Porter (1998), and WRAP (2013)

sign, symbol, or design, or a combination of them, intended to identify the goods and services of one seller or group of sellers and to differentiate them from those of competition" (p. 30). The five characteristics can be summed up under the term brand elements (Keller 2013). Besides, in practice a brand is considered as a collection of objects that create awareness, reputation, or prominence among consumers on the market (Keller 2013).

In the context of this seminar paper branded apparel is recognizable and can clearly be defined and associated with one certain brand ["name, term, sign, symbol, or design, or a combination of them" (AMA)].

10.2.3 Sustainability in the Fashion Industry

For the concept of sustainability or sustainable development different definitions and perceptions exist. "The scope is broad, ranging from environmental protection to conservation of resources, habitat preservation, biodiversity, recyclable/pollution-free products, sustainable (in the sense of stable) operations, and fair working conditions" (D'heur 2015, p. 1). A widely accepted definition of sustainable development is the description, published by the Brundlandt Commission: "Sustainable development is development that meets the needs of the present generation without compromising the ability of future generations to meet their own needs. An economy where natural resources are used only to the extent that they can regenerate" (Brundtland 1987).

In order to adopt measures and business practices that support the sustainable development of a company, firms introduce Corporate Social Responsibility (CSR) programs (D'heur 2015). The EU Commission was the first institution that determined sustainability to be deep-seated with corporate responsibility by defining CSR as "the responsibility of enterprises for their impacts on society" (Europäische Kommission 2011, p. 7).

Especially the fashion industry, more precisely fast fashion companies, is often accused for their unsustainable business operations that have a negative impact on social systems and the environment. Toxic materials, child labor, low wages, the violation of human rights, raw material exploitation, or the high level of textile disposal on the landfills are concerns on the environmental impacts (Choi et al. 2015).

As the discussion on environmental impact of textile production and distribution processes gains more attention and the number of skeptical consumers and stakeholders increases, more and more fashion companies, including fast fashion models, develop their own CSR strategies. As a result they take new consumption modes and value chain models into consideration. Depending on the overall business model, the fashion companies choose their degree of integration of sustainable development measures and the focus area (Pal 2014).

Textile recycling is categorized as a sustainable fashion practice and approaches the problem of textile disposal. It can be further split into upcycling, downcycling, and reusing (Chan et al. 2015; Choi et al. 2015). The fast fashion retailer H&M clarifies that textiles represent 5 % of the landfill produced in the USA and 95 % of the discarded textiles are still recyclable (H&M Hennes and Mauritz AB 2013). These numbers show that a major part of the, in the fashion industry created, value is simply discarded and textile recycling is still a less covered topic.

In the course of the discussion about extended producer responsibility (EPR) conventional fashion companies started screening their "business models from a more holistic perspective in an attempt to incorporate complete life cycle strategies into their practices" (Kant Hvass 2014).

This seminar paper will deal with new concepts that can be classified as measures to enhance the sustainable development of a fashion company, fighting the loss of value due to textile disposal by especially focusing on the recycling technique of reusing.

10.2.4 Circular Economy as an Answer to Textile Waste

Fashion is characterized by fugacity, variations, and seasonality. Especially, fast fashion retailers focus on an increasing number of collections per year, promoting quickly changing trends, making the consumer following the market movements in order to buy as often and much as possible. At the same time fast fashion retailers are able to offer their assortments at a low price because they source the products in low-cost countries (Bly et al. 2015; Choi et al. 2015). On the other side of the coin the products are not durable due to the low quality provided. The consumer therefore tends to buy even more disposable fashion (Joy et al. 2012). Pushing disposable garments into the market contributes to one of the major problems in fashion and clothing industry: the textile waste (Claudio 2007). Whenever a fashion item is simply discarded, meaning thrown away in the landfill, all the materials and the energy used during the manufacturing, as well as the carbon emissions from the transport of the good along the supply chain, would be wasted (Choi et al. 2015). Especially when the clothing is cheap and not qualitative, consumers simply discard the old garments (Goworek et al. 2012). Another study divides the manners of disposal into four categories: discarding the used product by throwing it away, passing or selling the used product to a second hand shop or to another person, donating it to charity organizations, or loaning the item (Ha-Brookshire and Hodges 2009). A study conducted by the Waste & Resource Action Programme (WRAP 2012), focusing on the UK fashion consumption, reveals that a change in the supply, use, and disposal patterns might reduce and enhance the water, carbon, and waste footprint. They amplify that per year one-third of the whole UK textile consumption ends at the landfill, thus loosing its total value. If the textile waste would be sold or donated as second hand, or would be recycled, £ 140 million as additional income could be generated. The reselling or recycling of used garments could therefore function as a new revenue opportunity for retailers (WRAP 2012).

In 2013 the Ellen MacArthur Foundation published the "Towards the Circular Economy" report, which broaches the issue of the circular economy in the European Union. This concept represents the evolution from a linear, consumption-based take–make–dispose system that wastes resources and depends on a continuous material supply, to a "restorative" economic model that is designed to regenerate and reuse its resources as efficient as possible from the beginning by designing out all sources of waste. It could therefore be compared to a closed-loop system.

"By definition, a closed loop system is a societal system where products and their components are designed, manufactured, used and handled so as to circulate within society for as long as possible, with maximum usability, minimum adverse environmental impacts, minimum waste generation, and with the most efficient use of water, energy and other resources throughout their lifecycles" (Brismar 2015).

Further, elements that are included in the circular or closed-loop system are the use of renewable energy or the elimination of toxic, non-reusable materials. They claim that research efforts in the field of energy- and resource-efficient production are made but systematically eliminating waste and material leakage are exceptional. The foundation furthermore describes the limitations of the mostly applied "take-make-dispose" model. In this concept the company extracts raw materials and energy to produce a product, which is sold to the customer. The customer discards it when he or she wants to replace it with a new product and therefore all inputs are lost (Ellen MacArthur Foundation 2013).

Especially fashion manufacturers and retailers, managing a linear system, are dependent on the raw material supply, for example the cotton supply. Prices evolve to be volatile, the supply might be unstable due to extreme weather conditions or resource scarcities, and wages in Asia inflate (WRAP 2013).

As an answer to this problematic the Ellen MacArthur Foundation introduces the system of "four powers" that function as sources of value creation and increases material productivity within a circular economy model: the power of the inner circle, the power of circling longer, the power of cascaded use, and the power of pure circles (Ellen MacArthur Foundation 2013). The power of the inner circle represents the principle of minimizing the material usage within the value chain. The power of circling longer means maximizing the number of times a product or material is reused, remanufactured, or recycled, thus maximizing the number of consecutive cycles (Ellen MacArthur Foundation 2013). The power of cascaded use implies that the way of reusing a product or material within a value chain has to be diversified. The power of pure circles signifies that non-toxic, clean materials enhance the efficiency of reselling or reusing the products.

In literature possible, circular approaches to resolve the textile waste problematic as textile recycling, changes in the consumers disposal manners, or post-retail involvement by the fashion industry are seen (Kant Hvass 2014).

The seminar paper deals with the value chain of a branded second hand store. Second hand consumption echoes in the power of circling longer and the power of cascaded use (Ellen MacArthur Foundation 2013). Offering branded second hand apparel can therefore be understood as a possible approach for retailers to work in the direction of a circular economy system, by extending the loop, in order to avoid textile waste and as a way to take the post-retail responsibility of garments (Kant Hvass 2014).

10.2.5 Second Hand Clothing

10.2.5.1 Product Lifecycle Versus Useful Life

In literature the useful life of a product and the product lifecycle are differentiated. The useful life of a product expresses the period between acquiring a new product until it does not perform satisfactorily anymore. The product lifecycle has different meanings depending on whether one considers the manufacturers or the consumer's point of view. The consumer regards the period of usage between the purchasing and the discard or replacement of the product. For the manufacturer two concepts exist: the product lifecycle in marketing and in manufacturing. The marketing product lifecycle covers the time between the product introduction on the market and the withdrawal, when sales are declining. The manufacturing product lifecycle considers the same time span as the marketing perspective but also includes the post-production processes in the initial conception phase, such as front-end deals, design, or development (Murthy et al. 2008).

In the course of this seminar paper the product lifecycle from the consumers' point of view will be underlain. Normally, the lifecycle of a product is shorter than its useful life because consumers regularly replace used items by new products. This implies that the period of ownership or the product lifecycle is shorter than its useful life. Accordingly the user opts for a new product. These worn or used products still serve the purpose. As a result, the market for second hand products is created (Murthy et al. 2008). This phenomenon does also apply to the fashion consumption and therefore depicts the precondition for the second hand clothing market.

10.2.5.2 Second Hand Clothing Versus Vintage Clothing

In practice, the terms second hand and vintage are often used synonymously; however, in literature the meaning of each terminology is distinguished. Second hand clothing is defined as apparel that has already been used (Cervellon et al. 2012). In a further definition second hand is classified as a form of product recycling, which means that a used product is taken into a new stage of usage without changing the product design, only optionally refurbishing it (Morana 2006). Morana furthermore specifies it as the recycling form, named reusing. Morana determines that in this mode the product is reused with the same intended purpose as before. Complementary Chan et al. (2015) define reusing as "selling, exchanging, or giving away the used clothes or fashionable items" (p. 155). The recycling from described should not be confounded with reutilization where the raw materials of the used product are recycled for remanufacturing of new products (Morana 2006). The age of the clothing does not determine whether the piece of apparel is second hand or not (Cervellon et al. 2012).

Vintage clothing, on the other hand, is identified by the age of the clothing because the generally accepted definition says that it has to be manufactured between the 1920s and 1980s (Cervellon et al. 2012). Besides products that are produced before 1920 are defined as antique whereas clothing date from 1980 is classified as modern pieces (Cervellon et al. 2012). Moreover, it is described as "a rare and authentic piece that represents the style of a particular couturier or era" (Gerval 2008). According to Cervellon et al. (2012) the consumption mode of second hand and vintage merchandise is often confounded because "vintage pieces might be second hand and second hand pieces might be vintage, but [...] not all vintage pieces are used and not all second hand pieces are old" (p. 958).

10.2.5.3 Consumer Motivations for Buying Second Hand Clothing

Prior studies mainly show that the buying intentions for purchasing second hand apparel are from economic nature and are indirectly influenced by ecological consciousness (Cervellon et al. 2012; Guiot and Roux 2010; Williams and Paddock 2003). Cervellon et al. (2012) compare the motivations of female consumers to buy second hand or vintage fashion. Their results highlight that strong differences in customer profiles and motives exist. According to their findings consumers, buying vintage items, are motivated by nostalgia proneness, a high fashion involvement, the need for uniqueness as well as the fact that they enjoy the feeling of treasure hunting when buying vintage clothing. The higher the level of education, the more a person is willing to invest in vintage pieces.

To the contrary Cervellon et al. (2012) state that a high number of second hand shoppers are influenced by frugality and value consciousness. The study conducted by Williams and Paddock (2003) also proves that economic motives are the main drivers for second hand shopping. Both studies at the same time reveal that the level of income or education does not necessarily have an impact on second hand buying behavior. Williams and Paddock (2003) deepen this statement by highlighting that besides price-conscious households also affluent households engage in consume second hand apparel for social, recreational reasons. Guiot and Roux (2010) find that seeking for uniqueness is another driver for second hand consumption. Moreover, "[e]co-consciousness is related to the intention to purchase second hand pieces through the mediating effect of bargain hunting" explain Cervellon et al. (2012). The consumers see the importance in eco-friendly consumption and disposal patterns nevertheless not all of them do directly link second hand consumption to sustainable buying practices due to their lack of knowledge (Cervellon et al. 2012). Research on young fashion consumer's disposal habits, conducted by Morgan and Birtwistle (2009), supports the finding mentioned by revealing that this customer group also values sustainable consumption practices but their habits are not environmentally friendly. Furthermore, they explain that this customer group is willing to recycle and reuse products that are more qualitative and durable (Morgan and Birtwistle 2009). Another study shows that only consumers with strong environmental attitudes and background knowledge about sustainable clothing consumption opt for second hand clothing and are generally motivated to use alternative textile acquisition and disposal mechanisms (Connell and Kozar 2014).

In their comprehensive study Roux and Guiot (2008) define four general categories of second hand shoppers: the polymorphous enthusiasts, the thrifty critics, the nostalgic hedonists, and the regular specialist shoppers (p. 365–366). The polymorphous enthusiasts consume most second hand products through various second hand distribution channels. The thrifty critics are interested in the low prices of second hand products and want to save money by reusing and recycling. The nostalgic hedonists are rather focused on recreational motives, especially nostalgic dimensions when looking for and purchasing second hand pieces, as for example jewelry. The regular specialist shoppers make most of their second hand purchases in permanent stores, searching for certain product categories.

Cervellon et al. (2012) point out that it is necessary for fashion companies to clearly communicate the positive ecological value of second hand consumption in order to profit from the potential of the existing ecologically conscious customers. The study valuing our clothes (WRAP 2012), focusing the UK market, adds evidence on customer's motivation to consume more second hand fashion. It becomes clear that two-thirds of the participants buy or receive second hand apparel and they are motivated to consume more under the condition that a wider and better assortment range is available.

10.2.5.4 Distribution Formats for Second Hand Clothing

Second hand apparel is distributed through different formats. In their study about the environmental benefits from reusing clothes, conducted in 2010, Farrant, Olsen, and Wangel describe that the quality of the used item determines how and where the second hand product will be distributed. According to them products that are in good condition are often sold in the country where it is taken back from the first user via various distribution formats, generally in Western Europe or North America, whereas low-quality garments, where people can for example drop off in public collection containers, are exported to developing countries, as for example Africa. In these countries the used garments are sold to wholesalers who resell the bales to small retailers that distribute the products on informal markets. The Sub-Saharan area is the biggest second hand destination worldwide. The rest is sold to recycling companies or ends in the landfill.

As the seminar paper focuses on the distribution of branded second hand items in branded stores, the retail formats existing in the Western European and North American markets, "the countr[ies] of collection" (Farrant et al. 2010), will be further deepened for exemplifying the existing second hand fashion market in these core regions.

Roux and Guiot (2008) defined second hand buying as "the acquisition of used objects through often specific modes and places of exchange" (p. 66). Second hand buying is further classified as an environmentally sustainable form of clothing acquisition (Connell and Kozar 2014). Originally, the distribution of second hand products was concentrated on local formats where used products were sold at low prices. Mainly price-conscious clientele was attracted (Williams and Paddock 2003; Williams and Windebank 2000). According to Chan et al. (2015) second hand fashion became a fashion trend between the 1960s and 1970s. Since then an informal demand is growing and used clothes are sold and exchanged on "second hand markets, flea markets, swap meets and garage sales" (Roux and Guiot 2008) or with family and friends (Connell and Kozar 2014).

At the same time formal distribution channels like permanent second hand stores, offering mono-brand or multi-brand assortments, vintage clothing shops, and charity or thrift shops evolve. These formal distribution formats have a fixed, permanent location as common stationary retail stores (Chan et al. 2015; Connell and Kozar 2014).

An example for such a formal distribution type represents the Swedish clothing company Filippa K. They engage their customers to return their old garments and sell them in their own-branded second hand store in Stockholm in order to extend the product lifecycle and profit from the remaining value (Filippa 2014).

Consignment models, where customers can provide their used items, in a retail store or on an online platform, and receive a part of the revenue when the article is sold, also exist (Connell and Kozar 2014).

In Charity shops like Oxfam, consumers have the possibility to discard the clothing they do not want or need anymore. Afterwards the donated, qualitative products that are in good condition are resold at low prices to people in need. For the further utilization or the appropriate disposal of the used, unsalable items, Oxfam cooperates with local partners who are members of the umbrella organization “Fairwertung.” These companies take care of the appropriate arrangements. All the revenues generated by the Oxfam shops are reinvested in social or environmental projects of the “Oxfam e.V.” (Oxfam Deutschland Shops gGmbH 2016; Pal 2014).

The marketplace by ASOS, the German online exchange platform Kleiderkreisel, or eBay are examples for the distribution format of Internet platforms where mainly C2C business for second hand clothing takes place. The demand for such models is growing (asos.com Ltd 2016; Chan et al. 2015; eBay International AG 2016; Kleiderkreisel GmbH 2016).

Especially channels like formal second hand stores, informal second hand markets, as well as the exchange of worn garments on digital platforms in the Internet gain in importance due to the growing formal demand from buyers, the increasing offers, and accessibility of second hand garments (Pal 2014; Roux and Guiot 2008).

In addition to the mentioned sales channels today’s retailers develop innovative approaches in order to extent the lifecycle of clothes and to reduce textile waste. They offer for example upcycling or repair services, execute new retail modes like leasing or hiring garments, and, as already aforesaid, resell their used products in own-operated mono-brand retail stores or engage peer-to-peer commerce for second hand merchandise (WRAP 2013). Five new business models focusing on these strategies will be amplified in the following.

10.2.5.5 Innovative Business Models for Second Hand Clothing

In the course of the study “Evaluating the financial viability and resource implications for new business models in the clothing sector,” written by Buttle, Vyas and Spinks, published in by the Waste & Resource Action Programme (WRAP 2013), the practicability of five new business models that concentrate on lengthening the product lifecycle, on decreasing the amount of garments being discarded in landfills, as well as on downsizing the quantity of new clothes being produced, is assessed. They draw their conclusion from desk-based research, assumptions, and in-depth interviews with several businesses.

The first model is based on the perception that the majority of wasted textiles could be repaired and it addresses the problem that many consumers are not able to mend their clothing on their own as well as the fact that convenient repair services often are not accessible. In this concept the retailer or manufacturer offers in-store repair or upcycling services. Additionally, workshops where customers can learn how to upgrade their wardrobe themselves are provided. According to WRAP (2013) the provision of such services is increasing but the revenues are relatively low, especially due to the high cost base. As described in the study this model is likely to be adopted by rather local, small initiatives that function as potential partners for fashion retailers than by large providers.

Model two, a large-scale leasing service, is predicated on the fact that there exist several product categories that are only needed and used by consumers for a limited period of time, as for example baby clothes. The presence of such businesses is limited. The low demand and user acceptance for leasing garments might be a reason for this. Nevertheless, the German company "Lütte Leihen" experiments in this field. The results of the study show that this is the most effective model when considering utilization of the garment's value (WRAP 2013).

The third strategy assessed in the work deals with hiring services for clothes, which can be classified as a way of collaborative consumption. In practice, this model is often applied to formal, business and evening wear, as well as designer accessories, prevalently for high fashion items, encountering strong demand and user acceptance. The concept has a long history but in the recent years the online business revived its dynamics by making it possible to hire up-market fashion products to customers wherever they are situated. The garments can be rented for 10–15 % of the original retail price. This model might work financially; however, it produces waste and has the highest environmental impact (WRAP 2013).

The fourth model examined by WRAP in 2013 comprises businesses that resell branded pre-owned garments that they originally manufactured and sold, in a conventional retail store. Most of the times consumers receive incentives like vouchers for returning the used products to the retailer. According to WRAP "[t]he garments are … sent to a central warehouse for sorting, cleaning and re-styling and distributed back to flagship stores" (2013, p. 3). This model is valued as being commercially practicable the most due to the low-cost base, it generates reasonable profits, and the consumer interest/acceptance as well as waste savings is high. The study reveals that UK consumers buy second hand leisure clothes at least once per month.

Peer-to-peer exchange of second hand clothes constitutes Business model 5. Due to the fast development of social media and online platforms peer-to-peer trade models like eBay, "Kleiderkreisel" or the "marketplace" by ASOS, where vintage pieces are offered, evolved in the last years. Its cost base is relatively low. This does also apply to the revenues that are realized through a transaction rate the provider gains. Attracting users to contribute actively and trading high volumes, to actually generate profits, are the main challenges.

Business model four is evaluated to be most suitable for fashion brands that drive fashion trends and focusing on a sustainable strategy as well as producing in large scales (WRAP 2013). Therefore, the value chain model by Porter (1998) will be applied to a similar business model in this seminar paper.

10.2.5.6 Value Activities in the Second Hand Fashion Industry

When used garments that should be resold without changing the original purpose of the product are distributed through formats as described in Sects. 10.2.5.4 and 10.2.5.5, the systematic of the reverse logistics as well as a number of key value creating activities is performed (Pal 2015; Sas et al. 2015).

Companies establish reverse logistics systems in order to recollect, transport, recover, dispose properly, and resell products that are returned by customers. These products were originally manufactured by the collecting firm. The concerns are to operate the recollection and recovery processes efficiently in order to maximize the economic and environmental value of the used product and at the same time minimizing the costs of the reverse logistics operations (Sas et al. 2015).

When a fashion brand is involved in the product take-back, reselling, recycling, downcycling, and upcycling activities, these initiatives are called product stewardship (Kant Hvass 2014).

In the empirical study, conducted by Pal (2015), where key actors from the Swedish second hand clothing market were interviewed, the following value adding activities become apparent: collecting, sorting, refurbishing, as well as designing and redesigning. These value-generating activities can be classified as actions of product stewardship as well as components of a reverse logistics systematic.

Collect

If consumers are not motivated and attracted to bring their used clothing to a second hand store or charity shop, these distribution formats would not be able to offer a varied assortment and could therefore could not exist. Since the formal second hand retailers and charity organizations are dependent on returns or donations, the attraction of consumers returning their used clothes has to be ensured. The retailer, the manufacturer, or a third party can collect the used garments (Savaskan et al. 2004). Different in-store take-back schemes or direct donations to charities are executed by fashion brands like Levi's, Jack & Jones, or Patagonia (Kant Hvass 2014).

For the collection of the used garments a convenient location is essential (Ha-Brookshire and Hodges 2009). It is furthermore said that the collection of textiles through retail stores or collection containers "is only financially viable in areas with a high density of population" (Pal 2015, p. 547). Since 2013 H&M, in cooperation with the global recycling company I: Collect, offers a worldwide clothes collection initiative in its retail stores (Godelnik 2012). The customers have the possibility to return worn clothes that they do not need anymore to H&M. Big boxes are provided where the garments can be dropped and the customer receives a voucher for future purchases at H&M and to increase customer loyalty. As a business partner I: Collect assumes the following value-generating processes as sorting, recycling, and redistribution, for example in developing countries (Choi et al. 2015; Godelnik 2012; Kant Hvass 2014).

Another point to take into consideration when collecting second hand clothes is the fact that the consumers are often not sure whether the quality of the used garment is appropriate for reselling. The second hand retailers need to provide assistance in this area (Kant Hvass 2014).

Sort

In the value adding process of sorting the quality of the used garments is assessed and the further use is determined. The garments can be directly resold as second hand clothes to the consumer, the textiles can be recycled and be used for remanufacturing, or the garments can be exported to developing countries. Non-recyclable items have to be discarded on landfills (Kant Hvass 2014; Pal 2014, 2015).

Large fashion retailers that collect garments, mostly in lower quality, for the second hand clothing market, often cooperate with large scale, international sorters like the company I: Collect. These recycling companies manage the process of evaluation of the quality of the collected garments. Fashion companies can profit from their market knowledge as well as from the existing infrastructure (Pal 2015).

Brands that are reselling their collection in second hand stores, as for example the premium label Filippa K, have to sort the returned products manually and thoroughly. During this kind of sorting process the collected garments are classified depending on the remaining quality into three categories. Only the best maintained clothes are resold in own stores. The rest is either recycled for remanufacturing or given to charity organizations. For premium labels that reoffer the garments in their own-operated stores the sorting process is significant and value adding (Pal 2015).

Refurbish

When the collecting and sorting process is finalized those items that are classified as resalable are repaired, refurbished, or mended in order to regain the total functionality. In this step the product lifecycle and the useful life, as well as the durability, are extended. Moreover, the items are cleaned, often by professional laundry services. Nudie Jeans for example washes each returned jeans and sells the garments as "used own brand" in a special section in the store. For branded products neck labels, hang tags, or embroidery are added, depending on the marketing concept (Pal 2015).

Design and Redesign

In the stage of product design the end-of-use, especially if the clothing is mend to be resold and thereby reused, has to be taken into consideration (Dickson et al. 2009). A durable quality and a timeless design are necessary when the clothes should be resalable and multiple life cycles of products should be realized (Fletcher and Grose 2012; Kant Hvass 2014).

By redesigning the second hand clothing existing design characteristics of the used garment are changed and the redesign creates new value for the customer. Parts of the garment might be dis- and reassembled for extending the service life of the product. Adding pockets, buttons, rivets, or altering sizes can be forms of redesign. Nudie Jeans for example collaborated with designers and other creative who promoted the second life of the worn-out denim during the redesigning process (Pal 2015).

10.3 Discussion

Looking at the derivation of the key aspects in the field of the second hand clothing market from the literature review, it becomes clear that the distribution of second hand apparel is a method to evolve the existing value system in the fashion market toward a circular economic approach. This approach enables to profit from the remaining product value in the post-consumer phase, environmentally and economically. In today's fashion industry most of the created values are simply discarded by customers who consume clothing in high volumes and short product lifecycles. Due to the lack of knowledge about sustainable textile disposal options most of the products end up in the trash bin. The product lifecycle from the customer's point of view is much shorter than the useful life, because the clothing still serves its full functionality. This is the point where a new field of action and profit generation for fashion companies can be seen: the post-consumer product life. In the phase after the consumer has worn the product and decided to dispose it, reusing, or from a fashion brands perspective reselling, as a form of product recycling, represents a possible value adding business strategy. In the existing literature, the textile waste problematic, the export of low-quality second hand apparel to developing countries, and the reutilization of the textile materials for remanufacturing to extend the product lifecycle in the fashion industry are commonly known and researched practices. Nevertheless, the post-retail responsibility and thus the benefits and challenges, coming along with the business model of reselling own-branded garments in an own-operated second hand store, are not in the forefront of retail research and therefore not widely discussed. However, research clarifies that reselling of branded second hand garments can function as value preservation and value extension for fashion brands. These existing research findings make it worthwhile to look at this business model in particular from a value chain perspective by answering the following research question: What are activities to be integrated in the value chain of a fashion retailer in order to resell own-branded clothing in a self-operated second hand store?

The value chain model by Michael Porter in Sect. 10.2.1 serves as a framework and the distribution structure of Business model four (Sect. 10.2.5.5), portrayed by WRAP in 2013, is underlain to discuss the possible value adding activities, partially advised in literature, of a branded second hand store.

The findings in literature clarify that a prerequisite for operating a second hand business as a fashion brand is that the goods that are resold are highly qualitative and are made from durable materials (Kant Hvass 2014). The more popular a brand is, the more it is likely to be successfully sold in a second hand store (Chan et al. 2015). Therefore, the model will be applied to a premium fashion brand.

10.3.1 Primary Activities

The *Inbound logistics* have to enable the company to execute reverse logistics, meaning especially executing the process of collecting or taking back the used garments from the customers. The take-back points could be either located in the common retail stores or in the second hand store. The strategic collecting network has to ensure the collection of high-quality products in high volumes. The volumes are of prime importance because the assortment can only be planned when enough qualitative merchandise is available. Here, a further key value adding activity is the integration of a well-working transportation system that connects the collection activities with a central warehouse (WRAP 2013). Working together with a partner, specialized in the collection of clothes, can be advantageous because these companies have comprehensive market knowledge and experiences.

Outbound logistics enable a fashion company to distribute the products. In the case of branded second hand garments, a second hand concept store or a separated sales section (shop-in-shop) needs to be established as distribution channels. The location of the store is from significant importance because on the one hand it functions as the point of collection and on the other hand it is the location where the reselling takes place (Ha-Brookshire and Hodges 2009). According to Pal (2015) such business models are only financially viable in densely populated areas.

The *Operations*, including manual sorting, refurbishing, cleaning, mending, as well as repairing the garments, are value activities that have to be integrated in the value chain of a branded second hand store. "Product quality and product price are the most important basic requirements the consumers treasure on their purchase of second-hand clothing products. To be specific, second-hand clothing needs to be clean, tidy and with little differentiation (in these aspects) from the new clothes. Therefore, filtering and cleaning processes are essential" (Chan et al. 2015, p. 162). In this field of action it is essential for a brand to develop a deliberate system for determining the quality of the garments during the sorting procedures.

Depending on the size of the business "[i]ntegration with external organization [s], such as developing partnership[s] with professional recyclers and collectors, is crucial and helps ensure the UGC [Used Garment Collection] program associated reverse supply chain system is effective" (Choi et al. 2015, p. 194). Operational activities as well as activities in the field of *Inbound Logistics*, collaborations with experienced partners, like laundry services or textile collectors and sorters, are imaginable.

Often customers are not willing to purchase second hand apparel in stationary retail due to the perception the poorly organized merchandise in the stores (Connell and Kozar 2014). A catching *Marketing and Sales* strategy is therefore essential when reselling branded garments. Second hand stores are often perceived as being outmoded. An appealing product presentation, attracting visual merchandising ideas, as well as an innovative store design need to be ensured. Pal advises (2015) to focus on strong story telling across all communication and sales channels in order to get connected with the customers. In the second hand business it is even more important to focus on a strong customer relationship. The consumers should not only come and buy the products, but should rather be motivated to bring back their used garments. Most of the customers do not link second hand clothes to sustainable consumption. It is therefore important that the brand communicates the advantages of second hand products. Since literature shows that second hand buyers are motivated by economic incitements, the fashion brand needs to offer the second hand garments at a considerably lower price. Choi et al. (2015) investigated the effects on brand awareness and image building of used garment collection (UGC) schemes. They ascertained that there exists a correlation between UGC systems and the brand awareness and brand image. The study shows that fashion brands with strong CSR activities are perceived to be more honest and trustworthy (Choi et al. 2015). A garment collection scheme could therefore be further used as an effective marketing tool for gaining the customers trust.

In terms of *Service*, a clothing repair service (Chan et al. 2015), as Filippa K offers, could be a possible value adding activity. Furthermore, the brand could organize workshops, as Nudie Jeans does, where the customers are educated in sustainable clothing consumption measures or learn how to upgrade their used garments on their own.

10.3.2 Support Activities

Overall, the fashion brand needs to integrate the second hand business within the existing *Firm Infrastructure*. Only by integrating the post-retail activities in the superordinate value chain of the organization enable finding interfaces for collaboration between the different models.

Sales personnel with lacking knowledge about the sustainable aspects of alternative textile consumption mechanisms, who cannot provide information about the environment-related product features, is a barrier for customers to buy second hand apparel. They wish to be better informed by the salespeople (Connell and Kozar 2014). Therefore, in terms of *Human Resource Management*, it is essential for a second hand retailer to offer appropriate sales and product trainings for the sales staff.

A precondition for reoffering own-branded garments is ensuring durability, longevity, and quality of the products in the area of *Technology Development*. The fashion brand therefore needs to integrate the holistic approach combined with the thought of prolonging the product lifecycles, in the design process. Since second hand items are used and therefore not from the latest collections the used garment offered by the fashion brand will always lack in being completely trendy. At the same time fashion consumers ask for up-to-date assortments and fresh trends. It is well known that trends come and disappear. Thus, finding the right balance between simplicity and timelessness of the styles and at the same time meeting the present zeitgeist will be a value adding activity that needs to be integrated in the value chain. The design teams have to ensure right in the beginning of the creation process that the items are still wearable and fashionable in the second or third ownership stage.

The *Procurement* activities come along with the *Inbound Logistics*. The *procurement* works through convenient take-back schemes that function as enablers of making the customers return their used, branded garments. As seen in further examples, the company should offer vouchers of, for example, 15 % for future purchases, the second hand or latest collection section, or cash coupon (Choi et al. 2015) for the customer. Chan et al. (2015) recommend a membership program for the procurement of used clothing. The firm comes into closer contact with the customers and might enhance their motivation to return worn, branded clothes.

When considering the activities and structures that arise, when a fashion brand implements a branded second hand store as an approach towards a circular economy, benefits as well as challenges can be identified.

The reuse of clothes "introduces a potential for reducing the environmental impacts generated over the overall life cycle of clothing" (Farrant et al. 2010). Consumers and other stakeholders of fashion companies develop an eco-conscious mindset and as a result create pressure on the prevailing clothing production and consumption systems. Even less-informed and scrutinizing customers show willingness to consume more sustainably. Facing the movement to sustainability, mainly driven by the social expectations of consumers in the fashion industry, reducing the environmental impact, by applying new consumption modes as second hand retailing, evolves to be a beneficial business strategy for fashion brands. Especially establishing branded second hand stores represents an effective measure because it attracts new customer groups due to the sustainable communication as well as due to lower prices offered. This form of CSR engagement enhances the overall brand image and might contribute to the sales of the new branded garments in the common retail stores as well, as the customers gain more trust in the brands. Moreover, this approach helps fashion companies to make use of the created value efficiently, by surpassing it to new owner, it makes them less susceptible to volatile raw material supply markets, and it is supposed to be a cost saving business approach.

Looking at the benefits described reselling branded garments represents a new way to achieve profit for fashion brands.

Nevertheless transforming the prevailing business model entails various challenges. First, offering second hand clothing must suit the overall brand strategy, as the existing core customer base must accept and understand the extension of the current business model. Especially premium or luxury brands, which provide value in form of prestige, sometimes through artificial scarcity, are only available for a solvent clientele. Opening the arms for more price-conscious customers could be unpleasant for the regular customers. The customer base must still be willing to invest in the new collections. Furthermore, the fashion brand needs to find the appropriate balance between the simplification, which is needed for reoffering styles and prolonging the product lifecycles, and the fashionable, trendy degree of its assortments, which should attract consumers to buy new pieces. Another challenge might represent the assurance of product durability, as the brands cannot directly influence how their customers handle their wardrobe. A further threat is the dependence on the regular customer returns, as the second hand assortment can solely be planned and filled with the worn clothes. In this case, high marketing and motivational efforts are in demand. In addition, the new business model might require working with external companies, because establishing the distribution network and the logistics background is a complex task. This might be challenging in terms of negotiations. However, collaborating with experienced partners might emerge as a strong benefit when starting in the second hand market.

10.4 Conclusion and Future Research

Considering the mentioned value adding activities that need to be integrated in the value chain of a fashion brand, when offering branded second hand garments in a self-operated store, one can derive that this model can be utilized as a differentiation strategy.

It gives the customer the possibility to consume sustainably with a clear conscience and enjoy high qualitative, branded products at reasonable prices. At the same time the strategy fights the textile waste problematic. The value provided to the consumers is therefore materialistic and idealistic in nature. As this research paper is limited to a desk-based research, here is the starting point where future empirical research through quantitative or qualitative approaches should continue, searching for answers to unsolved questions like How would the broad consumer landscape react towards such approaches? Would they recognize the differences between new goods and elaborately refurbished second hand garments? How much would they be willing to pay? Which unknown measures and incentives could be taken to make the customer return the worn garments to the brand? And at the end, would the consumers still be willing in paying the full price for new garments from premium brands?

Above all, it could be derived from literature and research that for being able to perform such kind of business model, a fashion brand needs to integrate holistic and complete product lifecycle strategies in its value chain and as a key enabler reverse logistics practices have to be established. The whole construct represents a challenging business approach to the industry. Therefore, fashion brands should start collaborating with advanced partners and profit from their market knowledge and infrastructure.

"Working together for a world without waste" put it clearly and concisely: "any fashion brand at the forefront of addressing sustainability issues and driving fashion trends, such as the recent increase in anything vintage, would be an ideal candidate for uptake of this model" (WRAP 2013).

References

asos.com Ltd (2016) ASOS Marketplace. Retrieved 2 Jan 2016, from https://marketplace.asos.com/

Bly S, Gwozdz W, Reisch LA (2015) Exit from the high street: an exploratory study of sustainable fashion consumption pioneers: sustainable fashion consumption pioneers study. Int J Consum Stud 39(2):125–135. doi:10.1111/ijcs.12159

Brismar A (2015) Ask the expert: what is a "closed loop"—and does the approach work?|Make it last. Retrieved from http://makeitlast.se/2015/03/13/ask-the-expert-what-is-a-closed-loop-and-does-the-approach-work/

Brundtland GH (1987) Report of the world commission on environment and development: our common future. United Nations, Oslo. Retrieved from http://www.un-documents.net/wced-ocf.htm

Cervellon M, Carey L, Harms T (2012) Something old, something used: Determinants of women's purchase of vintage fashion vs second-hand fashion. Int J Retail Distrib Manag 40(12):956–974. doi:10.1108/09590551211274946

Chan H-L, Choi T-M, Lok JC-Y (2015) Mass market second-hand clothing retail operations in Hong Kong: a case study. In: Choi T-M, Cheng TCE (eds) Sustainable fashion supply chain management. Springer, Cham, pp 155–165. doi:10.1007/978-3-319-12703-3_8

Choi T-M, Guo S, Ho SS-Y, Li W-Y (2015) Effects of used garment collection programs in fast-fashion brands. In: Choi T-M, Cheng TCE (eds) Sustainable fashion supply chain management. Springer, Berlin, pp 183–197. doi:10.1007/978-3-319-12703-3_10

Claudio L (2007) Waste couture: environmental impact of the clothing industry. Environ Health Perspect 115(9):A449–A454

Connell KYH, Kozar JM (2014) Environmentally sustainable clothing consumption: knowledge, attitudes, and behavior. In: Muthu SS (ed) Roadmap to sustainable textiles and clothing. Springer Singapore, Singapore, pp 41–61. doi:10.1007/978-981-287-110-7_2

D'heur M (2015) shared.value.chain: profitable growth through sustainable value creation. In: D'heur M (ed) Sustainable value chain management. Springer, Cham, pp 1–107. Retrieved from doi:10.1007/978-3-319-12142-0_1

Dickson MA, Loker S, Eckman M (2009) Social responsibility in the global apparel industry, 1st edn. Fairchild Books. http://www.bloomsbury.com/us/social-responsibility-in-the-global-apparel-industry-9781563675928/

eBay International AG (2016) eBay Kleinanzeigen. Retrieved 2 Jan 2016, from http://www.ebay-kleinanzeigen.de/

Ellen MacArthur Foundation (2013) Circular economy report—towards the circular economy vol. 1. Cowes. http://www.ellenmacarthurfoundation.org/publications/towards-the-circular-economy-vol-1-an-economic-and-business-rationale-for-an-accelerated-transition

Europäische Kommission (2011) Eine neue EU-Strategie (2011–2014) für die soziale Verantwortung der Unternehmen (CSR). Brüssel. http://eur-lex.europa.eu/LexUriServ/LexUriServ.do?uri=COM:2011:0681:FIN:DE:PDF

Farrant L, Olsen SI, Wangel A (2010) Environmental benefits from reusing clothes. Int J Life Cycle Assess 15(7):726–736. doi:10.1007/s11367-010-0197-y

Feller A, Shunk DD, Callarman DT (2006) Value chains versus supply chains. http://webcache.googleusercontent.com/search?q=cache:s_AThpqK_8AJ:www.ceibs.edu/knowledge/papers/images/20060317/2847.pdf+&cd=7&hl=de&ct=clnk&gl=de

Filippa K (2014) Filippa K Sustainability Report 2014. Stockholm. http://filippa-k.com/media/wysiwyg/Filippa%20K_sustainability_report_2014.pdf

Fletcher K, Grose L (2012) Fashion and sustainability: design for change. Laurence King Publishing, London

Gerval O (2008) Fashion: concept to catwalk, 1st edn. Bloomsbury Publishing PLC, London

Godelnik R (2012) H&M launches first global clothing collection recycling program. Retrieved 22 Dec 2015, from http://www.triplepundit.com/2012/12/sustainable-hms-new-global-clothes-collecting-initiative/

Goworek H, Fisher T, Cooper T, Woodward S, Hiller A (2012) The sustainable clothing market: an evaluation of potential strategies for UK retailers. Int J Retail Distrib Manag 40(12):935–955. doi:10.1108/09590551211274937

Guiot D, Roux D (2010) A second-hand shoppers' motivation scale: antecedents, consequences, and implications for retailers. J Retail 86(4):355–371. doi:10.1016/j.jretai.2010.08.002

Ha-Brookshire JE, Hodges NN (2009) Socially responsible consumer behavior?: exploring used clothing donation behavior. Cloth Text Res J 27(3):179–196. doi:10.1177/0887302X08327199

H&M Hennes & Mauritz AB (2013) Conscious action highlights 2012. Stockholm. http://about.hm.com/content/dam/hm/about/documents/en/CSR/reports/Conscious%20Actions%20Highlights%202012_en.pdf

Joy A, Sherry JF, Venkatesh A, Wang J, Chan R (2012) Fast fashion, sustainability, and the ethical appeal of luxury brands. Fash Theory J Dress Body Cult 16(3):273–296. doi:10.2752/175174112X13340749707123

Kant Hvass K (2014) Post-retail responsibility of garments—a fashion industry perspective. J Fash Market Manag Int J 18(4):413–430. doi:10.1108/JFMM-01-2013-0005

Keller KL (2013) Strategic brand management: building, measuring, and managing brand equity, 4 edn, global edn, Pearson international edn. Pearson, Boston

Kleiderkreisel GmbH (2016) Kleiderkreisel. Retrieved 2 Jan 2016, from http://www.kleiderkreisel.de/about

Morana R (2006) Management von closed-loop supply chains. Deutscher Universitätsverlag, Wiesbaden. doi:10.1007/978-3-8350-9061-3

Morgan LR, Birtwistle G (2009) An investigation of young fashion consumers' disposal habits. Int J Consum Stud 33(2):190–198. doi:10.1111/j.1470-6431.2009.00756.x

Murthy DNP, Østerås T, Rausand M (2008) Product reliability. Springer, London. doi:10.1007/978-1-84800-271-5

Oxfam Deutschland Shops gGmbH (2016) Oxfam Deutschland Shops gGmbH. Retrieved 5 Jan 2016, from https://shops.oxfam.de/

Pal R (2014) Sustainable business development through designing approaches for fashion value chains. In: Muthu SS (ed) Roadmap to sustainable textiles and clothing. Springer Singapore, Singapore, pp. 227–261. doi:10.1007/978-981-287-110-7_9

Pal, R. (2015). Applying 3DCE for value creation in second-hand clothing chain: a Swedish study. In: Reflections on supply chain research and practice, 544–553

Porter ME (1998) Competitive advantage: creating and sustaining superior performance, 1st edn. Free Press, New York

Roux D, Guiot D (2008) Measuring second-hand shopping motives, antecedents and consequences. Recherche et Applications En Marketing (English Edition) 23(4):63–91

Sas I, Thoney KA, Joines JA, King RE, Woolard R (2015) Reverse logistics of US carpet recycling. In: Choi T-M, Cheng TCE (eds) Sustainable fashion supply chain management. Springer, Cham, pp 3–30. doi:10.1007/978-3-319-12703-3_1

Savaskan RC, Bhattacharya S, Van Wassenhove LN (2004) Closed-loop supply chain models with product remanufacturing. Manag Sci 50(2):239–252. doi:10.1287/mnsc.1030.0186

Williams C, Paddock C (2003) Thc mcanings of informal and second-hand retail channels: some evidence from Leicester. Int Rev Retail Distrib Consum Res 13(3):317–336. doi:10.1080/0959396032000101372

Williams CC, Windebank J (2000) Beyond formal retailing and consumer services: an examination of how households acquire goods and services. J Retail Consum Serv 7 (3):129–136. doi:10.1016/S0969-6989(99)00016-8

WRAP (2012) Valuing our clothes: the evidence base. WRAP, Working together for a world without waste. http://www.wrap.org.uk/content/valuing-our-clothes

WRAP (2013) Evaluating the financial viability and resource implications for new business models in the clothing sector. WRAP, Working together for a world with-out waste. http://www.wrap.org.uk/sites/files/wrap/Clothing%20REBM%20Final%20Report%2005%2002%2013_0.pdf

Chapter 11
Case Study: Loop Extension of Filippa K

Jochen Strähle and Alexander Schnaidt

Abstract The purpose of this paper is to investigate the use of sustainable closed-loop supply chain of the fashion brand Filippa K. Information on green fashion has been gathered and a case study approach on the fashion retailer Filippa K conducted. Results show a switch in knowledge content between a fast fashion supply chain and a sustainable supply chain. Also there is an evolution in sustainability as companies, retailers, and manufactures suffer under pressure from the customers, governments, and the media. Sustainable fashion brands like Filippa K are interested in sharing precise knowledge on variety of aspects linked to the sustainable closed-loop supply chain. This research paper has been limited by less information and unexplored topics in the theme green fashion. This led to the personal critical disputation with the brand Filippa K.

Keywords Sustainability · Closed-loop supply chain management · Fast fashion · Recycle · Reuse · Consumer behavior

11.1 Introduction

The recent development of 'Fast Fashion' has had a significant impact on buyer behavior and is raising concerns about environmental sustainability within the fashion industry and its broader supply chain.

In order to understand the sustainability issues it is necessary to understand what the term 'sustainability' actually means. Sustainability is much more than the relationship to the environment, it's about the relationship with communities, institutions, and individuals. Furthermore, sustainability involves changing environmental dynamics that affect human livelihoods and wellbeing, with intersecting

J. Strähle (✉) · A. Schnaidt
School of Textiles and Design, Reutlingen University, Reutlingen, Germany
e-mail: jochen.straehle@reutlingen-university.de

A. Schnaidt
e-mail: AlexanderSchnaidt91@gmail.com

J. Strähle (ed.), *Green Fashion Retail*, Springer Series in Fashion Business,
DOI 10.1007/978-981-10-2440-5_11

ecological, economic, and socio-political dimensions, both globally and locally (Langenwater 2009).

Over the past decade, sustainability and social responsibility have become more important in the fashion industry with these issues often being brought to the forefront through increased community and societal awareness and subsequent media pressure (Emberley 1998). Additionally, governments from around the globe, particularly those from developed western countries have begun to focus on the proper treatment of garment workers in developing countries, and the end of garments life cycles due to the cost and increased scarcity of landfills, in addition to the production of harmful emissions and the depletion of nonrenewable resources. This combined increased pressure to collect and reuse old products from customers, to minimize emissions and recover the residual value of the waste requires companies, retailers, and manufactures to develop and market eco-fashion and better promote sustainable consumption (Henninger et al. 2015).

Many of the world's largest clothing manufacturing countries, such as Bangladesh, China, India, and Pakistan do not only have to cope with the rising demands of the apparel industry, but also deal with growing pollution issues due to population growth, and increased water scarcity due to climate change. Hence, there is a strong need to reevaluate the fashion supply chain (sourcingjournalonline.com 2015). Similarly, the same can be said for fast and affordable and trend-sensitive fashion, while typically highly profitable, it raises a range of ethical issues (Aspers and Skov 2006).

In order to reduce sustainability and ethical issues in an era of increased consumption of fashion, sustainable supply chains are needed. A sustainable supply chain is defined as "the integrated management of all forward processes in the supply chain (procurement, production, distribution) as well as the processes (collection, testing, recovery) that aim to reintroduce returned products, parts and materials into the supply chain" (Bloemhof et al. 2012). To sum up, it should be socially responsible, economically sustainable, and environmentally friendly coordinating forward and reverse operations (Henninger et al. 2015). Closed-loop supply chains include five key activities, besides the known forward process, such as sourcing manufacturing and distribution:

- Acquisition/Collection;
- Reverse logistics;
- Inspection and Disposition;
- Reprocessing (remanufacturing, repair, recycle);
- Remarking.

Related to those five activities, Fig. 11.1 shows the most relevant processes in a closed-loop supply chain.

Within the closed-loop supply chain a sales process is happening between distribution (market launching) and the usage by the final consumer because a lot of products are returned right after their sale without ever having being used. These types of returns are known as 'commercial returns' and can be for a variety of different kind of reasons, however, these are fundamentally different from

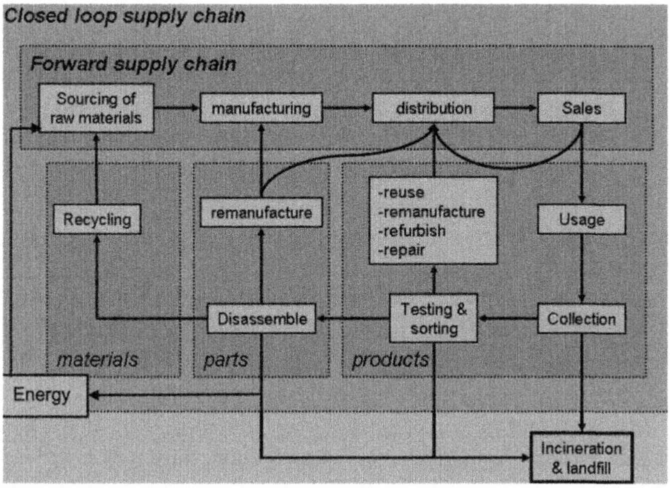

Fig. 11.1 The process of a closed-loop supply chain. Adapted from Bloemhof et al. (2012)

'end-of-use returns' and need to be taken care of differently as well. Hence, closed-loop supply chain management impacts sustainable products and services in three ways:

- Profit: Cheap resourcing through recovery of materials, party, and products give firms the chance to offer cheaper products with higher margins.
- Planet: Recovery of materials, parts, and products decrease the need for materials and energy and prevents landfill. Also reuse and repair use almost any resources while emissions are essentially lower than manufacturing.
- People: The number of jobs increases by the additional waste process and landfill.

The characteristics of the product and the circumstances in which the product is to be collected, establishes the optimal closed-loop supply chain configuration. For example, if the environmental impact is the usage phase, then it makes sense to design for environmental efficiency and safety in the usage period. The recovery options of the product depend on the economic endurance. In some cases, it is therefore better to recycle and replace the product with one that is environmentally more efficient and sometimes remanufacturing is the better solution. The positive environmental impact of product recovery balances the negative impact of more logistical activities. The point cost efficiency needs to be resolved through a better product design, strategy, and supply chain organization. Also energy has to be considered as a raw material. Thus in the forward supply chain energy sourcing occurs using fossil fuels (nonrenewable resources), solar energy and wind (renewable resources), and biofuels (recovered energy). In a nutshell, closing supply loops is about reusing and recycling products, parts, and materials and energy (Bloemhof et al. 2012).

The researcher and analyst Gwen Cunningham states "you cannot have infinite, unfettered growth and fast-fashion methods of consumption and production if you want to protect resources." From the sustainability side the fashion industry requires large amounts of natural resources and has to change how fashion is made because lots of these resources can be reduced, recycled, substituted, or eliminated (sourcingjournalonline.com 2015). But many companies are not using sustainable supply chain management because disposable fashion under the fast fashion concept is a trend that is also more profitable. The phrase 'fast fashion' refers to low-cost clothing collections that mimic current fashion trends (Joy et al. 2012). For example, Zara's fashion supply chain must be highly responsive to changes in the market and able to produce fashion in small quantities to satisfy the needs of the customers. The result is that new styles appear in the fashion market within a very short time period and the goal is to follow a final ready-to-sell process. In other words well produced and packaged products on the retail sales floor. The fashion concept leads to overconsumption because consumers only use the products for a short period of time before replacing them with new ones, but it leads to desired successful sales volume. Moreover, the fast fashion supply chain is known for involving hazardous materials in the production process, high volumes of pollutants and productions with low social responsibility. All in all, it can be said it shortens the valid lifecycle and generates more waste than regular clothing products (Henninger et al. 2015).

The fashion market needs to address many of these issues in order to approach the goal of a sustainable supply chain management (Henninger et al. 2015). One of the fashion brands in the industry that is notorious to use sustainable supply management and to make a big difference is called Filippa K. The Swedish fashion brand is focusing on sustainability and is trying to ensure the same quality and standards in working and production conditions throughout their supply chains. The designs are based around sophistication and simplicity. Considering the features of fashion products, Filippa K developed a system of a recycle- and reuse-based closed-loop supply chain to improve the sustainability of fashion products. This paper attempts to show by conceptually mapping the current situation, achievements, and commitments of Filippa K's sustainable closed-loop supply chain that the brand can survive against fast fashion retail companies and remain profitable whilst also being sustainable at the same time. Furthermore, it will give an understanding that consumers are accepting the model and this paper will highlight some of the challenges facing Filippa K's closed-loop supply chain.

11.2 About Filippa K

11.2.1 The Company

Filippa K is a Swedish clothing company, which is based in Stockholm, Sweden. The company was founded in 1993, by the fashion designer Filippa Knuttson and her husband Patrik Kihlborg. Filippa K has grown to be one of "the top leading fashion brands, with a strong position in long-lasting fashion" (Knutsson 2014). The successful journey of this iconic fashion brand started with the release of its very first product line, when Filippa K produced a small line of tight stretch jeans that immediately became very popular in Stockholm. The brand quickly grew and jerseys were soon added to its fashion line. After 20 years, Filippa K is still growing steadily and adding products. The brand describes itself as a "company with passion for design," which is linked to its strategy to make garments that are long lasting and stylish.

The company operates in 20 markets around the world through 50 brand stores, such as Belgium, Germany, Switzerland, and Scandinavia. In addition to that the fashion lines are represented by more than 600 retailers (Liberty, Bijenkorf, NK, Sockman, and Illum) and are also available via its e-commerce business.

Currently Filippa K has more than 350 employees, and owns seven subsidiary companies with its head office being based in Stockholm. Local offices are also located within Sweden, Norway, Finland, Denmark, the Netherlands, Belgium, and Germany. The enterprise offers fashion garments for men and woman, including shoes, bags, and accessories (Knutsson 2014).

11.2.1.1 The Vision

The Vision of Filippa K is "fashion where sustainability is the guide to growth." Planetary boundaries and ecosystems is the inspiration for the guide, with the brand wanting to be a part of the solution rather than adding global ecological problems related to sustainability.

Filippa Knuttsson, the founder of the brand Filippa K once said, "I set out to build a brand with substance and truth, not dependent on the superficial trends of the fashion industry." Hence, this remark refers to the overconsumption of textiles and describes the essence of Filippa K as a fashion brand that is not only wearable, but also aesthetically balanced as pieces will stand the test of time effortlessly, which is essential for garments of high quality and clean design. Filippa K believes "that design guided by conviction is the shortest path to fashion that is beautiful inside and out." In order to become more sustainable, Filipa K continued to develop garments that are personal, simple and long lasting. This is linked directly to the core values of the business, which aims to offer fashion that has style, simplicity, and is of high quality.

Through the development of a carefully curated wardrobe built on personal style and circular design with an integral approach to fashion business, the company describes the fashion industry and the subsequent intricate supply chain as challenging, long, and complex. And acknowledge that the production of clothing can often be responsible for the harming the natural environment and societies in developing countries where garments are often cheaply produced. Hence as a company they aim to be 100 % sustainable, through endeavoring to find every loop in the supply chain that is in the best interest of an environmental and social solution. In order to maintain their beliefs, Filippa K is trying to reduce any negative impact it makes by only choosing the best material; production practices and chooses to work in cooperation with experts who share the same ambitions. The main goal is not to produce more than needed or contribute to overconsumption in any other way (Knutsson 2014).

11.2.1.2 The Nine Planetary Boundaries

Filippa K calls out to respect a concept that is called the nine planetary boundaries (Filippa 2015a). Twenty-nine scientists developed the concept. They defined the limits of ecological development in the year 2009. Nine planetary boundaries is the central concept in an Earth system framework. The concept is designed to define a safe operating space for humanity, governments, international organizations, civil society, scientists, and the private sector as a precondition for sustainable development. The main driver of environmental change is humanity. Companies, organizations, enterprises, or others who cross these boundaries could risk negative environmental changes. However, some human actions have already crossed the boundaries, while other activities are in imminent danger of being crossed. The nine planetary boundaries include climate change, novel entities, stratospheric ozone depletion, atmospheric aerosol loading, ocean acidification, biochemical flows, freshwater use, land-system change, functional diversity, and biosphere integrity (functional diversity and genetic diversity). According to the following illustration, the green areas represent human activities that are within safe margins, the red areas show human activities that have exceeded safe margins, the yellow areas represent human activities that may or may not have exceeded safe margins and the gray areas with red questions marks point out human activities for which safe margins have not yet been determined (Rockström et al. 2009; Fig. 11.2).

Filippa K tries with their belief 'Long-Lasting Simplicity' to produce fashion that stays within these boundaries and is sustainable as possible (Filippa 2015a).

11.2.1.3 The Fair Wear Foundation

The brand Filippa K works together with the Fair Wear Foundation (FWF) to enhance labor conditions for garment workers. The FWF is an independent and nonprofit organization that also works with other companies and factories together to achieve this goal. The organization currently has 80 member companies that represent over 120 brands and they are active within 11 production countries across Asia, Europe, and Africa. The FWF helps to control the improvements made by the companies it works with, by helping to increase the effectiveness of the efforts made by the companies through the mutual sharing of expertise, and strengthening industrial relations and social dialog with garment workers and local communities. The overarching mission is to create good labor conditions for all garment workers all over the world, and prevent exploitation.

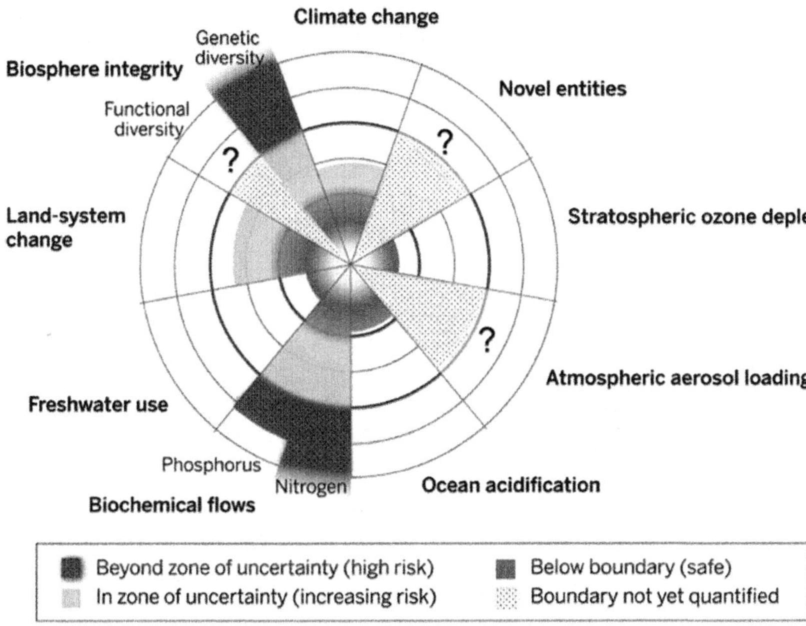

Fig. 11.2 Nine planetary boundaries. Adapted from Steffen et al. (2015) and Rockström et al. (2009)

This organization is not only open to companies that distribute products but also those who produce within the supply chain. Eight labor standards form the code of labor practices. To be a member of FWF you need to fulfill the following labor standards (Fear Wear Foundation 2015):

1. Employment is freely chosen;
2. There is no discrimination in employment;
3. No exploitation of child labor;
4. Freedom of association and the right to collective bargaining;
5. Payment of living wage;
6. No excessive working hours;
7. Save and healthy working conditions;
8. Legally binding employment relationship.

To secure good labor standards in the supply chain and help ensure compliance to their code of conduct, Filippa K has worked with the Fair Wear Foundation since 2008. The main goal is to maintain the human rights of workers and to ensure that suppliers support fair working conditions as part of their social responsibility, along with offering secure and healthy working conditions to their employees. Through this multi-stakeholder initiative factory audits are conducted and this helps Filippa K to consistently improving its action plans when needed. The FWF supports its member's development through an annual performance check, which analyses the processes and ways of working with social matters. In addition to that the organization supervises the development in several countries and conducts risk assessments to determine where to concentrate on social auditing. As a member of the Fair Wear Foundation, Filippa K is committed to audit 90 % of its suppliers in countries that are considered high risk for exploitation, a level that they overstepped for the last 2 years (Filippa 2015b). More information about auditing by Filippa K can be read in Sect. 11.2.4.1.

11.2.1.4 Initiatives and Organizations

Filippa K supports not only fair labor conditions, but is also a member of many initiatives and partner of organizations, and support a range of initiatives to stay in direct dialog. Through this approach they aim to increase their knowledge and awareness of new processes and systems that enable more sustainable production activities. In order to be socially active, they have built partnerships with a number of global nonprofit organizations.

For example, 'The Hunger Project' is an organization they support, despite not being of any direct value to the business way. However, the values of this project align well with Filippa K who's own belief is about creating sustainable relationships. The mission is to end hunger and poverty by pioneering sustainable, grassroots, women-centered strategies, and advocating for their widespread adoption in countries throughout the world, particular those in Africa, Asia, and Latin America. The Hunger Project is especially focused on ensuring that every person can fulfill a life of self-reliance and dignity (The Hunger Project 2015).

11.2.2 The Method: The Four R's

Filippa K describes the journey toward sustainability as a learning process. It is important to drive change and to learn from every step. Learning, determination, and hard work are the key to a better process that can, and will be adopted in every garment piece. The company is trying to explore every detail of sustainable fashion, develop ideas, and concept to find the most ecological solution. For that reason, they take one product group at the time and involve coworkers, suppliers, customers, and many experts to create better outcomes. Through the support of experts, Filippa K traveled across the world to get a better understanding of a sustainable life cycle evaluation—from cotton fields to the garbage can.

Additionally, they asked random people about their wardrobes to find out what a fashion conscious society really wants to wear, with the aim being to produce fashion garments that appeal to the taste of as many people as possible for a long period of time. In addition to live up their vision to offer "fashion where sustainability is the guide to growth," Filippa K invited 30 people of different nationalities from other parts of their supply chain and nongovernment organizations to a workshop in spring 2014. The idea was to get everyone motivated and especially committed and agree to the firm vision (Filippa 2015d).

11.2.2.1 Reduce

Filippa K is creating products with a minimal negative impact that has become part of the organizations mindset. For each season they select a few styles in order to be as sustainable as possible before they move on the next set of products. Also, they ensure that their fashion line is always clean and simple instead of excessive. Filippa K represents the opinion that a personal wardrobe should last and happen over time. The company tries with the help of the customers to select, maintain, and estimate the wardrobe. To reach this aim, Filippa K has to exanimate each aspect. It starts by choosing the materials for the season and ends with how the consumer will use and take care of the product until it reaches its end of life.

One of the main ideas is called ‘Circular Design.’ It is a concept inspired by ecosystems where nothing goes to waste and everything is reused or recycled. This means that Filippa K must continuously minimize the negative social and environmental impact it makes. Every product is created to be long lasting in style and of high quality.

While the brand overthinks, redesigns, and rebuilds the processes and structures it never compromises on its design. They truly believe in a circular design mindset where nothing goes to waste and every material can be reused or recycled. The production models are shifted from linear to circular and destroy harmful and wasteful activities (Filippa 2015d).

11.2.2.2 Repair

Filippa K likes to minimize its environmental footprint during the use phase of a garment and where possible extend a garments lifetime. In order to this, they have invested in educating and helping their customers to take better care of their garments so that they become the long-lasting sustainable products they were designed to be. In addition to this, in 2014 Filippa K introduced ecological washing products with Tangent GC (Ecological garment care company from Sweden). The products are called ‘Filippa K Fabric Wash’ (detergent) and ‘Filippa K Fabric Care’ (fabric softener). These products combine mild and efficient care to let the garments last as long as possible.

Furthermore, Filippa K tries to repair all products that become broken or worn with time, for example, a fallen button, to a rip or a tear in a blouse. This service is free for customers. In event that the product cannot be repaired, Filippa K will pass it onto a charity organization to ensure nothing goes to waste (Filippa 2015d).

11.2.2.3 Reuse

Filippa K opened a secondhand store in 2008 allowing customers to return used garments, giving these items a new lease on life by these products being passed on. The idea was to prove that the products stand the test of time, in style and quality.

Also, this demonstrated an alternative to the practice of shopping and removal. Filippa K built up a model that takes full responsibility for the entire fashion life cycle of the garments.

Besides creating a wardrobe that is simple instead of exorbitant, they have developed new ways for sustainable fashion consumption. This is why Filippa K is testing another model where people can lease the products rather than buying new items. The Filippa K Lease concept has existed since early 2015 and is currently in nine following Filippa K store countries including Sweden, Denmark, Belgium, Netherlands, Norway, and Finland.

Consumers have the option to test a wardrobe piece or use it only for a short time for a special event. The customer can lease any new product for 4 days at 20 % of the full price. If the product has been leased before to another customer, the cost is 15 % of the full price instead. Additional costs such as cleaning or fixing a garment is included in the price. Through this initiative Filippa K has shown another way to help with the environmental waste impact of fashion, and is also a great opportunity for keeping ones wardrobe curated and updated. The customer is allowed to temporarily own extra product pieces. This concept has helped the overall ambition to create products inspired by minimal footprint and long-lasting simplicity (Filippa 2015d).

Filippa K not only offers the Lease concept in their own stores, it also cooperates with organizations that serve clothing libraries, such as Lånegarderoben (The Closet Library) in Sweden and Lena—The Fashion Library in Holland. These companies require users to buy a membership that allows them to borrow clothes for a limited period of time (Lanegarderoben 2015). In the past few years, more and more clothing libraries appeared in industrialized countries. They were established for a reaction to clothing overconsumption and environmental concerns associated with the fashion industry. Especially in Sweden is the phenomenon spreading quickly (Esculapio 2015).

11.2.2.4 Recycle

The last method oft he four R's is Recycle. Filippa K believes that worn-out clothes still have value. They want the customers' old worn-out garments back to recycle them. For this reason, they developed a collecting system for Filippa K clothes in their stores in Sweden and Denmark in 2015. The concept collect from Filippa K is a step toward a circular economy. Every customer that returns a Filippa K product that is no longer used, to a local store receives a 15 % voucher to be used on the next purchase in a Filippa K store. Through that system the company has the ability to give clothes new life through secondhand stores. If this is not possible because the clothes are not in shape to sell anymore, they want garments to be recycled as materials for new products. In order to recycle the old products and improve the chances that textile fibers do not disappear in material loops Filippa K works with humanitarian organizations (Filippa 2015d).

In addition, Filippa K works together with Cirqle. Cirqle is an app that guides consumers to stores where used garments are collected in Sweden. The individual that has donated clothes gets rewarded through the app, e.g., a digital discount. The local partners for the clothes that are given away are:

- Sweden: Stadsmissionen
- Norway: Fretex
- Denmark: Røde kors
- Netherlands: Het Rode Kruis
- Belgium: Oxfam
- Finland: Fida International

11.2.3 The Front Runners

In 2013, Filippa K produced an organic cotton t-shirt with the help of experts from the Natural Step. The Natural Step is a nonprofit organization founded in Sweden in 1989 by scientist Karl-Henrik Robért. The Natural Step framework is a simple science-based tool for analyzing the complex issues associated with sustainable development (Robért 2015). They organized a sustainable life cycle assessment (SLCA) to produce a t-shirt. To makes sure, Filippa K followed the t-shirt all the way from the cotton fields, through the production phase, user phase, and finally to the end of its existence when the garments are burned and used as energy. The t-shirt turned out to be one of the best-selling products. After 2 years, the company launched the first three Filippa K front runners of long-lasting simplicity. These products were a jersey shirt, top, and a dress made of 100 % natural and renewable fibers.

Filippa K is following all steps of a product in a value chain. The most important decisions are taken during the design phase because they determine how sustainable things will be in the entire value chain. From stumbling on the design of the cotton t-shirt they encountered first-hand the challenges involved in sustainable production. This includes issues from the fabric itself, the label, the thread, the hangtag, the string, and the pin that holds the hangtag. They explored a myriad of raw materials and their effects on both people and the environment.

Also the brand discovered more efficient waste resources, chemical use, energy consumption, transportation and distribution, implications for people and communities, ecosystem impact, and a lot more.

The other aspect of a front runner was to figure out when it should become a wardrobe favorite, along with other important key factors, such as style, function, simplicity, and quality. Filippa K always emphases that style goes before trend because a product that is in this year could be out next year, which limits its longevity. As mentioned before, Filippa K knocked on people's doors to undertake a survey with the Swedish industrial design agency 'People People' of a wardrobe that is durable and beautiful for as many customers as possible (Filippa 2015c).

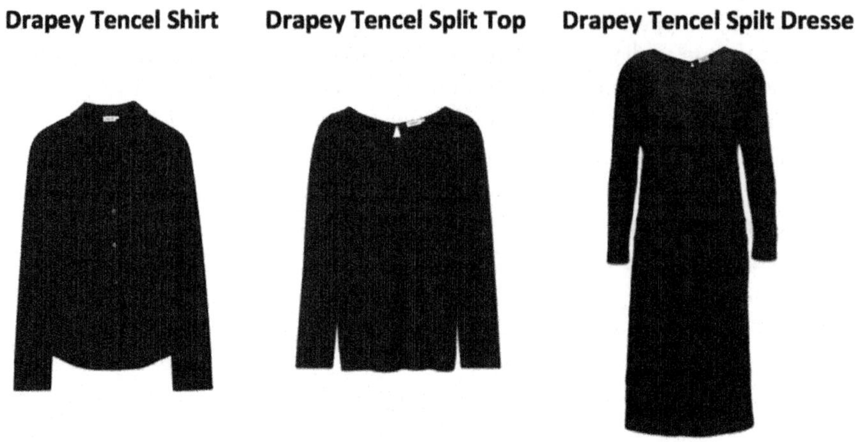

Fig. 11.3 Filippa K front runners. Own illustration adapted from Filippa (2015c)

In addition, the front runners are mostly black even so it is the least environmentally friendly color choice if seen to the production phase alone because people tend to combine this color with different garments (Filippa 2015c).

The Fig. 11.3 shows the three Filippa K front runners of the last season (Filippa 2015c):

Filippa K decided on 12 criteria that must be fulfilled to be a long lasting front runner. This measure was developed after having developed the knowledge to sustainable fashion first-hand.

11.2.3.1 Sustainable Materials

The first criterion is the choice of sustainable materials. Filippa K decides carefully upon which fabrics or threads to use for its products. For example, organic cotton with restrictions on chemical fertilizers and pesticides is a much better solution than conventional cotton. The disadvantage is that is still requires a huge amount of water during the cultivation and processing. Unlike polyester, cotton and polyamide are not as renewable as the materials are made from petroleum oil. This is one of the reasons why they have chosen the material tencel for the main thread, lining and fabric. Also tencel uses less chemicals, water and is coming from a closed-loop process where nearly all chemicals and water are reused again. For the button's Filippa K uses the sliced and dried palm trees from the company Corozo Buttons (Filippa 2015c). The firm manufactures genuine, high quality natural buttons. Some advantages are dry-clean washable, scratch and high temperature resistant, and 100 % natural (non toxic). They specialize in custom made tagua/corozo buttons for clients in big and small, available anywhere around the globe (Corozzobottons 2015).

11.2.3.2 Recycable

One criterion for the front runners product is recyclability. The difficulty in recycling a product is the separation of natural and synthetic fibers such as cotton an elastane. As just mentioned, Filippa K is using tencel not just because it is sustainable but also because it is a renewable cellulose-based material derived from wood fibers. When tencel eventually reaches the end of its product life it can be reused again by recycling into new cellulose fibers. To let the customers know that Filippa K is recycling them, they put a tag inside the garment that says: “Our clothes are made to last. To let it last even longer, pass it on to others if you don’t wish to wear it anymore. If worn out, we welcome you to return it to our Filippa K stores for recycling” (Filippa 2015c).

The recycling process from the brand Filippa K works in the following way: The old tencel clothes are brought from the Filippa K Stores including secondhand stores to a recycling company. Filippa K works together with the technology company called Re:newcell, which is focusing on developing sustainable processes for recycling garments in order to more efficiently use the resources that exist in the world today (Re:newcell 2015). The company brings the cotton to a factory that shreds the textiles to a porridge-like substance. The next step is to remove nonrecyclable pieces like buttons and zippers from the porridge. Now the porridge is broken down to a molecule level and turned into a fiber substance that can be used for thread. The best result is achieved by using pure cottons, but in general a mix of cotton can be used as well (Braw 2014).

Filippa K is also cooperating with other firms, such as Naturvårdsverket, which is a Swedish environmental protection agency. Naturvårdsverket is one of many stakeholders with responsibility for the environment. They have the task of monitoring, coordinating, and evaluating efforts involving other stakeholders to meet environmental objectives (Oscarsson 2015). They chose to work closely with this agency to develop more infrastructures and achieve better recycling methods (Filippa 2015c).

11.2.3.3 Transparent Supply Chain

Filippa K believes in cooperation through an open and honest dialog. This requires a solid commitment towards working efficiently and effectively. They set themselves goals, which they frequently measure to ensure that they keep in line with their expectations and those of their stakeholders. Filippa K represents the opinion that transparency in the supply chain is essential if they want to accomplish more sustainable products. Another important aspect is that Filippa K needs to ensure that their partners work in compliance with their ethical, social and environmental standards. This is the reason why Filippa K has traced every component of the front runners in the whole value chain from fiber extraction to the finished garment.

Filippa K knows exactly who makes the main fabric, lining, hangtags, buttons, zippers, and other attachments for their garments (Filippa 2014). The following suppliers and factories are used for the front runners:

- Tencel fibre: Lenzing—Austria
- Spinnery: Mercomahla—Portugal
- Cut and Sew: Caetano and Reistex—Portugal
- Corozo buttons: Nut—Eucador, Betussoni—Italy
- Tencel interlining: Wendler—Germany
- Tencel thread: Realfio—Portugal
- Freight forwarder: Garland—Portugal, First Cargo—Sweden
- Warehouse: DHL Texport—Sweden

11.2.3.4 Minimal Use of Resources

Filippa K uses as little resources as possible. One of the firm mottos is 'less is more.' This is especially important for the front runners. The main aim is to avoid wasteful processes. That is why Filippa K is continuously searching for practices that could potentially harm the environment in conjunction with its other stakeholders. They work closely with their supply chain to reduce the use of resources. The usage of energy and fossil fuel is extremely fundamental. Also they are trying to prevent the most water-intensive processes in all areas, especially in areas with narrow water supply. In addition, Filippa K has removed the safety pin on the products to save on metal resources. The plastic bags are made out of 50 % recycled material and the cardboard boxes are made from recycled paper (Filippa 2014).

11.2.3.5 Minimal Waste

Filippa K is calling out to the people to throw fewer clothes away and instead to reuse or recycle them. In the sewing factories, fabric waste is one of the major struggles during the production. This is why they optimized all the front runners to avoid waste from fabric cutting. The cut-offs that cannot be used for the product are either used to make new scarves or sent to the recyclers to become raw material for new garments (Filippa 2014).

11.2.3.6 Less Chemicals

Filippa K is limiting and trying to eliminate the long term the use of feasible harmful chemicals in the production of all garments. Nowadays, chemicals are basic ingredients for clothing production processes. This brings many challenges that can only be solved through the cooperation with others. This is why Filippa K has put in place a chemical restriction list that suppliers must follow. Also, they are part of several networks and initiatives, such as a chemical research group led by Swerea. The goal of Swerea is to create competitive advantages for the companies in product, process, and production development (Filippa 2014). They offer expertise in relation to materials properties and applications for textile materials (Swerea swedish research 2015). Also they are working with another agency called KEMi, which is working to reduce the risks to the environment and people from being harmed by chemical products (KEMI 2015).

11.2.3.7 Minimal Emissions

Today, pollution is a worldwide problem that is growing. Challenges that the fashion industry has to face are emissions to the air, ground, and water. In order to become more sustainable Filippa K has to narrow these difficulties down. The main problem for climate change is the use of fossil fuels in productions and especially in transportation. Filippa K wants their suppliers to invest in renewable energy sources as much as possible. One step in reducing the use of fossil fuels is by taking more trains and boats instead of trucks and planes. Filippa K is using them as well while still managing their lead times. In order to minimize emissions to the ground and water, Filippa K is cleaning the water and using sludge management. For example, the 'front runner dyer' uses an anaerobic water purifying plant to clean the water before it is released in Portugal (Filippa 2015c).

11.2.3.8 Respect for People

As already said, Filippa K clearly shows its respect for people by ensuring that all workers in the supply chain are treated safe, fair, and healthy. The working conditions and the wages are essential traits. They are expecting suppliers of all products to observe the code of conduct. This is based on the International Labor Organization criteria and is supported by the fair wear foundation (Filippa 2015c).

11.2.3.9 Respect for Animal Welfare

Filippa K is a long time member of the largest animal rights organization in Scandinavia, Djurens Rätt. This organization fights for animal rights in Sweden. For the company it is important that animals are treated with respect. In addition, animals should be protected from unnecessary stress and strain. None of the current front runners contain fur, leather, wool, angora wool, or feathers from animals. The overall goal is to only use fur and leather from cows, buffalos, goats, pigs, and sheep that have been raised for meet production (Filippa 2015c).

11.2.3.10 Long-Lasting in Design and Quality

It is important to start with decisions at the beginning of a value chain. Filippa K takes these decisions during the design phase to show how sustainable the products will be at the end of the value chain. This goes even back to the foundation of the company in 1993 when they decided to produce clothes that is all about long-lasting design and quality. The goal was to design clothes which were made to last. Long lasting also means to pass the garments on to friends or others if the customer wishes to not wear them anymore. Filippa K welcomes customers to return their garments to a Filippa K stores for usage within their own secondhand stores or in order to be recycled. Furthermore, the front runners are available in-store for at least 1 year (2 seasons). In the fashion industry 12 months is a short eternity (Filippa 2015c).

11.2.3.11 Perfect Fit and Comfort

Filippa K thinks that garments will only become ‘wardrobe favorites’ if they are comfortable, fit well, and are easy to laundry no matter how stylish they are. This is why staff at the company also wears front runners 1 month before they start the production to see if these products can handle all of life’s situations perfectly. The ambition is to ensure that these products live up to their desired standards (Filippa 2015c).

11.2.3.12 Financial Sound

Filippa K embraces the view that a product is not sustainable if it is not financially available for the consumer. Ensuring that this can be achieved while also maintaining their own bottom line as well as that of their partners in the supply chain the company is successfully producing the front runners and a range of other products. To ensure efficient and stable production planning Filippa K focuses on a professional and long-term partnership with their suppliers. Trust is always a key point between these partnerships. If the business is profitable, Filippa K wants to invest more money in sustainable practices along the total supply chain (Filippa 2015c).

11.2.4 Overall Achievements in 2014

In 2014, Filippa K fully implemented a fiber tool. They categorized fibers into fiber classes. One is the most sustainable class and five is the one that harms the environment the most. They increased the use of class one and two fibers to 44 % of their total fiber use. Also, the use of conventional cotton whilst in decline is slowly being replaced by the greater usage of organic cotton, lyocell and linen. More than half of their total collection contains a better sustainable fiber as the majority fiber.

Fig. 11.4 Fiber percentage share from 2012 to 2014. Adapted from Filippa (2014)

FIBRE	2014	2013	2012
Cotton	25,4 %	26,3 %	31,3 %
All Wool	25,4 %	16,1 %	17,4 %
Organic Cotton	9,3 %	7,9 %	8,4 %
Viscose	9,3 %	9 %	8,4 %
Polyester	8,3 %	6,9 %	3,5 %
Lyocell	8,2 %	6,7 %	5,6 %
Polyamide	5,5 %	5,8 %	5,2 %
All Leather	5 %	6,1 %	7,4 %
Silk	3,7 %	3,2 %	5,2 %
Other Fibres (less than 0.7%)	2,5 %	4,4 %	3,1 %
Linen Natural Retting	2,1 %	2 %	1,6 %
Elastane (Lycra and Spandex)	1,7 %	2 %	1,4 %
Acetate	1,5 %	2,7 %	1,5 %
Triacetate	1,3 %		
Mohair	1 %		

In addition, the set target to offer at least one product in every product group in a more sustainable way. They achieved the target in all product groups, except for the denim products due to the minimum quantities for purchasing organic cotton are very high. Filippa K said it was not possible from both a logistical and financial point of view (Filippa 2014).

The Fig. 11.4 (Filippa 2014) shows the approximately fiber share that is used to produce the garments over 2012, 2013 and 2014.

11.2.4.1 Production

In 2014, 62 % of production was placed in Europe with 55 % in other countries, as shown in Fig. 11.5 (Filippa 2014). These countries were classified as low risk countries by FWF.

The Fair Wear Foundation membership enables Filippa K to monitor the development of suppliers in other countries and conduct risk assessments to determine whether social auditing should be focused in. As a member of FWF the company itself is monitoring at least 90 % of its own suppliers. "For 2014, 98 % of our suppliers were monitored" (Filippa 2014). The audits were done in knitting factories and sewing factories in China and India. The most common findings from audits are issues relating to overtime and the payment of living wages. To solve overtime problems, the brand set up their production plans together with their suppliers so that the lead times can be discussed and the requested delivery times achieved. To handle payment of a living wage, Filippa K keeps an on-going dialog to accentuate the importance of this issue and FWF gives support to Filippa K's internal work and development. The code of conduct states that living wages should be paid to everyone. Furthermore, the ambition is to set a plan during 2015 on how they can intensify their work to make sure living wages for all workers in the supply chain (Filippa 2014).

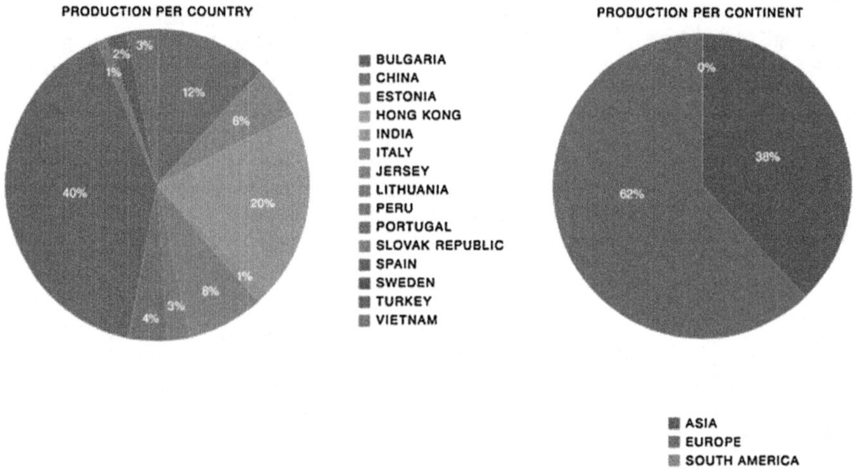

Fig. 11.5 Filippa K's production per country and continent in 2014. Adapted from Filippa (2014)

	2014	2013	2012	2011	2010
Turnover	662,5	635,2	572,9	524,0	425,4
Earnings Before Tax (EBT)	70,8	79,0	49,9	74,0	41,9
Tax Expense	16,0	18,2	11,9	21,8	10,0
Tax % of EBT	23%	23%	24%	29%	24%

Fig. 11.6 Numbers of filippa K's profit level. Adapted from Filippa (2014)

11.2.4.2 Profit Level

Filippa K achieved a profit level of 10 % Earnings Before Tax (EBT) and a turnover of $662,500,000 SEK for 2014. The brand complies with tax legislation in all seven subsidiaries countries. The company is profitable, pays taxes, and contributes to the society in which Filippa K operates. Figure 11.6 shows all the accounted tax expensed, turnover EBT, and tax % of EBT during the last 5 years (Filippa 2014).

11.2.4.3 Professional Long-Term Partnerships

One of Filippa K's key success factors is the long-term partnership with suppliers. The brand bought 66 % of their supplier volume from factories for which they have substantial leverage (10 % of the factory production capacity). In 2014, the objective was to deepen their collaboration with chosen key suppliers and to drive change in the area of sustainability over the years together. Also, the supplier base seems stable with 74 % of the purchase volume coming from factories that they have worked for more than 5 years. The same goes for the majority of the wholesale partners (Filippa 2014).

11.2.5 Overall Commitments Until 2030

According to their sustainability report in 2014, Filippa K aims to achieve the following (Filippa 2014):

1. Conscious design for a better future:
   - Only sustainable materials (FK class 1–2).
   - Only recyclable styles.
2. Sustainable sourcing and manufacturing:
   - Full transparency in their supply chain.
   - Only sustainable production processes.
3. Resource efficient business:
   - Accurate purchase precision in number of pieces produced.
   - Minimal footprint throughout our business.
4. People in our value chain are respected:
   - Total compliance with Filippa K's Code of Conduct based on Filippa K's values.
5. Long-term sustainable success:
   - Professional long-term partnerships.
   - Sustain a profit level of more than 10 % EBT.
   - Sustain a growth in comparable units.

11.3 Discussion

Filippa K developed a system of a recycle- and reuse-based closed-loop supply chain to improve the sustainability of fashion products. This service included maintenance, recycling, reverse logistics, and final waste disposal. Filippa K studied in detail their closed-loop supply chain and its effects on sustainability. Predominantly their focus is on material, production and logistics aspects. Filippa K's closed-loop supply chain is complex, involving many challenging stages and people. Filippa K triggers the product development process, including research and design. They are best placed to bring about change in the sustainable production of clothing through their choice of suppliers the design of their products and the control over the use of chemicals and energy in the production process and the final garment.

The ambition is to think in circular mindset—nothing goes to waste and should be reused or recycled to help the planet socially and environmentally. A Company like Filippa K is making a big sustainable impact and is inspiring other companies and individuals to make a difference, but are the consumers accepting this model?

Are they interested in paying more money for fairly produced, long-lasting garments that can be recycled and can be rented for a period of time? In the following paragraphs the author will discuss the positive and negative sides of their model and will end up with his personal conclusion towards Filippa K's sustainable closed-loop supply chain.

Filippa K's sustainability aspects are very important to the fashion supply chain due to current strong competition, intensive resource usage and labor conditions in some regions. Other firms that follow the trend and produce fast changing fashion and push low price strategies encourage consumers to buy more clothes, which they actually do not need. The result is called overconsumption, with some fashion.

Products even being unused, dumped, or just becoming waste. Filippa K's solution is to recycle and re-use the garments. Reusing clothes helps to prevent the production of new clothes and decreases the environmental burden of the life cycle of clothing (Farrant et al. 2010). Therefore, Filippa K collects used clothing products and resells them as secondhand clothes. They are not just concentrating in the energy consumption aspects, but also in the potential benefits of recycled clothes and clothes that can be reused.

Filippa K secondhand stores are another form of environmentally sustainable clothing acquisition. The customer has the chance to return an old Filippa K garments that he or she does not want anymore and obtain a 15 % discount for the next purchase in store. If the garment is not sellable, the customer has the chance to do something good and can still give it to a selected humanitarian organization. People that tend to go to secondhand stores rather buy 'need products' instead of 'want products' and are even willing to pay up to 25 % more for an organic cotton shirt compared to a conventional cotton shirt. Also, through secondhand clothing stores like Filippa K's, it is possible to recycle clothing and keep it out of landfills (Connell and Kozar 2014).

Filippa K's rental service has a big positive affect on sustainability. Trough rental services, sustainability can be improved through various processes and materials in the fashion supply chain. Some consumers that are shopping at fast fashion outlets prefer not to wear clothes for a long time because they prefer to wear different clothes as often as possible. Therefore, consumers have the chance to just rent a product, wear it a couple times and return it. Instead of paying for something and getting rid of it with no value when they are done—customers have the ability to extend the value. It is efficient and it is green. One person's trash is another person's treasure. However, the service is not always better than procurement. The total system costs are apparently better in the procurement model than the rental model, because consumers will use the product carefully. They think the garments are their own property. Another negative aspect is that customers might be scared to make a stain on it or tear the garment and the result is to pay for it. Furthermore, it is quite expensive because it costs 20 % of the original price, but you are only allowed to lease it for 4 days (Hu et al. 2014). In my opinion, basic garments will be less interesting than outfits for a certain events because these are mostly more expensive. Unfortunately, there is no information given about how profitable the rental service is.

Moreover, are the costumers even adopting Filippa K's model or do they rather buy clothes from fast fashion companies? How profitable is Filippa K's business? Due to the numbers given on the Filippa K Sustainability Report 2014 and recent studies about consumer behavior and attitudes regarding to sustainable clothes from Connell and Kozar (2014), the author came to the following view. Nowadays, people are concerned about the natural environment and rightly believe that it should be protected.

For example, participants in a study have not only a general concern about the environment, but also believe that company's like Filippa K or the government should encourage individuals in behaviors that protect and improve the natural environment. They should be influenced in purchasing clothing that is sustainably produced. However, a variety of studies have shown that even when consumers are aware of how clothing is produced, the distribution process, and consumption impact on the environment, people will still buy the lower priced clothing.

These consumers represent the opinion that sustainability in clothing production is important but less stylish, very counterculture in style, not well-fitting generally uncomfortable and too expensive. Some even have negative attitudes towards characteristics and attributes of environmentally sustainable clothing. For example, they avoid purchasing secondhand garments because of a perception that the merchandise is badly organized and there is not a lot of variety of products.

Furthermore, there is still a lack of knowledge about the environmentally sustainable clothing consumption and awareness on how clothing production affects the natural environment (Connell and Kozar 2014). Why would these people buy at secondhand stores or even think about recycling? Filippa K is trying to avoid this lack of knowledge through best marketing strategies and media outlets by informing consumers about environmental issues in associations and practices of sustainable versus mainstream firms. They constantly work to improve their internal processes, communicate their work on their website and in their sustainability report in order to raise greater awareness in our society. In addition, they work with several organizations, and support initiatives and networks to help find the best potential sources in conjunction with its suppliers. Filippa K is trying to challenge not only itself but also the fashion industry by participating in several projects, seminars, and conferences. At these events, the brand raises questions and talks about recent activities and possible future scenarios to spread awareness in the field. For example, each year they visit the political week in Almedalen, Sweden (Filippa 2014). Through practices with collaboration and partnerships they gain knowledge to detect better sustainable products and processes. Pressure from consumers who also believe that some suppliers will not act not responsibility; creates both benefits and difficulties in their attempts to search for ideal environmental performance requirements (Pereseina et al. 2012).

Another point to consider is the challenges and conflicts in a closed-loop sustainable supply chain. The major challenge is the cost increase because costs and revenues are still the main drivers in the development of a supply chain (Pereseina et al. 2012). Filippa K's relationship through collaboration in regards to sustainable goal can be cost intense. For example, collaborative waste reduction, beneficial solutions to production and service problems, and innovation that are social and environmental friendly are very cost effective.

Filippa K's operationalization of sustainable development is also an important point to keep in mind. Continuously, the company has to look for tools, systems, partnerships, suppliers, etc. to find the best sustainable processes. This absorbs a lot of time and effort. It is becoming increasingly difficult for Filippa K to manage and monitor their supply chain activities and ensure quality standards along the chain. Also, sometimes employees could lose their motivation if results do not show up how they were wished them to be. The worst case would be the fear of change because people tend to think that change is connected to difficulties of interpretation. Another challenge to face for Filippa K could be the change of mindset and culture around the globe. There exist different governmental decisions, consumer demands, and harsh strategies formulated by organizations that can be hard to change because supply chain processes and the way logistics influences society and the environment. There are many difficulties in the negotiation of contracts with suppliers and organizations, such as the choice of fuel, routing vehicles, trade-offs between environmental effects, and delivery times as well (Pereseina et al. 2012).

A further aspect to think about is how sustainable a closed-loop supply chain really is? Are there gaps that we forgot? For example what about transportation? Surprisingly the company does provide a lot of information to stakeholders about how they undertake their transportation. With Filippa K admitting they have limited opportunity to affect current air emissions with transportation being handled by third party providers (Filippa 2014).

Filippa K has a store with newly produced garments and a store where u can buy garments that are secondhand or rent new garments, but what is mostly forgotten is the transportation cycle gets longer as a result of this. First, the garments have to come from the production country to the Filippa K stores. After the product is used from a customer the garment needs to be brought to the secondhand store. The store personal has to bring it to the laundry salon or to a place where it gets recycled. After that, it has to be picked up and brought to the secondhand store again. Same with the renting service, people drive there and the garments need to be prepared over and over. This means a lot of transmission for our air. On the other side, a lot of transmission could be saved because no long-distance transport is needed to move the finished garments from the factories located in low-labor-cost countries to the consumer in a developed country. This adds to the overall quantity of nonrenewable fuel consumed. To sum up, Filippa K has to find the best solutions to reduce the transmission for our air transportation. A possible solution could be a pick-up service that collects the garments and brings the new or reuse products to the customer. This way, more consumers would also adopt Filippa K's service because some customer's do not have the time or are too comfortable to turn in or buy the garments.

As we can see, there are still some challenges that need to be confronted by companies like Filippa K, but analyzing these challenges and conflicts can add to logistics sustainability, and allows Filippa K to improve their approach to sustainable supply chain management. It is important to learn from every step of the journey and drive change. But even more important is to inform the world about fast fashion products and to influence the consumer's opinion, request, and perspective. In addition, members (institutions, organization, companies, manufactures) play a role in educating consumers and sustaining the development of the green fashion sector by supporting them with information or subsidies. Also it is necessary to design and produce garments that are personal, long lasting and optically considerable for the consumers. Otherwise Filippa K's model will make it stiff to buy the fashion products. It is interplay of several challenges, but after this extensive research and analysis, the author came to the conclusion that the brand is and will be successful, because Filippa K is seeing these challenges and confronting them together with their stakeholders. They are overlapping breakdowns and setting commitments for a better future, a future where Filippa K helps the planet and is profitable at the same.

11.4 Conclusion

In this paper, the author focused on introducing the brand Filippa K and the associated vision and sustainable operation structure. The brand proves that it cannot only innovative and be sustainable, but also produce approachable great fashion. In order to promote sustainability, developing a closed-loop supply chain is an important strategy for Filippa K in the fashion industry. Also, the author wanted to point out that the consumer's awareness of sustainability, regarding environmental and social issues is increasing.

The author first studied the structure of a sustainable fashion supply chain and revealed how Filippa K works in the corresponding stage of the sustainable supply chain. Also it was observed that Filippa K has implemented sustainable programs like collecting, recycling, and leasing. The brands sells new, recycled and secondhand product in their stores. Eco-Materials like tencel is used, the productions of the garments are as sustainable as possible, green distribution approaches with fewer emissions and energy used, and green retailing and education to customers is promoted. The results of the study reveal that Filippa K focuses significantly on supplier compliance with their code of conduct, employing further monitoring and auditing activities to prevent production problems in developing countries, improve overall supply chain performance and set sustainability criteria for their suppliers.

Furthermore, the author discussed the implications of the fast fashion industries in comparison to sustainable supply chain management. The results were that fast fashion companies are notorious for their excessive contribution of pollution and waste to our environment. The practices and trends of poor designing, raw material selection, processing, manufacturing, and disposing of apparel garments are the major contributing factors to these issues. Moreover, the production is mostly in low-cost labor countries. In the fast fashion industry, companies take advantage of low input costs through procuring their materials from developing countries. Also, it was talked about consumer behavior toward sustainable supply chain management and the fast fashion industry. The findings were that some consumers are still not aware of the negative effects through fast fashion products and some of those who are aware support for different kind of reason the fast fashion industry.

The author would like to finish this paper with the manifest of Filippa K adapted from their sustainability report in 2014 (Filippa 2014): “If we want fashion to stay relevant and aesthetic, inside and out. We need to be personal, simple, and long lasting. Ecosystems are our inspiration. Planetary boundaries are key, not limitation. Sustainability is our guide to growth. We are dedicated to a carefully curated wardrobe built on personal style and circular design with a holistic approach to business. How we do something is how we do everything. What we do is long lasting.”

These expressive words or mission statement can serve as an orientation for all sustainable fashion companies. Finally, the author is convinced that companies, such as Filippa K with it is closed-loop supply chain are the future of green fashion.

References

Aspers P, Skov L (2006) Encounters in the global fashion business, vol 54. Current Sociology, London

Bloemhof J, Van der Laan E, Wassenhove LV, Zuifwijk R (2012) A sustainable future for closed loop supply chains. https://rzuidwijk.files.wordpress.com/2012/11/26-chap-15-bloemhof-et-al-rz.pdf

Braw E (2014) Sweden reveals world’s first garment made entirely from recycled cotton. http://www.theguardian.com/sustainable-business/sustainable-fashion-blog/sweden-recycled-cotton-technology-fashion-composting

Connell KYH, Kozar JM (2014) Environmentally sustainable clothing consumption: knowledge, attitudes, and behavior. Manhatten

Corozzobottons (2015) Why corozo? http://www.corozobuttons.com

Emberley V (1998) Venus and furs: the cultural politics of fur. I.B. Tauris & Co, London

Esculapio A (2015) Check out some fashion: clothing libraries in Sweden. http://sds.parsons.edu/designdialogues/?post_type=article&p=723

Farrant L, Olsen S, Wangel A (2010) A environmental benefits from reusing clothes. http://www.mdpi.com/2071-1050/6/10/7063/htm

Fear Wear Foundation (2015) Labour standards. http://www.fairwear.org/488/labour-standards/1.-employment-is-freely-chosen/

Filippa K (2014) Filippa K sustainability report 2014. Stockholm. http://filippa-k.com/media/wysiwyg/Filippa%20K_sustainability_report_2014.pdf

Filippa K (2015a) Long lasting simplicity. http://www.filippa-k.com/se/filippak-world/our-belief/

Filippa K (2015b) The fair wear foundation. http://www.filippa-k.com/se/filippak-world/fair-wear-foundation/

Filippa K (2015c) The front runners. http://www.filippa-k.com/se/filippak-world/our-front-runners/

Filippa K (2015d) The method. http://www.filippa-k.com/se/filippak-world/our-method/

Henninger C, Alevizou P, Oates C, Cheng R (2015) Sustainable supply chain management in the slow-fashion industry. In: Sustainable fashion supply chain management: from sourcing to retailing. Springer, Heidelberg (pp. 129–155)
Hu Z-H, Li Q, Chen X-J, Wang Y-F (2014) Sustainable rent-based closed-loop supply chain for fashion products. Sustainability 6(10):7063–7088. doi:10.3390/su6107063
Joy A, Sherry JF, Venkatesh A, Wang J, Chan R (2012) Fast fashion, sustainability, and the ethical appeal of luxury brands. Fash Theory Journal Dress Body Cult 16(3):273–296. doi:10.2752/175174112X13340749707123
KEMI SCA (2015) Chemicals in the everday environment. http://www.kemi.se/en
Knutsson F (2014) About Filippa K. http://www.filippa-k.com/se/filippak-world/our-vision/
Lanegarderoben (2015) HUR GAR DET TILL? http://www.lanegarderoben.se
Langenwater G (2009) Planet first. Industrial Management
Oscarsson B (2015) About the Swedish Environmental Protections agency. http://www.swedishepa.se/About-us/
Pereseina V, Jensen L-M, Hertz S (2012) Challenges and conflicts in sustainable supply chain management. http://www.diva-portal.org/smash/get/diva2:690851/FULLTEXT01.pdf
Re:newcell (2015) Recycle textiles from waste to fribres. http://renewcell.se
Robért K-H (2015) Our mission. http://www.thenaturalstep.org/about-us/
Rockström J, Steffen WL et al (2009) Planetry boundaries: exploring the safe operating space for humanity. http://www.stockholmresilience.org/download/18.8615c78125078c8d3380002197/ES-2009-3180.pdf
sourcingjournalonline.com (2015) Are closed loop textiles the future of fashion? http://static1.squarespace.com/static/55d4d0e1e4b00f3b16cfd99d/t/56244d2be4b059c47b4f5166/1445219627335/closed-loop-textiles-article.pdf
Steffen W, Richardson K, Rockström J, Cornell SE, Fetzer I, Bennett EM, Sörlin S (2015) Planetary boundaries: guiding human development on a changing planet. Sci, 347 (6223):1259855. http://doi.org/10.1126/science.1259855
Swerea Swedish Research (2015) Swerea IVF. http://www.swerea.se/en/ivf
The Hunger Project (2015) Vision, mission & principles. http://www.thp.org/who-we-are/mission/

Chapter 12
The Role of Social Media for a Sustainable Consumption

Jochen Strähle and Chantal Gräff

Abstract This study focuses on the different roles of social media for the promotion of a sustainable lifestyle, behaviour and consumption, especially with regard to the typically non-ethical fashion industry. Research findings include eight roles of social media influencing a sustainable consumption contrary to prior research naming one to five impacts. Results show that social media educates and engages the young and ethically interested target group besides increasing supply chain transparency and brand or theme awareness. Furthermore, social media provides a platform for organisations' relationship management and social interaction since users get empowered to share experiences which leads to a higher level of trust.

Keywords Social media · Sustainability · Sustainable consumption · Fashion · E-WOM

12.1 Introduction

With the rising global usage of the Internet, social media has become one of the most powerful innovations in the twenty-first century. In social media, users generate content by sharing information in a community. Therefore, social media can be a powerful resource for sustainability communication engaging users in sustainability discussions besides promoting sustainability-related information. It is a consequence of the digital age that information and value exchange has to be taken place online, preferably on social media platforms (Diaz 2015; Johnson 2014; Zolkepli and Kamarulzaman 2015).

J. Strähle (✉) · C. Gräff
School of Textiles and Design, Reutlingen University, Reutlingen, Germany
e-mail: jochen.straehle@reutlingen-university.de

C. Gräff
e-mail: chantal_graeff@web.de

J. Strähle (ed.), *Green Fashion Retail*, Springer Series in Fashion Business,
DOI 10.1007/978-981-10-2440-5_12

Prior research has focused on the different types and the general role of social media besides referring to various marketing tools for sustainable consumption. But the special role of social media for a sustainable consumption has only been mentioned without naming these roles in specific. Especially with regard to the fashion industry, social media's influences and advantages have hardly been explored. This research will unite and examine all aspects of sustainability marketing and social media communication to detect the exact position of social media in ethical communication. Besides, green organisations' key-ambitions for using social media as a communication tool will be figured out.

The paper is organised as follows. The first part shows general information about social media and sustainability communication to provide an overview about the topic. Then, the green target group will be introduced because of the importance to address the right users with targeted messages. Afterwards, the special influence of users' online reviews and recommendations on other users' actions will be explained. This paper also lists five challenges companies have to succeed to implement targeted customer approach in social media. Thereupon, research results for advantages and key-ambitions of a social media sustainability communication will be developed. The last point of the literature research shows examples of different social media platforms supporting previous findings. At least, the research results will be included, compared and linked in the discussion to present all compiled roles of social media for a sustainable fashion consumption as well as resulting effects in detail.

The purpose of this paper is to show that companies without a social media presence are missing an opportunity to spread their sustainability messages to the society. Social media has the power to influence users' decisions, shopping behaviour and moreover their whole world view. Therefore, these new online technologies can change the population to think and act in a more sustainable and ethical way.

12.2 General Information About Social Media

12.2.1 Definition

With the rising global usage of the Internet and the introduction of Web 2.0, new technologies started to allow social interaction between individuals, communities and the society with user-generated content (Zolkepli and Kamarulzaman 2015). Based on Internet technology applications, social media tools have been established which has become one of the most powerful innovations in the twenty-first century (Diaz 2015). Social media transforms passive individuals into active users by supporting social interaction (Zolkepli and Kamarulzaman 2015). Today, these interactive platforms are integrated in user's everyday life due to the possibility to communicate boundless via smartphones (Ka-Yan Ng et al. 2015). The most influencing social media tools like Facebook, actively used by more than one million monthly Internet users, Instagram, Twitter and LinkedIn position social media as the most frequented online sites on global scale (Zolkepli and Kamarulzaman 2015).

12.2.2 Six Types

Since social media has become an important marketing tool for individuals and companies, it is important to know about the different social media categories. Currently, there are six types of social media, but the most known and simultaneously highly influential social media tools are social networks (Grahl 2015).

Social networks like Facebook, Instagram, LinkedIn, Twitter, Pinterest and Google+ connect people with similar interests and backgrounds who upload user-generated information. Normally, social network users create profiles, get the opportunity to found groups and interact in various ways (Grahl 2015). Besides, companies are able to deliver content to specific target groups in these networks (Ka-Yan Ng et al. 2015). At the moment, visual social networks like Instagram or Snapchat are becoming more popular due to the fact that an image has a multiple higher chance to be shared than text information (Diaz 2015; Fig. 12.1).

Other social media tools with high influences are forums and especially blogs. Forums and blogs provide a platform for users to post and respond to information and messages. Forums like Yahoo! normally include a high number of users communicating about different topics whereas blogs are often designed by one user and centred about one specific or several connected subjects. Since 2010, the number of blogs, especially in the field of fashion and lifestyle, increased a lot and with that, the popularity and influence of blogs increased (Grahl 2015). Blogs are often told as online journals with content ordered chronologically or by categories (Diaz 2015).

Besides social networks as well as blogs and forums, there are four types of social media mostly known and used in the US, but with the opportunity to get more globally important in future. First, there are bookmarking sites like StumbleUpon. These sites recommend web links to their users which can save, tag and share these links to users with similar interests. Then, in social news like Reddit, users can post news and the community decides by voting which news will be displayed the most prominently. A next social media type is media sharing like YouTube or Flickr. These websites allow people to share pictures and videos as

Fig. 12.1 Examples of social network platforms. Adapted from Tan (2015)

well as the opportunity to create profiles and comment the uploaded media. The last category is microblogging whereat Twitter is the most commonly used site to post short updates (Grahl 2015).

To complete, there are several other tools belonging to social media. For example, wikis like Wikipedia are websites allowing people to add and edit information to a database. Then, podcasts such as Apple iTunes provide audio and videos through subscription services. In addition, rating and review sites like Yelp or music sharing sites like Spotify among to social media (Grahl 2015). Although there are several different types of social media, they can overlap among the services. A good example is Facebook which is a social network with microblogging features because of the status updates (Grahl 2015).

12.2.3 General Role

To realise the role of social media for a sustainable consumption, the general role for individuals, brand communication and the fashion industry has to be clarified.

It is proven that social media has quickly revolutionised the communication and interaction between individuals as well as between companies and costumers. Individuals use social media in multiple ways as they are creators, conversationalists, critics, collectors, joiners and spectators (Ka-Yan Ng et al. 2015). Users can co-create news by sharing and filtering out irrelevant information on their own opinion. This form of bottom-up news production has been established by social media. Another point is that individuals have to multi-task when receiving information and news from different sources simultaneously. On top, many users express their opinions and personal content openly and worldwide. Another character of social media is the interaction and connection of individuals with similar needs, tastes and backgrounds which helps information publishers to deliver content to the right consumers (Pentina and Tarafdar 2014). Users can be engaged in social interactions but there is also the possibility for individuals to share information on platforms like Wikipedia or to offer valuable advice by rating. Thus, these electronic networks enable members to share information and experiences quickly on a global scale (Hajli 2014).

As described before, social media has not only revolutionised the interactivity between individuals but also the ways companies share information. Since the rise of social networks, the marketing and public relation departments have to include these new dynamic communication channels in their marketing plans (Johnson 2014). Nearly, all companies nowadays have adopted social media as an effective method of corporate communication (Reilly and Hynan 2014). But companies have to consider the differences of social media and traditional media communication. While traditional marketing tools publish information to the masses without a direct consumer reaction, social media allows a direct consumer contact, called two-way interaction. This is why digital media replaced many traditional print media over the last decade (Reilly and Hynan 2014). The low cost of social media marketing is another benefit. Carrascosa et al. (2015) talk about negligible costs to set up marketing in social media compared to traditional print media. In addition, individual users become information channels themselves so that the ownership barrier of traditional marketing tools is conquered.

Another point is the "ability to reach consumers worldwide" (Reilly and Hynan 2014, p. 3). Although most of social media users are younger than 30 years old, the majority of adults have at least one profile on a social media site. This explains why Facebook already reaches one million monthly users and measures several millions of interactions, likes and comments daily (Reilly and Hynan 2014). Furthermore, "social media is an ideal environment for building brand communities" (Habibi et al. 2014) which can influence the consumers' trust in brands. Online forums, reviews, ratings, recommendations and online communities produce electronic word-of-mouth between consumers. Because of the high faith in other users' recommendations, these social media interactions can support brand trust and reduce the risk for customers (Hajli 2014). But todays customers also expect brands to act like trusted friends and "engage with a new sense of intimacy" (Briggs 2010). Precisely because anonymity gets reduced by social media, authenticity has to be shown in a real company story and the valuation of customers (Briggs 2010).

All in all, social media communities support the customer relationship management of companies by providing opportunities to interact, creating trust and becoming more attractive to consumers (Hajli 2014). But "perhaps the most influential upshot of social media is consumer empowerment; now consumers have a strong voice and firms are not in control of the conversations among consumers" (Fournier and Avery 2011). Consequential, marketers have to note every customer and include their opinions in new products and advertising (Habibi et al. 2014). Especially in the fashion industry, users' norms, values and preferences communicated in social media are highly influential to trends. Users often share pictures and style-related information expecting feedback on their style choices. Thereby, social media plays an important role in trend building and users' self-realisation (Wolny and Mueller 2013).

12.3 General Communication of Sustainable Consumption

After analysing the general role of Social Media, it is important to identify the general communication of sustainability, ethical brands and green products to finally include all aspects in the role of social media for a sustainable consumption.

Reilly and Hynan (2014) underlines that an organisation has to use different marketing tools like Corporate Social Responsibility reports, sustainability reports, annual reports, social media or TV ads to communicate corporate sustainability. He adds that "through content, style and tone, corporate communications provide insight regarding the values underlying an organisation's culture" (Reilly and

Hynan 2014, p. 3). Uncontested is the fact that public's environmental and social awareness has to be educated and increased through a variety of ways and means (Prothero et al. 2011). Finney (2014) also strengths the use of various channels of information with the statement that green consumers are more interested in environmental claims and could be supported in their interest through a comprehensive, coordinated communication process. In addition, he underlines that the availability and access of information sources influence the decision-making process so that "the type of purchase will determine how involved a consumer is and how much time they devote to the analysis of available information" (Oates et al. 2008 as cited in Finney 2014, p. 10).

To give consumers the information they need to make purchase decisions, green products and sustainable corporations have to be marketed successfully (Rex and Baumann 2007). "The media is widely acknowledged to play an important part in gaining public support for sustainability initiatives" (Seip et al. 2006 as cited in Kolandai-Matchett 2009, p. 1) because of media's capacity to educate and inform a multitude of people within a short space of time. Meanwhile, the success of environmental protection depends on public support. For example, information campaigns could be used to participate consumers in the design and implementation process by asking about their needs and wishes (Kolandai-Matchett 2009).

Furthermore, Rex and Baumann (2007) highlight the challenge to demonstrate environmental product qualities. Green consumers have to be informed about the benefits of ecolabels and green products, but meanwhile, green products have to appeal to customers like non-green products. Moreover, a company has to be clear if they want to inform all consumers about the green dimension or if they want to focus on already green consumers (Rex and Baumann 2007). Whatever consumer group a company focuses on, it is a fact that communication technologies influence the reshaping of beliefs and values surrounding consumption activities (Stanik 2010). Lee et al. (2012) points up the capability to reshape consumers' values by educating about the goal of green consumption which is to benefit the environment. She is persuaded of the dependence between an increase of consumers' awareness of environment and a global rising demand for green products. Communicating sustainability with consumers and communities and thereby changing values, companies' environmental consciousness gets strengthened (Lee et al. 2012). Furthermore, ethical products can support the Corporate Social Responsibility by changing consumers' values or improving the corporate image. She concludes with the importance of companies' voluntary promoting of a green consumption society to shape a global greener consciousness (Lee et al. 2012).

During the last years, many companies recognised this importance and especially the rising demand for green products so that many sustainable companies emerged and existing organisations launched sustainable collections. Since then, messages about sustainability are all over lifestyle media and a social change of consumption has started (Lundahl 2014). Lundahl (2014) calls the active involvement of a market actor the "fashionalisation of sustainable consumption" (Lundahl 2014, p. 1).

This fashionalisation led to social innovations with high potential for promoting sustainable consumption (Jaeger-Erben et al. 2015). Jaeger-Erben et al. (2015) allude Do-It-Yourself-Workshops and swapping or sharing communities. These innovations work against the inefficient product usage, the mass consumption. Especially, those social innovations can be supported by new social settings like the Internet or social media (Jaeger-Erben et al. 2015).

12.4 Social Media Communication of Sustainable Consumption

12.4.1 Green Target Group

Organisations have to be aware of the green target group to address the right consumers with targeted content in their social media presence (Reilly and Hynan 2014). In the following, the social media target group living an ethical lifestyle is described and different types of these consumers are extended.

As described before, social media in general speaks to a younger audience; although most of the adults have at least one profile on a social media site by today (Reilly and Hynan 2014). Certainly, new media are a normal part of daily living for most of young people who spend several hours per day on social media platforms. As a result, young adults have a high technical affinity, more and foremost with social media (Vaterlaus et al. 2015). Lucy Atkinson (2013) suggests that those with a high sense of agility for social media are more open to use and enjoy trying new technical phenomenon instead of being scared.

The sustainable target group in social media is mainly a community of younger green consumers who are already interested in sustainability and ethical manufacturing. These green consumers use to be more interested in information about the organisation's ethical stance (Finney 2014). In social media, consumers self-select into lifestyle groups which helps advertisers to create targeted advertisements based on consumer needs and wishes. Organisations easily reach the green target group searching online sustainability communities or sustainability-related posts (Minton et al. 2012).

But it is indicated that organisations should not address the 'green consumer' but understand all consumers at an individual level. In general, companies can distinguish between three different groups of consumers living a sustainable lifestyle. The first group is Selectors who focus on one selected aspect in their lives to consume sustainable. The second group is Translators who are green in some aspects of their lives. If the promoted products or values are of interest for Translators, they can get deeply involved and make material changes in their own and even in other's lives. But if a green product is not interesting for Translators, they are going to ignore all marketing communication about this product. Only for Exceptors, sustainability plays an important role in all aspects of life. The lifestyle is against mass consumerism and mainstream society. Exceptors are seeking the communication with other Exceptors in networks, but this group of green consumers is really hard to reach through mainstream marketing because of their critical attitude towards the society (McDonald et al. 2012). In conclusion, this typology perfectly shows the dependence between social circles, social media usage and marketing success. But all types have in common, that social media let users be part of a friendship circle, keep updated and informed and get different points of view (Zolkepli and Kamarulzaman 2015).

Zolkepli and Kamarulzaman (2015) describe different motives for individuals to use social media. At first, users follow personal needs—trendiness, enjoyment, entertainment and interactivity. Trendiness is considered as a lifestyle like the wish to be involved in latest trends. Furthermore, enjoyment stands for pleasure and happiness while using social media and is a motivation tool for users to communicate in online discussions and share pictures or experiences. Entertainment satisfies users' needs for emotional release and gives a way to escape from pressure. The last personal need is interactivity which is operationalised as the exchange of messages in the era of two-way communication. Besides personal needs, individuals follow social needs in social media usage. Both personal and social needs are positively influenced by the innovation characteristics of social media. Social needs describe the intention to be part of a group. This belongingness is important for users to avoid feelings of loneliness, but also pressures users to use certain technology. Another key driver for using social media is the interaction with friends and other users. Being a member of a group of friends and networks, called companionship, is another motivator. Besides, the option to escape reality in social relationships in the virtual world motivates users to engage in social media. At last, innovation is another mediator as innovative technologies get users new experiences (Zolkepli and Kamarulzaman 2015).

Lundahl (2014) adds that consumers like to show their selflessness through promoting their sustainable consumption; and more and foremost that they can afford to spend more money for green products. Because of these outcomes, users see social media as a medium it is worth spending time for. The multitude of spend user time explains the high amount of social network marketing (Ka-Yan Ng et al. 2015).

12.4.2 Special Role of E-WOM

A main reason why organisations use social media to communicate with users is the impact on consumers' trust. The social interaction between consumers in communities or through recommendations and reviews develop social support for a brand, value or product. Good feedback, high ratings and positive comments therefore influence the level of trust and the reduction of risk. This leads to a higher intention to buy. Therefore, social media has empowered consumers who support business through generating content (Hajli 2014). "Social media provide an unparalleled platform for consumers to publicize their personal evaluations of purchased products and thus facilitate word-of-mouth communication" (Chen et al. 2011, p. 1). Thus, WOM, the abbreviation of word-of-mouth communication, is communicated as the evaluation of products and services between consumers. As WOM is an overall term for experience sharing and information dissemination, E-WOM is the online version of WOM and can be translated with electronic word-of-mouth communication (Ka-Yan Ng et al. 2015).

Wolny and Mueller (2013) suggest that E-WOM is a part of the basic human needs of being helpful and giving advice. Therefore, E-WOM is seen as the human impulse to transmit experiences, and thereby, supporting purchase decisions of others (Wolny and Mueller 2013). Compatible with this proposition, a majority of social media users is inclined to try new products based on other users' online suggestions. Users with a good experience of E-WOM recommended items are likely to write online reviews as well (Ka-Yan Ng et al. 2015). In addition, customers are more likely to share recommendations to less available products because their experiences will be more unique and more important as for common products (Phang et al. 2013). Phang et al. (2013) add that companies should use social media to encourage a high level of user participation in content generation. Through enabling greater user interactions, the level of activity on social media networks and the dimension of E-WOM can be influenced (Phang et al. 2013).

The electronic word-of-mouth communication can be positive or negative and, in rare cases, neutral (Ka-Yan Ng et al. 2015). The main motivation behind positive word-of-mouth is to share expertise with others (Chen et al. 2011). Ka-Yan Ng et al. (2015) adds that positive WOM includes great new experiences or recommendations a user is willing to share with others. Another incentive behind positive WOM is the social confirmation as well as the self-approval by demonstrating great purchase decisions and through the altruistic sharing of expertise (Chen et al. 2011). "Negative WOM can be broadly defined as undesirable behaviours, such as unpleasant shopping experiences, complaints, or product or service disparagement" (Ka-Yan Ng et al. 2015, p. 53). The motivations behind negative word-of-mouth are sharing dissatisfaction and seeking vengeance (Chen et al. 2011). Paradoxically, Wolny and Mueller (2013) suggest that a few negative comments can positively influence the attitude by improving E-WOM credibility.

Consequentially, social media mediated WOM is an important influence on consumers' behaviours. Ka-Yan Ng explains that the influence of WOM is higher as radio, newspaper and magazine advertising. Thus, E-WOM affects consumers' purchase behaviour through the influence between users' intentions, attitudes and behaviours. This fact can be explained with the high effort and time spent on social media platforms. In addition, the trust in E-WOM, even recommendations of unknown users, is higher than in other advertisements (Ka-Yan Ng et al. 2015). Ka-Yan Ng et al. (2015) also found out that fashion apparel is the most E-WOM influenced product category because fashion products reflect users' social status besides their values. The high level of social visibility is a reason for the high engagement in sharing information and opinions about fashion products.

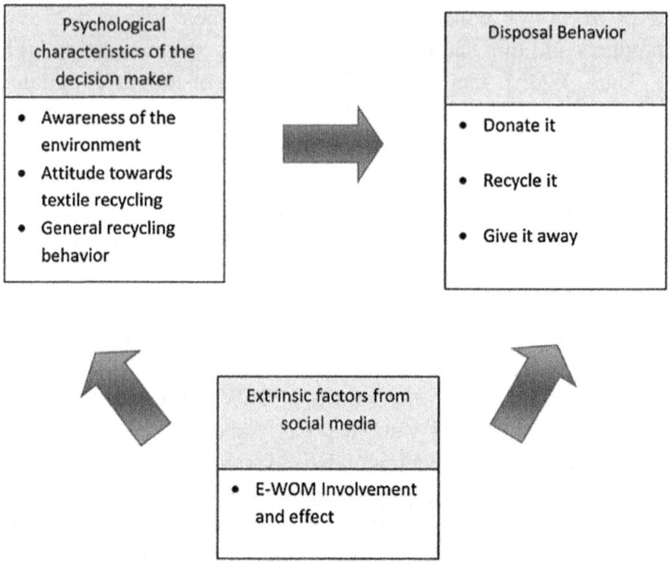

Fig. 12.2 The influence of E-WOM on young consumers' disposal behaviour. Adapted from Ka-Yan Ng et al. (2015)

To show the whole extent of E-WOM, Ka-Yan Ng et al. (2015) introduces the process of consumer decision-making which is heavily influenced by E-WOM (Ka-Yan Ng et al. 2015). The first step in the process is the recognition of a need or problem which often appears when interacting in social media networks. Then, the first contact point between consumers and organisations is the search for information which is distinguished into internal search, the memory, and external search including word-of-mouth and company information. Furthermore, social media is a rating source providing useful information and comments of users to enable comparison in the evaluation of alternatives. In the stage of purchase decision, E-WOM can persuade users to buy at a certain organisation because of the trust in recommendations. After using the product or service, customers can use social media to share their personal experiences in positive or negative comments in the post-purchase behaviour. This E-WOM, in turn, can influence other users in their purchase decisions (Ka-Yan Ng et al. 2015).

The effect of E-WOM as an extrinsic factor from social media is shown in the following Fig. 12.5. E-WOM influences young consumers' disposal behaviour in a sustainable context. The green decision maker is said to be aware of the environment, has an attitude towards general recycling and especially textile recycling. These psychological characteristics affect the disposal behaviour which is donating, recycling and give away instead of throw away (Ka-Yan Ng et al. 2015; Fig. 12.2).

12.4.3 Challenges

Social media innovations do not only easy the way of communication, they also challenge companies and users with several difficulties. These challenges are specified below.

12.4.3.1 Fast Changes

Before internet technologies took over society and marketing, people and organisations had time to get accustomed with innovations. Nowadays, the challenge for organisations and individuals is the intense pace of innovations because social media services can be developed within days. This speed gives no time to adapt to a new technology before the next innovation appears (Dennett 2015). The best and as well challenging solution is choosing few networks suitable for individual's or company's aims. Within this limited number of platforms, the company's message has to be consistent to show authenticity (Reilly and Hynan 2014).

12.4.3.2 Platform Profiles

The creation of a social media profile is a challenge itself. Companies should clarify that the profile is the official account to prevent misinformation and confusion among users. Then, content should be updated frequently, with a sense of social media etiquette and at its best in a unique company language. Besides, questions of consumers should be answered as soon as possible (Reilly and Hynan 2014).

12.4.3.3 Messages

Especially in promoting environmentally friendly products, marketers should use explicit messages to ensure the understanding among consumers. Therefore, the organisation's sustainability report should get broke down to easily sharable social media postings which pick up the sustainability highlights. Thus, social media challenges organisations to include information of a 140-page report into 140-symbols status updates. Due to the fact that the inclusion of too much information in fashion advertising can effectuate disinterest or even distraction, organisations should create variety through videos or pictures (Johnson 2014; Yan et al. 2012). In summary, companies should avoid information overloads, boring performances and misinformation (Carrascosa et al. 2015).

12.4.3.4 Cross-Cultural Differences

Cross-cultural issues like value differences and varied consumer behaviour need to be taken into account when addressing different cultures through one social media performance. At first, the social media attitude varies in the daily time spent on social media and the platforms themselves differ among countries. Then, the demand and willingness to pay a premium for green goods vary by the grade of development in countries. Thus, marketers have to choose between multiple profiles for different cultures or one general profile. Another point is the creation of individual advertising plans and the usage of different social media platforms for the different countries to tell the right messages to the right target group (Minton et al. 2012).

12.4.3.5 Negative Comments

The most obvious challenge for companies is the interactivity of social media. Users "are free to comment upon corporate messages—negatively as well as positively" (Reilly and Hynan 2014, p. 10). Negative comments or reviews have a much higher effect on consumers as positive ones. Besides, consumers trust other consumers' recommendations so that social media can negate brand trust (Habibi et al. 2014). Particularly because of the fast sharing of company missteps, organisations have to attend to comments quickly to counter negative information. A company protocol for social media strategy can support a quick handling in case of adverse user comments (Reilly and Hynan 2014). But any negative opinions can influence product improvements and companies' attitudes in social media in a positive way (Johnson 2014). In conclusion, the real challenge is to find the right balance between "empowering customers to spread the word about their brand through viral networks whilst still controlling the company's own core strategic marketing goals" (Wolny and Mueller 2013).

12.4.4 Advantages

The role of social media in a sustainable context is connected with impacts on consumers' behaviours, decision-making and values (Johnson 2014). "Social Media provides an efficient means for not only sharing sustainability-related information but also engaging stakeholders in sustainability discussions" (Johnson 2014). Johnson (2014) adds that the most important links between social media and sustainability reporting are user engagement and transparency. Thus, companies without a social media presence are missing an opportunity to ensure that the society receives their sustainability messages. It is a consequence of the digital age that information exchange has to be taken place online, preferably on social media platforms (Johnson 2014). Besides, social media is a powerful communication tool as it allows online communities to have access to requested information without having to read a long and complicated sustainability report. In contrast, social media messages are short and easy to understand. Plus, users have the possibility to participate in the organisation's sustainability performance through discussing and commenting (Johnson 2014).

Due to the fact that the Internet is the primary source for green product information, organisations have to understand the consumer receptiveness to sustainability marketing to address the right values and wishes. Therefore, advertisers can create campaigns letting users feel like doing something for a sustainable future. It is concluded that social media can make consumers a part in changing the society into a more sustainable, green community (Minton et al. 2012).

Besides, the platforms may be used for many other reasons. Of course, companies use targeted social media marketing to increase product demand by developing new products or promoting the brand or special offers. But social media is also a good tool for building customer relationships by answering user questions like career opportunities. In addition, consumers can get educated about the company's initiatives for a sustainable consumption and other environmental projects or economic issues (Minton et al. 2012; Reilly and Hynan 2014). Another positive part of social media communication is the similar power between all actors on platforms. All participants publish across hierarchies so that decision-makers, experts and end users have independent and similar conversation power (Fraoua et al. 2014). Fraoua et al. (2014) believe that social media are platforms for reproducing the usual public conversations between marketer and consumers as well as between individuals.

Danielle Vermeer (2013) summarises the roles of social media for promoting sustainable fashion in five stages which seem to her to be the most important roles.

At first, she mentions education since social media educates consumers through sharing producers' stories, manufacturing countries, production facilities and the makers' working environment. Then, empowerment is named due to the fact that social media empowers consumers to communicate their experiences in reviews and recommendations which influence other users' purchase decisions. In addition, the third role is the connection between fashion producers and consumers in social media which in turn results in a higher trust level. Furthermore, transparency is mentioned because companies' supply chains get more transparent through information published on social networks. At least, she adds a new point, the reduction of waste. Through social media, waste can be reduced with the help of crowdsourcing technologies. These tools help fashion companies to limit the production to the items consumers really want (Vermeer 2013).

12.4.5 Key-Ambitions

The detected roles and advantages support companies in implementing their missions and aims. The key-ambition of using social media as a tool to communicate sustainable consumption is the inspiration of people to engage in sustainability discussions, to start thinking and to take actions for a sustainable change in society. Organisations want to spread their sustainability mission to make a statement and to find people following the same principles. Reilly and Hynan (2014) emphasise the aim of green firms to be an inspirational source for people who get engaged in taking small actions. Together, these little individual actions are able to effect a global change (Reilly and Hynan 2014). Other key missions are the involvement of the population to "debate and to fit a new solution encouraging all population to get part of the policies adopted" (Fraoua et al. 2014, p. 1) as well as the dissemination of green information with the aim to fix the importance of sustainability in consumer's mind (Fraoua et al. 2014). Based on the social media advantage of a two-way information flow, building trust between consumers and organisations is the last big mission to create a sense of social and personal investment of consumers (Barris-O'Neill and Schuitema 2016).

12.5 Social Media Examples Supporting Sustainability

Each social communication medium provides different messages to users (Minton et al. 2012). In the following, different examples of social media platforms with a sustainable or ethical fashion background will be introduced; starting with mobile applications through highly influential blogs and the most common social networking platforms Facebook and Instagram to swapping and sharing communities. Often, media platforms are combined, for example if blogger have social networking profiles or communities include blogs.

12.5.1 Mobile Applications

There are many different types of ethical apps, whereof the following are prime examples.

The Fair TRACE TOOL is an app for iPhone and Android using barcode scanning to support a global supply chain transparency. The app shows users the story of their products' production, a real name and face of the makers, working conditions and the impact of the purchase. In addition, users can support the Fair Trace Tool campaign which is creating a fund for fair trade start-ups (Sarahana 2012).

Closet Swap is an app for iPhones collecting fashion items on users' Facebook pictures. Then, friends with the same app can borrow or swap these items so that second-hand usage of fashion apparel is supported and users get deflected of buying new items (Sarahana 2012). Besides the fashion swapping, the app shows vintage and second-hand stores in the user's surrounding and provides information about the production and consumption of ethical fashion (Fehrenbacher 2012).

Fig. 12.3 Structure of the app "Free2Work". Adapted from Mustafa (2013)

Instead is an app for IPhone as well. Via Instead, everyday purchases can be switched into donations. For example, if users are willing to buy a new dress, they are persisted to think about the better money usage in charities. Then, the user can donate right through the phone app (Sarahana 2012).

Free2Work is a free application for iPhones and Android Phones providing information about production circumstances. Users can scan a product's barcode to get information and ratings of the brand's usage of forced and child labour (Mustafa 2013; Fig. 12.3).

12.5.2 Blogs

Blogs have an immense influence on followers' behaviours. Bloggers spread their opinions and experiences to a high amount of people who trust on the bloggers' messages (Vermeer 2013). Besides, bloggers constantly share fashion advices and information about different topics they are wondering about. This is how trends, for example the trend to wear second-hand clothing, are co-created by consumers (Wolny and Mueller 2013).

A good example for the promotion of sustainable fashion apparel consumption through fashion blogs is the 'Fashion Detox' challenge. In 2015, over 70 American bloggers have been asked to refrain from buying clothes for 10 weeks and to share their opinions and experiences on their blogs. In the end, the challenge raised the bloggers' creativity in many ways. The influencer implemented do-it-yourself projects like repairing old clothes or sewing new garments, tried new combinations and wore some old items for the first time. Several bloggers also began to borrow and swap fashion apparel with friends to get the desired look (Ruppert-Stroescu et al. 2015).

In the following, some blogs will be introduced which focus on waste reduction, sustainable social change and education of the society in the context of fashion.

Fashion activists in New York City created the blog Ecouterre to promote ethical fashion. The aim is provocation, education and inspiration of consumers through sharing trends and innovations in the sustainable fashion context (Cadwell and Cadwell 2015).

In the blog Trash is for Tossers, Lauren Singer documents her zero waste journey in New York City. Her aim is to show her followers that a zero waste lifestyle is possible for anyone. For example, she propagates that second hand is the solution besides the question which clothes are really needed (Singer 2016).

Life + Style + Justice was launched by Hanna Theisen in the US featuring social enterprises and their work for a sustainable world. On one side, she shares her experiences in exciting and thoughtful articles and on the other hand, she is hosting events to show other consumers the possibilities of an ethical lifestyle (Cadwell and Cadwell 2015).

Eco Fashion Talk has been established in New York City by Sass Brown, a proponent of heritage craft skills to recycle and reuse and other forms of slow fashion. The blog shows links, information and videos about slow fashion to inspire consumers of an ethical lifestyle (Cadwell and Cadwell 2015; Fig. 12.4).

The Good Wardrobe is an online style-sharing community with the aim to push against fast fashion. The blog provides a platform for slow fashion with services

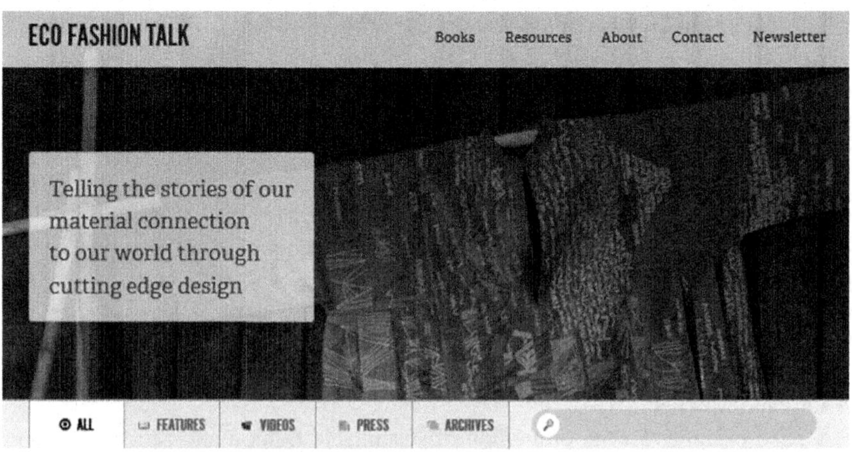

Fig. 12.4 An example of the blog "Eco Fashion Talk". Adapted from Brown (2016)

Fig. 12.5 An example of the blog "The Guardian". Adapted from Cadwell and Cadwell (2015)

that prolong the life of the wardrobe like sewing classes or sharing the know-how to repair old clothes (Yates and Alley 2015).

The Guardian is an award-winning news source and one of the most known blogs worldwide for sustainability questions. In the blog, tough topics around ethical manufacturing and a sustainable lifestyle are discussed in a trending performance (Cadwell and Cadwell 2015; Fig. 12.5).

12.5.3 *Instagram*

Normally, Instagram accounts are connected to sustainable blogs or ethical online shops. Due to the fact that only pictures are shared on Instagram, this medium has a high influence on the company's lifestyle appearance and can help to get a higher customer reach.

An example for a brand using Instagram as Lifestyle medium and online shop is "Vintage Lover CT". The brand from Cape Town, capital of South Africa, uses Instagram to post pictures of their second-hand and vintage-inspired collection and names the prices. Interested consumers can comment their wishes or order items via Email. Besides, inspirational pictures of customers in their clothes or atmospheric photos are shared to get a stylish and modern attitude (Emmanuel 2015).

12.5.4 *Facebook*

Like Instagram, Facebook is used to support the brand image with lifestyle media and short information. There are a lot of Facebook profiles featuring a sustainable society and ethical brands, for example the Centre for Sustainable Fashion, the Sustainable Fashion Initiative, Sustainable Fashion Regulars' Table or Sustainable Brands. What all of these profiles have in common is the fact that they are the link between users and communities or brands because of the transition to their blogs or homepages (Zolkepli and Kamarulzaman 2015).

12.5.5 Swapping and Sharing Communities

Swapping and sharing communities enable second-hand shopping and wardrobe updates from home (Emmanuel 2015). These communities rate among social innovations promoting transformation towards a more sustainable society besides a sustainable development. The collaborative consumption type "often responds to the negative impacts of mass consumption and the inefficient consumption of products" (Jaeger-Erben et al. 2015, p. 8), the so-called throwaway society. On homepages, items can be adjusted of all users to find people who want to swap their clothes or share items (Jaeger-Erben et al. 2015).

Examples of online Swapping and Sharing communities are the German pages Kleiderkorb and Kleiderkreisel. Both pages enable users to swap, sell or give away their clothes. In addition, Kleiderkorb has a forum for shopping ideas, user questions, homepage information or trends as well as a community page where users can introduce themselves and their profiles and a blog with updates and trend information. Kleiderkreisel has a big community with eleven millions of members worldwide and more than two billions of adjusted items. The advantage of Kleiderkreisel is the mobile App which enables a swapping and sharing anytime and anywhere (Janauskas 2016; Kayabas 2016).

12.6 Discussion

In the age of social media, everything is about the online interaction between individuals or organisations and consumers. This research provides detailed insights in sustainable development through social media. Considering the target group of mainly young users living an ethical lifestyle, findings show eight roles of social media for sustainable fashion consumption. In contrast to previous research which only considered one to five different roles, this paper presents eight roles combining all research findings. In detail, these roles are publicity, transparency, education, engagement, customer relationship management support, social interaction, trust and empowerment.

The first role of social media, publicity, can be seen as the common role of marketing, namely to raise brand awareness, to higher the attention of sustainability themes and to increase product demand. By promoting the brand, special offers, lifestyle or new products, social media supports common marketing aims.

Especially, short and easy messages or creative campaigns addressing the right values of the green target group contribute to organisations' prominence. Since Internet is the primary source for green product information, performing on various social networking platforms can raise the awareness level positively. In conclusion, social media platforms provide opportunities to become more popular and more attractive universally.

Raising the transparency of fashion supply chains is one of the main roles of social media. Especially in the typically non-transparent and unsustainable fashion industry, information published on social networks, in blogs or applications can increase consumers' interest, knowledge and trust. A sustainable consumption gets more interesting for the society once the advantages of sustainable consumption and the differences between ethical and unethical supply chains are demonstrated or, at best, visualised in videos or pictures. Good examples for social media supporting supply chain transparency are the apps Free2Work and The Fair TRACE TOOL. Both applications use barcode scanning to provide product information about production, working conditions, makers or child labour to support a global supply chain transparency in fashion industry.

Matching with the transparency aspect, sharing producers' supply chain also supports the education of consumers. Users can get educated about the company's initiatives for a sustainable consumption and other environmental projects or economic issues. Thus, social media allows online communities to have access to sustainability information without having to read complicated and long company reports. Thereby, social media shows the potential to affect individual's world view, political knowledge and values. The aim of this education in social media is to fix the importance of sustainable consumption in consumer's mind, especially in the theme of fast fashion. Moreover, social media has the role to change the society into a more sustainable, green and ethical living community. Examples supporting this society change are the three blogs The Guardian, Ecouterre and Eco Fashion Talk. These blogs share links, information, videos and pictures about sustainable lifestyles, ethical manufacturing, slow fashion, sustainability trends and innovations with the aim of provocation, education and inspiration. Considering the high amount of blogs about sustainable fashion consumption, education is already paying off due to the fact that many bloggers are interested in writing about this theme and readers like to read these blogs with an educational background.

If transparency and education are given, the engagement of the community is the next role of social media. On one hand, organisations want to apply people following the same sustainability principles and engage these in sustainability discussions. On the other hand, social media inspires the community to start thinking and to take small actions that add up to a big change in society. A perfect example for the already proceeded change in society is the app Instead because users can switch everyday purchases into donations for charities right through their phone app. Besides, the blogger Hanna Theisen demonstrates on her blog Life + Style + Justice that engaged individuals can support a sustainable consumption through hosting events like swapping parties.

Another outcome of the society change through education is the increasing requirement of waste reduction. For instance, crowdsourcing technologies support a production based on real consumer needs to counter mass production and overproduction. Or Lauren Singer shows on her blog Trash is for Tossers how to live a zero waste life in New York City, for example by wearing second-hand clothes.

Social media is also a good tool for building customer relationships because organisations can connect with individuals. A quick and satisfying answer of users' questions like career opportunities, store opening hours or product demands heavily supports a good customer relationship management (CRM). The company feels touchable and personal to consumers for the first time because of the ability of direct contact. Examples for the CRM support are companies' Facebook profiles which enable consumers and fashion producers or companies to communicate directly.

Social interaction is the first role of Social Media based on E-WOM, the electronic word-of-mouth. Individuals join online communities searching social support and friendship online. Thus, the interconnectivity between individuals has expanded with the aid of social media. A good example for a social media tool that enables social interaction is the online style-sharing community The Good Wardrobe. The community shares knowledge and tips to push against fast fashion with services that prolong the lifetime of clothes. Social media enables a similar power between all users which means that users have the same independent conversation power as decision-makers or experts. On social media platforms, usual public conversations between marketer and individuals or between individuals can be reproduced online. This is why social media empowers consumers who can generate content, offer valuable advice and share experiences to others. Users have the possibility to discuss and comment social media postings of companies which enables consumers to participate in the organisation's sustainability performance. Examples for consumer empowerment are fashion swapping and sharing communities which promote second-hand usage and recycling. In these communities, users share tips and tricks, describe their experiences or discuss about sustainable companies.

The described empowerment and social interaction in social media has a high impact on consumers' trust. One source of trust is the social support between users who become familiar with one another. Communities, reviews and recommendations are likely to establish a trust in products, services or companies. Positive comments, high ratings and good feedback influence the level of trust in organisations and reduce the risk for consumers due to the ability of comparison. For instance, Rating platforms are tools to raise the trust in companies because users' trust in E-WOM, even experiences of unknown people, is usually higher than in other advertisements. These impacts of E-WOM strongly influence consumers' purchasing decisions. Besides interconnectivity, the two-way information flow between consumers and companies create social and personal investment and with that build a basis of trust. Facebook for example supports the trust of consumers in companies by giving personal feedback.

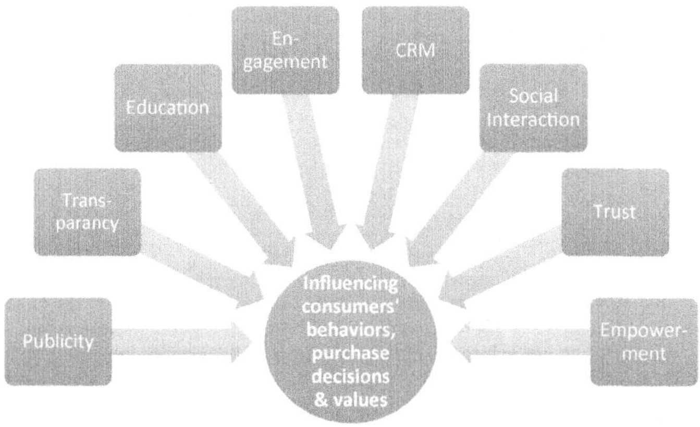

Fig. 12.6 The roles of social media for a sustainable consumption. Own illustration

In conclusion, social media plays an important role in changing the society into a more ethical and sustainable thinking community due to the fact that consumers' behaviours, purchase decisions, decision-making and values are connected with the impact of social media on users.

The researches' findings are illustrated in Fig. 12.6.

The research of this paper is mainly build on latest scientific literature. Due to fast changes in technologies, only resources of the year 2012 and younger are really meaningful for the topic social media. This paper can ease the way of future research in the field of social media and the upcoming trend topic Fashion Sustainability. More and foremost, the paper gives a clear and easy overview about the connection of the topics sustainability and social media.

In addition, this work raises interesting questions. For example, whether social media provides the appropriate platform to introduce sustainability news or is it only a tool for sustainability promotion. Should green companies, the innovators in sustainable consumption, be the first users of new social media technologies as well to show a general fast improvement? In the end, this paper shows that themes around social media and sustainability will never be completely explored due to ever changing technologies and innovations.

12.7 Conclusion

Social media plays a sensitive role in sustainable consumption because it provides a platform for sustainability communication. On one hand, organisations can reach and involve the right target group with creative and understandable content. On the other hand, the sustainability message gets spread through the interconnectivity of users and their exchange of experiences and advices. This research proves the necessity for organisations to be present on appropriate social media platforms to find consumers following the same sustainability ambitions as well as to gain new customers willing to take little steps which help to get to a global sustainable change.

References

Atkinson L (2013) Smart shoppers? Using QR codes and "green" smartphone apps to mobilize sustainable consumption in the retail environment. Int J Consum Stud 37(4):387–392

Barris-O'Neill D, Schuitema G (2016) Online engagement for sustainable ener-gy projects: a systematic review and framework for integration. Renew Sustain Energy Rev 54:1611–1621. doi:10.1016/j.rser.2015.10.084

Briggs T (2010) Social media's second act: toward sustainable brand engagement. Des Manag Rev 21(1):46–53

Brown (2016) Eco fashion talk. http://www.ecofashiontalk.com/

Cadwell A, Cadwell B (2015) The good trade. http://www.thegoodtrade.com/thoughts/sustainable-living-blogs

Carrascosa JM, Cuevas R, Gonzalez R, Azcorra A, Garcia D (2015) Quanti-fying the economic and cultural biases of social media through trending topics. PLoS ONE 10(7):1–14. doi:10.1371/journal.pone.0134407

Chen Y, Fay S, Wang Q (2011) The role of marketing in social media: how online consumer reviews evolve. J Interact Mark 25(2):85–94. doi:10.1016/j.intmar.2011.01.003

Dennett DC (2015) Our transparant future. Sci Am 312:3

Diaz A (2015) Defining social media. https://www.boundless.com/

Emmanuel C (2015) 5 Secondhand social media accounts to fol-low and shop. CapeTown ETC. http://www.capetownetc.com/blog/5-secondhand-social-media-accounts-to-follow-and-shop/

Fehrenbacher J (2012) "Closet Swap" iPhone App encourages you to trade clothes with friends. Ecouterre. http://www.ecouterre.com/closet-swap-iphone-app-encourages-you-to-trade-clothes-with-friends/

Finney SL (2014) Encouraging sustainable consumption: an exploration of con-sumer behaviour. Mark Rev 14(2):189–203

Fournier S, Avery J (2011) The uninvited brand. Bus Horizons 54(3):193–207. http://doi.org/10.1016/j.bushor.2011.01.001

Fraoua K, Bourret C, Sotto E (2014) Developing sustainable process in water economy using social media. Int Conf Hum Comput Interact. doi:10.1007/978-3-319-07854-0_3

Grahl T (2015) The 6 types of social media. http://timgrahl.com/

Habibi MR, Laroche M, Richard M-O (2014) The roles of brand community and community engagement in building brand trust on social media. Comput Hum Behav 37:152–161. doi:10.1016/j.chb.2014.04.016

Hajli MN (2014) A study of the impact of social media on consumers. Int J Mark Res 56(3):387–404. doi:10.2501/IJMR-2014-025

Jaeger-Erben M, Rückert-John J, Schäfer M (2015) Sustainable consumption through social innovation: a typology of innovations for sustainable consumption practices. J Clean Prod. doi:10.1016/j.jclepro.2015.07.042

Janauskas J (2016) kleiderkreisel.de. http://www.kleiderkreisel.de/

Johnson LA (2014) Connecting sustainability reporting and social media. Natl Provis 228(1):30–35

Kayabas Y (2016) Kleiderkorb. http://www.kleiderkorb.de/home

Ka-Yan Ng N, Chow P-S, Choi T-M (2015) Impacts of social media mediated electronic words of mouth on young consumers' disposal of fashion apparel: a review and proposed model.

Sustainable fashion supply chain management: from sourcing to retailing. Springer, New York, pp 47–58
Kolandai-Matchett K (2009) Mediated communication of "sustainable consumption" in the alternative media: a case study exploring a message framing strategy. Int J Consum Stud 33 (2):113–125. doi:10.1111/j.1470-6431.2009.00754.x
Lee N, Choi YJ, Youn C, Lee Y (2012) Does green Fashion Retailing make consumers more eco-friendly?: the influence of Green Fashion products and campaigns on green consciousness and behavior. Cloth Text Res J 30(1):67–82
Lundahl O (2014) Fashionalising sustainable consumption in lifestyle media. Adv Consum Res 42:343–347
McDonald S, Oates CJ, Alevizou PJ, Young W, Hwang K (2012) Individual strategies for sustainable consumption. J Mark Manag 28(3/4):445–468
Minton E, Lee C, Orth U, Kim C-H, Kahle L (2012) Sustainable marketing and social media. J Advert 41(4):69–84. doi:10.2753/JOA0091-3367410405
Mustafa A (2013) Fashion ComPassion. http://www.fashioncompassion.co.uk
Pentina I, Tarafdar M (2014) From "information" to "knowing": exploring the role of social media in contemporary news consumption. Comput Hum Behav 35:211–223. doi:10.1016/j.chb.2014.02.045
Phang CW, Zhang C, Sutanto J (2013) The influence of user interaction and participation in social media on the consumption intention of niche products. Inf Manag 50(8):661–672. doi:10.1016/j.im.2013.07.001
Prothero A, Dobscha S, Freund J, Kilbourne WE, Luchs MG (2011) Sustainable consumption: opportunities for consumer research and public policy. J Public Policy Mark 30(1):31–38
Reilly A, Hynan K (2014) Corporate communication, sustainability, and social media: It's not easy (really) being green. Elsevier Inc, Amsterdam. doi:10.1016/j.bushor.2014.07.008
Rex E, Baumann H (2007) Beyond ecolabels: what green marketing can learn from conventional marketing. J Clean Prod 15(6):567–576. doi:10.1016/j.jclepro.2006.05.013
Ruppert-Stroescu M, LeHew MLA, Connell KYH, Armstrong CM (2015) Creativity and sustainable fashion apparel consumption: the fashion detox. Clothing Text Res J 33(3): 167–182. doi:10.1177/0887302X15579990
Sarahana (2012) 9 Ethical shopping apps and plug-ins. http://www.shopethica.com/features/ethical-shopping-apps
Singer L (2016) Trash is for Tossers. http://www.trashisfortossers.com/p/about.html
Stanik L (2010) Computing our way to paradise? The role of internet and communication technologies in sustainable consumptions and globalization. In: Rattle R (ed). AltaMira Press
Tan D (2015) The 6 types of social media « SEOPressor—Best SEO Wordpress Plugin. http://seopressor.com/social-media-marketing/types-of-social-media/
Vaterlaus JM, Patten E, Roche C, Young JA (2015) #Gettinghealthy: The perceived influence of social media on young adult health behaviors. Comput Hum Behav 45:151–157. doi:10.1016/j.chb.2014.12.013
Vermeer DL (2013) 5 Ways technology can promote sustainable fashion. http://daniellelvermeer.com/blog/technology-sustainable-fashion
Wolny J, Mueller C (2013) Analysis of fashion consumers' motives to engage in electronic word-of-mouth communication through social media platforms. J Mark Manag 29(5/6):562–583. doi:10.1080/0267257X.2013.778324
Yan R-N, Hyllegard K, Blaesi L (2012) Marketing eco-fashion: the influence of brand name and message explicitness. J Mark Commun 18(2):151–168
Yates L, Alley M (2015) The good wardrobe| Style.Shared. http://www.thegoodwardrobe.com/
Zolkepli IA, Kamarulzaman Y (2015) Social media adoption: The role of media needs and innovation characteristics. Comput Hum Behav 43:189–209. doi:10.1016/j.chb.2014.10.050

Chapter 13
Potentials of a Fashion fTRACE App

Jochen Strähle and Marie Caterina Sfameni

Abstract The purpose of this paper is to identify the potential of a Fashion fTRACE (ffTRACE) application that gives transparent insight on the supply chain of a fashion item. The research methodology applied to this purpose is a literature review examining academic references. The key findings of this paper are that information plays a major role in the consumer decision process and is therefore beneficial to the demand for sustainable products. Given the right information content in a transparent, credible and understandable way is important. It is found that the functions of such an application would be able to satisfy this consumer demand and therefore has the potential to raise the sales of a sustainable company as well as increase the brand's awareness and improve its image. While mainly indicating the potentials of the ffTRACE application, their relevance is not examined in this paper.

Keywords Consumer decision making · Sustainable information · Sustainable fashion · Consumer behavior · Mobile application

13.1 The Importance of Information Within Sustainable Consumption

When considering sustainable fashion we can no longer talk about a niche market. Consumers around the globe are becoming more aware of the importance of sustainability every day (Keller et al. 2014). But contrary to the increasing concern for the environment, consumers do not purchase accordingly. Although the number of sustainable products in retail is rising, the sales are not increasing significantly (Gleim et al. 2013). Sartory (2014) claims that the market share of sustainable

J. Strähle (✉) · M.C. Sfameni
School of Textiles and Design, Reutlingen University, Reutlingen, Germany
e-mail: jochen.straehle@reutlingen-university.de

M.C. Sfameni
e-mail: Marie.sfameni@gmail.com

J. Strähle (ed.), *Green Fashion Retail*, Springer Series in Fashion Business,
DOI 10.1007/978-981-10-2440-5_13

fashion is still very small. Rex and Baumann (2007) state that there is a general lack of information on sustainable products and therefore consumers make unsustainable purchase decisions. Furthermore, 65 % of the people worldwide would buy more sustainable products if they had more basic knowledge and knew more about the positive impact of these products (Accenture 2014). Meanwhile, the numbers and types of information distribution channels are expanding rapidly. For example, the number of smartphones is increasing steadily. By 2020, 70 % of the world population will be using a smartphone (Scholz 2015).

Regarding the rising number of smartphones and the lack of information on sustainable clothing, the question arises whether an application giving the information needed to make a sustainable purchase decision could be beneficial for the demand of sustainable clothing. The potentials of such an application will be determined in this paper.

The method chosen to give an insight is a literature review that will present the state-of-the-art in research regarding this topic. More than 35 academic references are cited. Additionally, an analysis of possible functions of a fashion application will be presented, which was derived from a comparison with a similar working application in the food sector. The presentation of the literature findings starts in Sect. 13.2 where the role of information within the consumer decision process is described. The process will be illustrated and the limits of the information search will be examined by looking at the lack of information, possible sources available, and the information demand. In Sect. 13.3 the relevance of the application use while shopping will be examined and compared to the already existing fTRACE app for food. Possible functions of a fashion application, referred to as ffTRACE, will be pointed out. After a comparison of these functions the potentials of the fashion fTRACE app will be presented in Sect. 13.4. The paper closes with a conclusion in Sect. 13.5 that points out the results and limitations of this work, and gives recommendations about further research topics.

13.2 The Role of Information Within the Consumer Decision Process

The demand determines the offer. Becoming aware of the environmental impact of any purchasing decision, the consumer realizes that his choice has a big influence on the market. By rejecting products that are not sustainable the consumer can use this power to support sustainable progress (European Commission 2006). By choosing sustainably responsible products the consumer can exert his power by supporting sustainable companies. Consequentially, making the right choices is the way consumers can influence and advance the sustainable development (Moisander et al. 2010).

To understand how a decision is made and what factors influence this decision toward being sustainable or not, first the consumer decision process model will be presented. Next, the information search which is part of the decision making process will be examined.

13.2.1 The Consumer Decision Process Model

What step does a consumer undertake before buying a product? Kroeber-Riel and Gröppel-Klein (2013) defined 7 steps as the purchase decision process. This process gives a rough overview on how people prepare for a purchase. In reality of course those steps overlap or mix sometimes, but this model gives insight into what is happening before a sustainable or unsustainable purchase decision is taken. The time spent on every single step depends not only on the character of the consumer but also on the characteristics of the purchased product (Yurchisin and Johnson 2010). The decision making process is usually more intensive if a very expensive product or a product with a special meaning to the person is bought (Fig. 13.1).

First, the consumer realizes that there is a need or a problem. Consumers feel like there could be some improvement in their life by acquiring something, typically by buying something. Because of this desire the person imagines an ideal state of her or himself. To reach this state, instead of remaining in the status quo, he or she needs to solve the problem. The solution is transferred to acquisition—the person needs to buy or somehow obtain something to reach her/his ideal state. To solve this problem in the second step one has to search for the information needed to make a purchase decision. While learning what kinds of products exist to solve the problem, one also needs to learn which of them are available and accessible. This

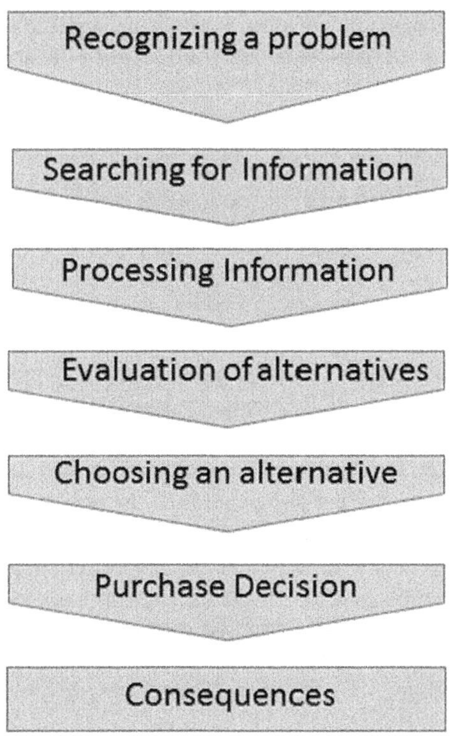

Fig. 13.1 Consumer decision making process. Adapted from Kroeber-Riel and Gröppel-Klein (2013)

additional information can be searched for internally (within one's own knowledge base) or externally (friends, advertising, Internet, etc.). In Sect. 13.2.2 the different sources will be named and explained. By processing all information found, the consumer starts to judge the alternatives provided. Depending on his values, he will choose the best alternative. Finally, the purchase decision is made and followed by the consequences of this purchase (Yurchisin and Johnson 2010).

There are many more purchase decision making models. The one from (Blackwell et al. 2006) for example includes more steps after the purchase decision such as use, care and storage, post-purchase evaluation, and disposal. Mainly, the steps before the purchase decision are the same as in the model of Kroeber-Riel and Gröppel-Klein (2013).

13.2.2 Relevance of Information

The attitude–behavior gap is a phenomenon that describes the fact that consumers are concerned about sustainability, but do not translate this concern into their purchasing behavior. When having a choice available they still do not choose the sustainable product and therefore do not act in a sustainable way (Young et al. 2010). Many authors talk about missing information as a reason for unsustainable purchase decision. To be able to evaluate the significance of this aspect and to prepare for deciding whether an application could help to close this gap, in this section the second and third steps of the consumer decision making process will be analyzed in detail. First, the lack of information will be presented, then possible information sources will be stated and finally the information demand of the consumer that supports sustainability will be pointed out.

13.2.2.1 Lack of Information

Rex and Baumann (2007) state that there is a general lack of information on sustainable products and therefore consumers make unsustainable purchase decisions. These findings are supported by this statistic made by Accenture (2014).

Figure 13.2 illustrates which parameters should be given so the interviewed person would buy more sustainable products. 64 % of the people worldwide would buy more sustainable products if they had more basic knowledge, and 70 % answered that knowing about the positive impact of these product would convince them to buy more sustainable products. 66 % mention that they do not really trust the claims of the company regarding sustainability and therefore do not purchase those items. 63 % of the interviewees claim they would buy more sustainable products if they were easier to find (Accenture 2014).

Fig. 13.2 Tangibles for purchasing more sustainable products. Adapted from Accenture (2014)

In summary, the figure states that there is a general lack of trustworthy information on sustainable products and therefore consumers make unsustainable purchase decisions. The environmental information needs to be visible and more efficient. Transparent and non-misleading information given to the consumer in the right way, influence his purchase decision enormously (McDonald et al. 2009). The increased awareness and knowledge about sustainability is therefore beneficial for the demand of sustainable products (Moisander et al. 2010). Also, Connell (2010) states that customers already prefer products that have sustainable content. He therefore suggests that more consumers are willing to choose sustainable products if there is credible information about them given.

One of the main hindrances is that consumers still need to actively search for green products and for information on their sustainability, which means they have to invest extra time to make a sustainable purchase decision.

Sheth et al. (2011) mention specifically that the consumers, as one of the principle stakeholders, are not given enough attention by the companies. They are not informed in the required way and therefore information that they might need is missing. Hence, the effectiveness of the companies' engagement on being sustainable is limited as the consumer does not know about it and therefore cannot actively support it. Furthermore, if the company does not have a sustainable value chain giving information to the consumers, it would not be beneficial to their sales figures (Moisander et al. 2010). Other reasons for lacking information are the consequences that arise when raising the general awareness of the consumer regarding sustainable consumption. Mass consumption and fast changing fashion trends, in general, are not sustainable. If a company teaches their consumers about sustainability the consumers might rethink their purchasing decisions and maybe change their behavior toward buying fashion. As this is not the intention of the retailer, he holds back information about sustainable fashion consumption to not lower his sales numbers (Strähle et al. 2015).

If information is missing one could say that consumers can still ask for it. This statement is true. But there is a phenomenon, described by Ehrich and Irwin (2005), called "willful ignorance" that shows that if information is missing many people do not ask for it, even when they would have used it to make their decision if it was available. If they like a product and there is no information on sustainability that they would normally take into account when making a decision, they do not want to ask for it as it might expose the product as unsustainable and therefore they can no longer buy it with a good conscience. They prefer to stay uninformed in order to buy the product without feeling guilty.

13.2.2.2 Sources of Information

When looking for information there are several sources available. To understand the characteristics of the information given by various sources, the different sources are presented below.

Internal information is the information that is already stored in the consumer's brain. This includes experiences with previous products and memory of previous purchases.

There are four main sources to obtain external information:

- Family, friends, and colleagues build the first source, the personal source.
- Marketer-controlled sources are advertisements, information from sales personnel, or other information sources at the point of sale (POS), as well as information found on the internet, as provided by the producer, fashion websites, and consumer reviews. Applications that provide information by the label itself are also part of this source category. The detailed use of these applications is described in Sect. 13.3.1.
- Perceived as the most neutral source of trade information are public sources like newspapers, magazines, or TV reports (Kroeber-Riel 2009).
- The last source, the objective source is a third-party-biased source that gives information such as magazines specialized on comparing products, not accepting advertisement or user reviews in, for example, online shops (Kroeber-Riel 2009). Also Oates et al. (2008) found that consumer that supports sustainability trusts more in independent sources such as green guides or groups like NGOs instead of trusting in the brand itself.

In general, consumers that support sustainability get more involved in the decision making process take longer time and are more critical toward the information given by the companies (Hassan et al. 2013).

13.2.2.3 Information Demand of the Consumer that Supports Sustainability

When talking about the information demand regarding sustainable products, one needs to consider two customer groups; those who did not yet know they want to purchase something sustainable and those who do know they want to.

Customers who are unaware of the importance of sustainability will not learn about the unsustainable practices behind some products by purchasing them (Allwood et al. 2006). Often they do not understand the link of the purchasing decision and the ethical problems occurring in the production process of this product. If there is no information on the product and then unsustainable fashion is purchased, there is no immediate consequence to this. One does not feel the bad working conditions on the skin and one cannot see the chemicals used in the production process on the garment. Information is the only way to make those background processes visible on the finished product. Therefore, information is fundamental for people who did not know the consequences of their purchase (Allwood et al. 2006).

On the other hand ecologically conscious consumers are extremely aware of the consequences occurring from their actions. They seek to only consume products that do not or at least almost do not harm the environment, other human beings or future generations. In addition to only purchasing sustainable products they also refuse buying unsustainable products to contribute to the environmental preservation (Akehurst et al. 2012). To continue to act as they wish, for them information is fundamental.

The influence of availability and access to information on the decision making process of these consumers is very high. They are very interested in the information given and are willing to dig deep into a company's organization and get informed via various channels (Oates et al. 2008). The information needed includes awareness of the product, the supplier, and the socio-environmental awareness. Sustainable consumers will additionally focus on aspects like durability, eco-performance, locality, responsibility, and corporate "identity" when making a purchase decision. Eventually, they will buy second hand, make, rent, or repair the item they need to act sustainable (Oates et al. 2008). In summary, consumers that already support sustainability as well as those not yet fully on-board with the concept of sustainability need information on the product to make an informed purchase decision.

Required Information Content

When receiving information, the consumer can only use this information if it fulfills certain requirements. The right content must be given in a credible, transparent, and understandable way.

To ensure the required information is given one needs to look at the definition of a sustainable company. Adapted from the three-pillar model created 1998 by the

Enquete Commission based on the definition of sustainability defined in the Brundtland report "our common future" from 1987, Gessner (2008) defined a sustainable business as one that realizes his business activities in harmony with the sustainable development while simultaneously satisfying the demand of the company itself and its stakeholders (UN world commission on Environment and Development 1987). This means ecological, economical, and socio-environmental aspects are respected equally in every part of the supply chain as well as in the main business (Jonker et al. 2010). Therefore, every raw material should be produced and harvested without damaging the environment, every worker involved in the process should be treated fairly, and every supporting function, such as transport or packaging, should be thought through and done with the least possible use of energy. The retail store should function as an energy and water saving unit, while the costs are kept in a range such that the company still can survive economically (Bailey and Baker 2014). To give full insight on the sustainability of an apparel item, the company needs to provide information on every single step of the supply chain.

Especially in fashion business the supply chain is extremely complex. For the production of one garment; for example, a printed T-shirt there are several supply chain steps necessary that often take place in different locations (Gardetti and Torres 2013). To trace the whole supply chain of a fashion item is still impossible for most of the clothing manufacturers as there are so many different steps from raw material to the ready-to-sell item (Bhaduri and Ha-Brookshire 2011). To illustrate the variety of production processes that all take place in different locations and the possible sustainability issues that can arise in each process step this figure from Gardetti and Torres (2013) is very illuminating (Fig. 13.3).

Furthermore, the companies who transport the item from location to location, the companies who produce little extras like buttons or zippers, and the companies who make the care and brand labels have to be named (Gardetti and Torres 2013). Only if a company succeeds in tracing every step of the supply chain and all the sustainability issues arising, the consumer gets full transparency on what he is buying.

Required Information Frame

Additionally, consumers often cannot trust the information that is given by the companies (Schäfer 2012a, b). To convince the consumer that supports sustainability, companies have to communicate about their business practices openly (Bhaduri and Ha-Brookshire 2011). Building up an environmentally and socially correct brand image, a company has to implement the required sustainability standards as well as to create a credible and authentic communication strategy, so their effort is recognized by the customers (Strähle and Köksal 2015). Next to being credible and reliable, the information has to be transparent. Transparency means reporting about the company's business practices in an open and honest way. The information given should be easy to understand and nothing should be hidden. The accessibility of the information plays a major role (Merriam-Webster 2010). Only if

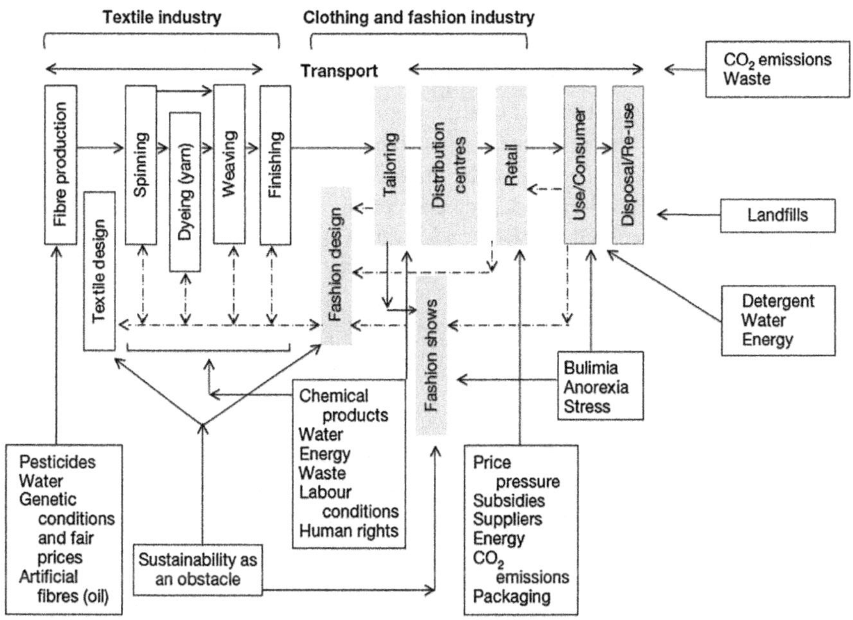

Fig. 13.3 Exemplary fashion supply chain including sustainability issues. Adapted from Gardetti and Torres (2013)

the information given is credible and transparent on every single supply chain step, the consumer can really use this information to make an informed sustainable purchase decision. He can only do that, if the information given is also understandable. In a study by Hassan et al. (2013) consumers emphasize the difficulty to determine the credibility as well as to understand it.

Eco Labels

An obvious way to get insight on the behavior of a company seems to be the labeling with eco labels. As a third party is involved who judges about which product fulfills all required standards and therefore gets the label, there is no chance of greenwashing by the company. Yet, the market penetration of eco-labeled products is still low. A reason could be that the link between information and label is not given. The labels are not clearly following national or international policies and therefore do not mean a lot to the consumer (Rubik and Frankl 2005). Furthermore, many customers do not even trust the companies who give certificates to the fashion products as those companies also need to be economical, profitable, and therefore need customers. Fashion labels (Bhaduri and Ha-Brookshire 2011).

Another possible way to inform consumers is the clear stating of explanations and information directly on the product label (Borin et al. 2011). Many of the consumers asked in a study by Bhaduri and Ha-Brookshire (2011) would prefer information at the POS and could imagine information on sustainability to be on hangtags or care labels directly on the product. When thinking about a purchase, one often tries on the garment and can then immediately get additional information on the product. Here the problem occurs that many consumers do not trust the information given by the companies themselves (Bhaduri and Ha-Brookshire 2011). Besides, most product packages are not big enough to carry all this information. The consumer would feel stressed and overstrained having to read everything, especially at the POS (Ehrich and Irwin 2005).

13.3 Are Applications a Solution?

13.3.1 Increasing Usage of Smartphones

The worldwide number of smartphones is increasing every minute. According to the mobility report of Erricson it will have doubled from the current level by year 2020. In 2014 there were an estimated 2.6 billion smartphones in use worldwide. By 2020 it is projected to be 6.1 billion smartphones. This means, that by 2020, 70 % of the world population will be using a smartphone. In Fig. 13.4 one can see that the western European area as well as the North American one is already covered by smartphones, which explains the smaller growth rate of smartphone subscriptions in these areas. Remarkable is the huge growth that is predicted for the Asian Pacific Area (Scholz 2015).

Meanwhile, Yahoo published new developments regarding the usage of applications. They analyzed 1.8 billion smartphones users and determined that by 2014 1.4 billion people used applications. In 2015 all of them, 1.8 billion people, did so which equates a growth of app users of 30 % (Scholz 2015).

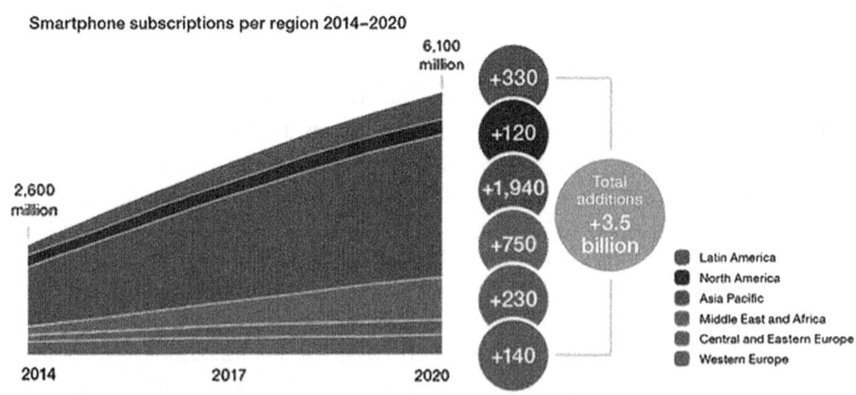

Fig. 13.4 Smartphone subscriptions per region 2014–2020. Adapted from Scholz (2015)

Fig. 13.5 Information searched for online. Adapted from AGOF e.V. (2015)

What the people use the applications for is indicated by a study from AGOF about German smartphone users starting at the age of 14. 78 % of them use the smartphone to search for information via websites like Google or Bing. The only application used more (78 %) is for making phone calls. The conclusion is that apps are mainly used for general phone functions and information gathering. Which kind of information the users look for is shown in the following Fig. 13.5. 25.5 % of the polled users search for information about female clothing. Search for information about male clothing is at almost the same rate (AGOF e.V. 2015). But why do so many people use the smartphone to search for information? Limitations such as the small screen or limited data volumes would suggest using a desktop or laptop computer at home to look for detailed information.

A reasonable assumption is that most people cannot anticipate that they will be needing information on some specific thing while out shopping. Often they come into situations where they wish for information on a specific topic, but the information needed is not given. Especially, missing product information can be a hindering factor during the decision–purchasing phase. So people choose the one way that is always available, anywhere and anytime to get this missing information, their smartphone, which they always carry with them and which always provides internet access (Scholz 2015). More than 90 % of the consumers use their smartphones while shopping at the POS, according to a survey from SessionM. While more than approximately 54 % use their devices to compare prices, others search for product information (48.4 %) and 42 % search for reviews (Taylor 2015).

These figures show that already many people are using smartphones and that the amount of them that use applications on the smartphones is almost equally high.

That they mostly use applications for searching information is giving a hint on how important the role of applications is during the "information searching" step in the consumer decision making process model. One application that delivers information on food will be presented as an example in the next subsection.

13.3.2 Functions of the fTRACE App

The fTRACE is an application that gives access to the platform fTRACE where manufacturers and retailers provide information on the origins, the production processes, and the quality of their brand's products. The platform is operated by the company GS1 German which is mainly known for the EAN barcode. The aim of fTRACE is to track the whole supply chain of the product and give all required information to the consumer in an easy and understandable way, while focusing on credibility and transparency.

By scanning barcodes on meat, poultry, fish, fruits, and vegetables or entering their number on the fTRACE Website, the consumer gets information directly from the manufacturer of these products. Next to the manufacturer company's name, there is information about the harvest or slaughter date, the packing date, and the name of the packing company. Furthermore, the location of the manufacturer is shown on a map and eco labels are named, as well as explained. Furthermore, the manufacturer has the possibility to give more information than is legally required. He can upload photos of his plantation or lakes and make videos from his stables or the harvest. He can even suggest recipes to cook with the scanned product. Coming directly from the manufacturer the information is believed to be trustworthy. Of course he can also state his quality standards and certificates of quality checks (GS1 Germany 2015).

To get an fTRACE code, the manufacturer of the product has to join fTRACE by making a contract with them. fTRACE and GS1 demand certain quality standards certificates that have to be provided by the manufacturer (Liegl 2015). Then he can enter the information about his products either via the fTRACE Website or, if it is a bigger company that has an own ERP system, the information can directly be transferred from this system to fTRACE. If the fTRACE-coded products are a huge amount, the manufacturing company can think about getting EPCIS software that combines the companies' information with fTRACE.

At the moment fTRACE has about 300 partners in the meat, fish, fruit, and vegetable sector. About 30,000 products have an fTRACE code on them and in April 2014 about 12,500 users daily visited the fTRACE application or website to scan a product (Hennecke 2015).

In comparison to how many people buy these products daily, this is still a small number. But it is increasing steadily and the users who are using it are mostly confident about it. Of course, right now not enough products do have a code to scan. The consumers wish for a platform that combines and collects the information on all the products they use, so they do not have to search long for this information (Gontek 2015).

13.3.3 Functions of a Fashion fTRACE App

Many people are already used to purchase food with eco or fair trade labels on them or even to using applications like fTRACE. As one can imagine the meaning of these labels for our food, the link within the fashion industry is not so obvious (Beard 2008). As the Lifestyle of Health and Sustainability (LOHAS) becomes more popular, it is a big trend to care about the conditions of one's own body and to be careful what to eat. As one puts food directly in his body one is very concerned about production processes and substitutes added. But in the end it is just as important what touches the body, specifically the skin from the outside. Fabrics containing dangerous chemicals or colors can also harm the body as well as the workers producing the fabric (Schäfer 2012a, b). That is also why the fashion world is changing. Some experts go so far to say that fashion as known today will not exist much longer as the consumer is changing his values regarding clothing. They want clean and fair products that do not harm the environment or workers. Hence, fashion companies that act sustainably have a competitive advantage in today's fashion world. Not black, but "closing the loop" and "saving resources" will be the new trend (Nina Piatscheck 2015).

Now ffTRACE as a new application to help fashion companies to communicate about the company's efforts to follow this trend requested by the customers will be defined.

Companies that want to cooperate with fTRACE have to agree on a contract with GS1 and have to fulfill certain sustainable standards. The company has to provide all needed information on their products and substitutions to get a barcode which they put on the clothing to make this information accessible to customers. The manufacturers themselves can also add additional information.

If the fashion fTRACE application works similar to the fTrace app in the food sector the following information possibilities would be achieved: By giving the standard information needed to fill the ffTRACE information requirements the company already gives much more information to the customer than normally available without the use of an application like ffTRACE. Information on every single production location is saved in the application. The company's names, the working conditions, the harvesting or producing times, the packaging procedure, and the date, when the product was sent to the next supply chain location, would be available to the application user. It is possible to add photos and videos of every location, showing the workers there or the production process. So an insight on the different supply chain steps can be given. Additionally, one could include knowledge about the working wages in the different manufacturer locations.

Furthermore, a list of achieved eco or fair trade labels is given and the different labels are explained according to which standards they fulfill. Another possible function of the ffTRACE app could be information about how to handle the product. Information on how to wash it without wasting too much water and energy or the possibility to advertise eco-friendly washing liquids that one can also buy in the store would make sure to provide instructions on how to act sustainable, even beyond the purchase of a sustainable product.

Instead of giving additional information on recipes, like in the food fTRACE application, the ffTRACE app could suggest outfit combinations or matching sustainable accessories. If there are some self-help projects occurring in one of the involved manufacturer locations, the application could refer to them and help collecting donations. Another possible function could be that the app saves information on every item that was bought after the code was scanned. If after some years the T-shirt is not wearable anymore, one can recall the shop where it was bought and get information about recycling options. Maybe the store itself can take the product back or suggest a special point where the garment can be donated. A similar recycling program is now available for sending batteries back to the manufacturer.

13.4 Potentials of a Fashion fTRACE App

The primary potential of the fashion fTRACE (ffTRACE) application is that it will actually be used by consumer. In 5 years seven out of ten people will have a smartphone. Almost every smartphone user takes advantage of using applications and 78 % today already use those applications to search for product information. Women's and menswear are in the top ten most searched for subjects via applications like Google or Bing. 90 % of the smartphone users already use their smartphone at the POS, and half of their searches are for information on the products. Right now they still need to use search engines to get the required information since the information is scattered on the label's website, the department stores website, or independent websites, for example, NGO's. As sustainable conscious consumers are willing to take some time for the purchase decision making process and especially for the information search step, they will take the time to use the application. When used, it is much more effective than the previous searching for information all over the Internet. The previous extra time needed to search for required information can be reduced and the consumer does not have to invest extra effort anymore.

In an ideal state, all clothes carry an ffTRACE code, so all information is united in one place and the consumer does not even have to think about where to look for it. Of course not all labels will join immediately, but as there is no such a platform right now it is a first step to collect information from different labels and producers into one place. In comparison to having to look for information on various websites, where the level of information and credibility can differ enormously, a platform with collected information is much more user-friendly. As the fTRACE application for food has 12,500 visitors daily (April 2014), growing the fTRACE application for fashion might be used extensively as well.

As the application links with the barcode on the apparel item itself, the information is directly accessible for the consumer when thinking about a purchase. Information at the POS is what many customers desire. As the packaging of the product does not suffice for all the information content, and anyway the consumer would be stressed by feeling the need to read all the information, the ffTRACE is a chance to provide all information available while not taking space on the packaging, which gives further product design options. Moreover, the consumer can exactly choose which information he/she is interested in and does not have to read everything, while still he could read everything if he wanted. By providing the information via the ffTRACE application, it is always accessible for most consumers.

To check for information will be much faster than before and the possibility to do so directly at the POS, before having to make the decision without information is a huge advantage. Furthermore, people no longer can "willfully ignore" information on sustainability. Previously, sometimes they did not ask for it, although they would have taken in into account if it had been available from the start. With the ffTRACE application it is easily available and cannot be ignored.

When a customer uses the applications he is served with all needed information. Content-wise, the ffTRACE application requires minimum information from the fashion labels, the producers, or manufacturers. This minimum information is already much more than the one usually found on the product or care label. As it includes basic information on every supply chain part, the consumer can find out which companies took part in the production process, where they are located, when the product was produced, which raw materials were used, how the working conditions are, as well as where it was send afterward and who cared about the packaging and transporting.

Being able to deliver further information about how to care for the garment, how to wash it, or what to do with it if it is no longer needed, might meet another need of the consumer that supports sustainability. He does not only care about the information upfront during the purchasing decision, but he also wants to know how sustainable the item is in use. Durability and recyclability are also aspects he wants to know about.

Considering every supply chain step the whole process gets transparent. As no information or production location can be hidden, the information covers everything the consumer that supports sustainability wants to know. As transparency means honest and understandable information, the application tries to make the information easy traceable. By showing the different production locations on a Google maps card, the consumers get an impression where the facilities are located in the world and how far the product has traveled to now hang in front of him in the store.

Because of the possibility to add photos or even videos of the production places, showing the working conditions, the cotton plantation, or the harvesting process for example, the company reduces the "distance" between the workers involved in the production process and the consumers. Many western consumers cannot imagine the production conditions in faraway countries and might imagine them quite primitive. By giving insights with photos or videos, these prejudices might be determined and trust could be build up. Material like this also counts as proof for many consumers that the information given is credible, since they can see the places and the conditions on the picture.

As there is the chance to present eco and fair trade labels within the application, more consumers might pay attention to them. Often the understanding of the meaning of eco labels is fragmentary and therefore they are not trusted. By having the possibility to list the standards required to reach certain labels and explaining them, transparency on the labels and their meaning could be created.

In summary, the application is capable of giving all needed information in a transparent, credible, and understandable way to the consumer. Hence, the application could help to fight the general lack of trustworthy information on sustainable products.

Since today not many companies are able to track every step of their own supply chain and create such a transparency on all the conditions while still looking good, being able to deliver this information via the ffTRACE application could be a great competitive advantage. Consumers nowadays want to know about the conditions under which their garments are produced and want to be informed about them. The ffTRACE barcode could soon become a quality seal regarding sustainable consumption. For this reason, it will also get easier to refuse unsustainable items. As the code becomes a proof for sustainable business practices, not having a code becomes an indicator for unfulfilled sustainable engagement of a company.

The application as an information source is very likely to be accepted by the consumer. Although it is a partly marketer-controlled source, since the manufacturer himself gives the information on his processes, there is still an objective third-party involved. When agreeing on a contract with GS1 Germany, the production companies need to prove that they have certificates for certain quality standards. Those certificates can also be uploaded in the application so the consumer himself has full insight on what companies produced this fashion item. This possibility is a further step in building up trust and credibility.

From the figure in Sect. 13.2.2.1, it is known that people would buy more sustainable products if they had more basic knowledge, knew more about the positive impact of the products, could trust the claim by the companies, and find the sustainable products with less effort. Transparent information is beneficial for increasing the demand for sustainable clothing, and costumers can support sustainable companies more effectively when they are informed about the efforts undertaken by the companies. As the ffTRACE application delivers all these factors, it is possible that the application leads to higher sales numbers for those labels providing a ffTRACE code on their items. The brand image could be improved and the general brand awareness can be raised.

As the attitude–behavior gap might also exist because of the lacking information, maybe this gap could be closed and people who are already concerned about the environment will in future translate this concern into their purchasing behavior.

One more potential of the application, if it becomes a sign for sustainable quality, is the pressure it might put on labels that do not yet have a sustainable supply chain or are not even capable of tracing their whole supply chain. If the supply chain of a company is sustainably exemplary, the company will certainly take advantage of the ffTRACE application to make this effort visible to their customers. Therefore, supply chains that do not use the application to give a transparent insight, as this insight might rather drive the consumer away from the product, might be confronted with decreasing sale and feel a pressure to create a value chain that people can look at and still buy having a good conscience.

13.5 Conclusion

After highlighting the growing concern for the environment and in general sustainability, the attitude behavior gap in the fashion market place is presented. It was found that consumers do not translate their concerns into their purchasing decision and therefore the market share of sustainable products is still low. After stating the possible correlation of unsustainable purchase decisions and missing product information in Sect. 13.1, a closer look at the role of information within the consumer decision making process is taken in Sect. 13.2. Based on the purchase decision making process model described in Sect. 13.2.1, the lack of information is presented in Sect. 13.2.2.1. Giving further information on information sources and the information demand of the consumer that supports sustainability it gets clear that one of the main hindrances for purchasing sustainable products is the missing basic knowledge as well as the missing information on the positive impacts of sustainable products. It is found that information has to be easy accessible, it needs to deliver the right content in a way the consumer can use it, and it has to be credible, transparent, and easy to understand.

By giving an overview of the worldwide use of applications in Sect. 13.3.1, the possibility of using a computer/smartphone application to provide the needed information is analyzed. It becomes clear that applications used to search for information are already widely spread, especially for information search on fashion products. By presenting the fTRACE app for food, possible functions of fashion fTRACE applications are indicated.

The knowledge gained is used to identify the potentials of the ffTRACE application that delivers full transparency on the supply chain to every consumer. The main potentials found are the delivery of information required by the consumer in the way he wishes to obtain it—credible, transparent, and understandable. Therefore, the application has the potential to meet the customer's information need while in the purchase decision making process, everywhere and anytime. With the right information an informed and therefore sustainable purchase decision can be made. For the company this means increasing sales numbers and higher profit as well as improvement of brand image and awareness. Furthermore, the applications can function as a sustainably quality seal and put pressure on other companies to become more sustainable in order to receive this quality sign as well.

13.6 Limitations and Future Research

While the potentials of the ffTRACE application were analyzed in detail, the relevance of these potentials was not specifically focused on. The question arises how many people will actually use the applications and therefore experience the potential that it has. Another unknown is whether many fashion labels and manufacturers will already have established a sufficient sustainable baseline to join the platform and promote their sustainability. The effort arising from collecting, uploading, and sharing this amount of information might be more expensive than the higher profit occurring from increased demand caused by the information given to the consumer. If not many fashion companies take part, the platform will not unite all products available and therefore there will be only one more minor source one can go took when searching for information and not the only one place where it can be found.

Due to the attitude–behavior gap, the evaluation of the real consumer demand is limited. One reason for the consumer behavior that does not correlate with the consumer's attitude might be the missing information. There can still be more reasons for this attitude–behavior gap. Because of the low quantity of literature findings regarding the demand of the consumer that supports sustainability and the also limited quality and structure of the data, since only a small sample of customers was studied, the results regarding the consumer behavior and his requirements for making a sustainable purchase decision are not complete. Also, international differences in country or culture patterns were not specifically considered in the existing studies.

To overcome the limitations of this research one could conduct quantitative studies exploring the consumer behavior when purchasing sustainable products. Finding out if there are more reasons explaining the attitude–behavior gap than missing information and trust could give a hint on further potentials that could be created by functions of the ffTRACE application. Furthermore, next to analyzing the potentials of the ffTRACE application, one could try to find out how far those potentials will be used and therefore how big their relevance is. An analysis of how the consumer acts at the POS and which information he mainly reads when opening the application and scanning a code could be beneficial for the development of the application.

References

Accenture (2014) Kriterien, die die Kaufbereitschaft ethisch einwandfreier Pro-dukte erhöhen 2014| Umfrage. Retrieved 2 December 2015, from http://de.statista.com/statistik/daten/studie/350714/umfrage/umfrage-zu-anreizen-haeufiger-ethisch-einwandfreie-produkte-zu-kaufen/

AGOF e.V. (2015) Mobile facts 2014—III. Retrieved from http://www.agof.de/download/Downloads_Mobile_Facts/Downloads_Mobile_Facts_2014/Downloads_Mobile_Facts_2014_III/mf%202014-III%20AGOF_mobile_facts_2014-III.pdf?f97c52

Akehurst G, Afonso C, Martins Gonçalves H (2012) Re-examining green purchase behaviour and the green consumer profile: new evidences. Management Decision 50(5):972–988. doi:10.1108/00251741211227726

Allwood J, Laursen S, Rodriguez C, Bocken N (2006) Well dressed? The present and future sustainability of clothing and textiles in the United Kingdom. University of Cambridge, Cambridge. Retrieved from http://www.ifm.eng.cam.ac.uk/uploads/Resources/Other_Reports/UK_textiles.pdf

Bailey S, Baker J (2014) Visual merchandising for fashion. Bloomsbury Academic, London

Beard ND (2008) The branding of ethical fashion and the consumer: a luxury niche or mass-market reality? Fashion Theory 12(4):447–467. doi:10.2752/175174108X346931

Bhaduri G, Ha-Brookshire JE (2011) Do transparent business practices pay? Exploration of transparency and consumer purchase intention. Cloth Text Res J 29(2):135–150

Blackwell RD, Miniard PW, Engel JF (2006) Consumer behavior, 10th edn. Thomson Business and Economics, Mason

Borin N, Cerf Douglas C, Krishnan R (2011) Consumer effects of environmental impact in product labeling. J Consum Mark 28(1):76–86. doi:10.1108/07363761111101976

Connell KYH (2010) Internal and external barriers to eco-conscious apparel ac-quisition. Int J Consum Stud 34(3):279–286. doi:10.1111/j.1470-6431.2010.00865.x

Ehrich KR, Irwin JR (2005) Willful ignorance in the request for product attribute information. J Mark Res 42(3):266–277. doi:10.1509/jmkr.2005.42.3.266

European Commission (2006) Consumer choice. Retrieved 16 December 2015, from http://ec.europa.eu/environment/archives/youth/waste/waste_consumer_en.html

Gardetti MÁ, Torres AL (eds) (2013) Sustainability in fashion and textiles: values, design, production and consumption. Greenleaf Publishing, Sheffield

Gessner C (2008) Unternehmerische Nachhaltigkeitsstrategien: Konzeption und Evaluation. Lang, Frankfurt am Main

Gleim MR, Smith JS, Andrews D, Cronin J, Joseph J (2013) Against the green: a multi-method examination of the barriers to green consumption. J Retail 89:44–61. doi:10.1016/j.jretai.2012.10.001

Gontek F (2015) Verbraucher-App: Transparenz mit Tücken. sueddeut-sche.de. Retrieved from http://www.sueddeutsche.de/wirtschaft/verbraucher-app-transparenz-mit-tuecken-1.2553394

GS1 Germany (2015) fTRACE. Retrieved 10 December 2015, from http://www.ftrace.com/en/gb/about

Hassan L, Shaw D, Shiu E, Walsh G, Parry S (2013) Uncertainty in ethical consumer choice: a conceptual model. J Consum Behav 12(3):182

Hennecke A (2015) Wie schmeckt unsere Zukunft? http://www.tagesspiegel.de/advertorials/ots/die-lebensmittelwirtschaft-e-v-footage-und-film-material-des-vereins-die-lebensmittelwirtschaft-im-vorfeld-der-anuga-wie-schmeckt-unsere-zukunft/12412280.html

Jonker J, Stark W, Tewes S (2010) Corporate Social Responsibility und nach-haltige Entwicklung: Einführung, Strategie und Glossar, 2011th edn. Springer, Berlin

Keller C, Magnus K-H, Hedrich S, Nava P, Tochtermann T (2014) Succeeding in tomorrow's global fashion market. Retrieved 19 November 2015, from http://www.mckinseyonmarketingandsales.com/succeeding-in-tomorrows-global-fashion-market

Kroeber-Riel (2009) Konsumentenverhalten (9., überarb., aktualisierte und, erg edn. Vahlen, München

Kroeber-Riel W, Gröppel-Klein A (2013) Konsumentenverhalten. Vahlen, München
Liegl P (2015) Ecosio—innovating EDI. Retrieved 10 December 2015, from https://ecosio.com/de/blog/2014/09/04/Rueckverfolgung-von-Lebensmitteln-mit-Hilfe-von-fTRACE/
McDonald S, Oates C, Thyne M, Alevizou P, McMorland L-A (2009) Comparing sustainable consumption patterns across product sectors. Int J Consum Stud 33(2):137–145. doi:10.1111/j.1470-6431.2009.00755.x
Merriam-Webster (2010) Definition of TRANSPARENT. Retrieved 22 December 2015, from http://www.merriam-webster.com/dictionary/transparent
Moisander J, Markkula A, Eräranta K (2010) Construction of consumer choice in the market: challenges for environmental policy. Int J Consum Stud 34(1):73–79. doi:10.1111/j.1470-6431.2009.00821.x
Nina Piatscheck (2015) Less is mode 37:60–61
Oates C, McDonald S, Alevizou P, Hwang K, Young W, McMorland L-A (2008) Marketing sustainability: use of information sources and degrees of voluntary simplicity. J Mark Commun 14(5):351–365. doi:10.1080/13527260701869148
Rex E, Baumann H (2007) Beyond ecolabels: what green marketing can learn from conventional marketing. J Clean Prod 15(6):567–576. http://doi.org/10.1016/j.jclepro.2006.05.013
Rubik F, Frankl P (eds) (2005) The future of eco-labelling: making environmental product information systems effective. Greenleaf Publishing, Sheffield
Sartory M (2014) Rat für Nachhaltige Entwicklung : Faire Textilien: „Es bringt nichts, nur eine Liste abzuhaken" - Interview mit Maren Sartory von Fairtrade Deutschland. Retrieved from http://www.nachhaltigkeitsrat.de/index.php?id=8138
Schäfer S (2012a) Nachhaltige Kleidung: Das Interesse an nachhaltig produ-zierter Kleidung wächst. Retrieved 5 December 2015, from http://www.zeit.de/zeit-wissen/2013/01/Nachhaltige-Kleidung
Schäfer S (2012b) Nachhaltige Kleidung: Unsere zweite Haut. Retrieved 10 December 2015, from http://www.zeit.de/zeit-wissen/2013/01/Nachhaltige-Kleidung
Scholz H (2015) Studie: Zahl der Smartphones verdoppelt sich auf 6,1 Milliar-den. Retrieved 10 December 2015, from http://www.mobile-zeitgeist.com/2015/08/10/studie-zahl-der-smartphones-verdoppelt-sich-auf-61-milliarden/
Sheth J, Sethia N, Srinivas S (2011) Mindful consumption: a customer-centric approach to sustainability. J of the Acad Market Sci 39(1):21–39. http://doi.org/10.1007/s11747-010-0216-3
Strähle J, Köksal D (2015) Impact of brand- and country image on the perception of sustainability in the fashion business. Int J Bus Commer 5(1):25–58
Strähle J, Will C, Freise M (2015) Communication of sustainability at European fashion online shops. Int J Econ Commer Manag 3(7). http://ijecm.co.uk/wp-content/uploads/2015/07/375.pdf
Taylor G (2015) More than 90 % of consumers use smartphones While shopping in stores. Retrieved 10 December 2015, from http://www.retailtouchpoints.com/topics/mobile/more-than-90-of-consumers-use-smartphones-while-shopping-in-stores
UN World Commission on Environment and Development (1987) Brundtland report. https://en.wikisource.org/wiki/Brundtland_Report
Young W, Hwang K, McDonald S, Oates CJ (2010) Sustainable consumption: green consumer behaviour when purchasing products. Sustain Dev 18(1):20–31. doi:10.1002/sd.394
Yurchisin J, Johnson KKP (2010) Fashion and the consumer. Berg, Oxford

Chapter 14
Case Study: Total Transparency at Honestby.com

Jochen Strähle and Laura Merz

Abstract Since there is no denying that transparency is increasingly central to corporate sustainability, the purpose of this paper is a case study on a company's attempt to be fully transparent, hence, picking up the existent scholarly conversation about uncompromising supply chain transparency. Literature so far was found to be fairly limited, but, following a trend, has been rising in numbers over recent years. Addressing these shortcomings in the methodology, an in-depth literature review about the multiple dimensions of supply chain transparency has been performed and links within supply networks stressed. On this basis, a case study by exemplary illustrating the fashion label Honest by has been drafted and the effort to become the world's first 100 % transparent company further examined. Findings are discussed whether more supply chain transparency is desirable in any case, obstacles listed and an outlook for this kind of business model has been drawn. The research is clearly limited by the amount of scholarly literature concerning Honest by in particular. Out of this reason, magazines and journal entries are used as reference as well. Only with the extension of the topic itself to supply chain transparency and the literature review beforehand, the paper gained its necessary academic standard. Concerning implications, it needs to be mentioned that even though Honest by demonstrates to be fully transparent, it was not possible to find any public information about the degree of supplier relationship. In particular, concerning the applied control mechanisms used to exert influence and to balance out the power gradient between company and suppliers.

Keywords Sustainability supply chain management · Transparency · Corporate responsibility · Supply network · Social desirability

---

J. Strähle (✉) · L. Merz
School of Textiles and Design, Reutlingen University, Reutlingen, Germany
e-mail: jochen.straehle@reutlingen-university.de

L. Merz
e-mail: lauramerz@web.de

J. Strähle (ed.), *Green Fashion Retail*, Springer Series in Fashion Business,
DOI 10.1007/978-981-10-2440-5_14

Abbreviations

CEO	Chief Executive Officer
CSR	Corporate Social Responsibility
FFDS	The Future Fashion Designer Scholarship
GOTS	Global Organic Textile Standard
IVN	Internationaler Verband der Naturtextilwirtschaft
JOCA	Japan Overseas Cooperative Association1010
LCCS	Low-cost Country Sourcing
MoMu	ModeMuseum
NGOs	Non Governmental Organisations
PLM	Product Lifecycle Management
PR	Public Relations
QM	Quality Management
SISP	Sebastian Indian Social Projects
SSC	Sustainable Supply Chain
SSCM	Sustainable Supply Chain Management
VAT	Value-added Tax

14.1 Introduction

"Try honesty. If you don't do it now, you'll have to adapt when complete transparency becomes mandatory. I think it's wiser to be a leader who's ahead of the rest rather than being behind because it will become mandatory one day. Buying a mystery will be an absurd concept soon. There's no luxury in a riddle", says Bruno Pieters, founder of Honest by BoF (2015). Transparency is not a term easily defined. When looking for a proper definition it seems to be relevant in which context transparency is used. Concerning economy, it is defined as the minimum degree of disclosure to which agreements, dealings, practices and transactions are open to all for verification (Business Dictionary 2015; Mol 2015). Obviously, already the definition leaves room for interpretation. Corporate transparency can take various forms such as sustainability reports (Hahn and Kühnen 2013), environmental product declarations (Schau and Fet 2008) and sustainability certifications (Bartley 2007). With purchased materials and components accounting for a growing share of companies' total expenditure and sustainability impact (Seuring and Gold 2013) transparency requirements have increasingly extended beyond internal corporate boundaries into supply chains (Mol 2015). This has given rise to the notion of supply chain transparency (Egels-Zandén et al. 2015). State-of-the-art of corporate sustainability within the fashion industry is code of conducts, for example, which are used by businesses as means for voluntary supply chain commitments. However, they can only be seen as a way to make these commitments more meaningful. In many cases, they are solely the reaction for deduction of so-called name-and-shame campaigns by activists or NGOs and are therefore highly suspected to exist out of green washing purposes only. Hence, corporations are acting upon a recent trend since there lies an increased societal emphasis on transparency in consumerism. It is a clear indication that not only the offer on the part of companies counts, but on a balanced scale the demand by customers. Bruno Pieters is deeply convinced that when it comes to driving forward ethics and sustainability in fashion the consumer has all the power (Borromeo 2013). In order to give the consumer the possibility to make a more informed decision, the first step of disclosing the exact origins of the products is decisive here. Taking the reproach of green washing into account, many cases exist in which companies defined supply chain transparency in an inconsistent manner. This way of wangling statements is not as conspicuous in the first place, because transparency can be applied in many dimensions depending on the field of application as well. As a result, companies are often being claimed to be either transparent or non-transparent. The framing is again critical because this radical approach ignores the fact that companies might make use of various means of transparency on different dimension levels. Several conceptual studies have analysed supply chain transparency, (e.g. Mol 2015) but few empirical studies have examined companies' attempts to be transparent in practice within the fashion industry in particular. A case study on one of the pioneers for this type of expertise can serve as a prime example for others. First, a conceptual theory building is used to introduce the concept of sustainability to the field of supply chain management. Digging deeper a literature review on multidimensional definition of transparency is performed to offer the needed theoretical background. Further it is demonstrated how the relationships among all actors of a value chain are in context with each other. Power, control and conflict (Gadde 2010) describe the different dimensions of a certain relationship atmosphere. This basis applied and later discussed on the case of Honest by gives a clear picture of the company`s business model. The paper is concluded by giving an outlook about the sociable desirability of today, but keeping the attitude–behaviour gap in mind. In doing this, the paper contributes to the sustainability and supply chain literature and is meant to provide a starting point for future empirical studies of supply chain transparency.

14.2 Sustainable Supply Chain Management

Until recently, SSCM was understood to be facing issues such as social and environmental factors in a subordinate order of preferences on a corporate agenda. On the contrary, it is perceived as the most pressing responsibility today. Sustainability with the integration of social, environmental, but also economical responsibilities centred in the triple bottom line has been set on the same level of business disciplines such as operations for instance (Fig. 14.1).

As the business term implies due to sheer definition it is the overall goal to generate profit. As a natural consequence, companies which engage in SSCM practices are concerned about attaining enough economic success in comparison to

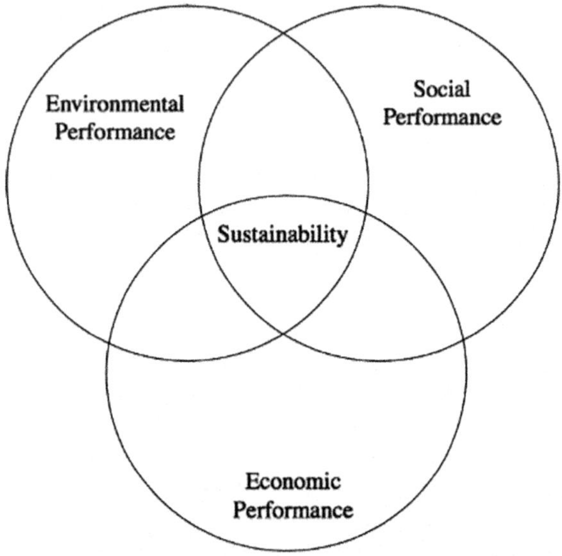

Fig. 14.1 Sustainability and the triple bottom line. Adapted from Carter and Rogers (2008)

businesses which concentrate solely on economic performance. This research question can be found in various literature and is, again naturally, curiously followed by the world of economy. The key to resolving this debate is the recognition that the way to run a business has changed dramatically in recent years. Profit making alone is proven to not be successful on long-term, but sustainable economising. Not only from an economical point of view, but also by governments companies are taken into responsibility for their impact of doing business (D'Amato et al. 2009). To start off at the very basics of a producing company, the concept of sustainability is crucial to be applied to the field of supply chain management in order to ensure sustainable production processes. Thereby, the ability to provide transparency is directly proportionate to a company's supply chain complexity (The Parker Avery Group 2015). As a rule of thumb, it can logically be said that smaller and younger companies rather know their suppliers in a much more detailed fashion. Bigger companies tend to develop such a complexity due to their sheer size that it becomes hard to keep an overview. Even more, transparency is of particular relevance here and it has been found to be an outstanding position to impact activities as well. Simple steps such as the reduction of packaging and the use of more fuel efficient transportation are just two examples among many. Not only costs can often be reduced, but reputation improved at the very same time. Corporate responsibility has become the latest and furthermore the most widespread fashion trend, at least in the Western part of the world. This can be observed since start-ups tout made in Europe in their crowdfunding campaigns and social media platforms are plastered with the world's social and environmental disasters caused by the fashion industry, commented with sharp disgust. Too, consumers more and more realise that they are strong influencers and that every single purchase is a choice with which they have an impact on a brand's equity. They increasingly push companies towards providing more focus and commitment to transparency (The Parker Avery Group 2015). Today's consumer seems to be savvier than ever and understand that corporate supply chains were originally designed to be anything but transparent (Ramsey 2014). Measuring and testing the social, environmental and commercial impacts of supply chain choices is now part of the public domain and ethical fashion is noticeably rewarded and celebrated (The Parker Avery Group 2015). "If Vivienne is so concerned about the ethics of fashion, why doesn't she have a transparent supply chain?" At a podium discussion this very question was addressed to Brigitte Stepputtis who works as head of couture at Vivienne Westwood. The reply was "Change is happening, but it's a process. Sourcing ethically will be difficult until governments legislate for that process. Also one of the most important issues for us is to be able to work sustainably but to scale. We have a global brand. It is easier for smaller lines to be more transparent because they don't have the scale that we have" (Borromeo 2013). The brand Vivienne Westwood represents itself as political, cultural and rebellious. It acknowledges the fact that climate change is the most urgent problem the human race has ever faced (Westwood 2015). And this is the reason why this statement can serve as an exaggerated example for general defensive arguments repeatedly found across the media. Responsibility is shifted to be a political affair and blame is found in various disputed issues. Even more transparency is particularly important to customers, as they lack the background details and knowledge available of the company's internal information. It can be said that transparency represents a way to transfer power from the company to its stakeholders (Egels-Zandén et al. 2015). Through this approach the information asymmetry between these two actors can be reduced and as a result it allows consumers to make a more informed evaluation of the company's offer. Due to the demand on the part of the customer transparency is en vogue and exploited as key to numerous problem-solving solutions. The general emphasis on transparency is mirrored across literature as earlier discussed (Carter and Rogers 2008). There it is connected to desirable characteristics such as accountability, legitimacy and trust (Martinez and Crowther 2008). Out of obvious reasons, it is a logic result that corporations strive to call transparency within their list of expertise. Corporate initiatives around supply chain transparency have therefore become just as essential as a brands' logos, mission statement and value proposition (Ramsey 2014). The overall perceptions that stake are high rises and that companies which do not resolve their supply chain issues are part of the problem. However, especially well-known brands are aware how crucial it became to take a closer look at the conditions within their own supply chain since any incident of indiscretion will immediately be echoed and amplified throughout consumer social media channels around the world. Businesses, too, seem to have understood that SSCM is decisive when keeping an eye to the future. Out of this reason, especially bigger companies with tarnished supply chain legacies are at the forefront of taking the lead (The Parker Avery Group 2015). An attempt in trying to keep up with the promises for the companies' strategic goals mostly includes commitment to innovation and the desire to embed sustainability best practices in the company's structure. By assessing the global impact and redefining corporate performance to meet new standards for excellence, fashion and retail brands can play a role in designing the future. This is the trend that will define the industry as a whole (Ramsey 2014).

14.3 Dimensions of Supply Chain Transparency

However, the transformation towards SSCM is long-winded and might therefore be misused for hastily marketing purposes. Companies are pressured because of the rising demand for this issue not only by their customers, but also by other stakeholders such as NGOs and politics (Carter and Rogers 2008). Headlines during recent years have proven that corporate disclosed information cannot be trusted blindly because companies tend to focus on the more successful examples while not being transparent on some of their CSR hot spots (Toppinen and Korhonen-Kurki 2013). Unfortunately, it seems like that commitment and progress of companies are as equally rewarded as actual impact (Ramsey 2014). It is cheered for every valiant effort at transparency which makes it easy for companies to perform green washing (The Parker Avery Group 2015). The claim for supply chain transparency lies therefore in addressing this deceptiveness by disclosure. The level of disclosure can therefore be used as a means for judgement of companies' supply chain activities. On the other hand, it is an option for companies to underline their legitimacy (Carter and Rogers 2008). As MacLean and Rebernak (2007) put that there is no better way to build trust among stakeholders than through transparency. Statistics, however, show that mistrust among consumers is widespread, but clearly state that consumers are more willing to purchase products from transparent companies (Bhaduri and Ha-Brookshire 2011). The question rises why so little companies have embraced transparency although these facts speak for themselves. However, as mentioned above, the judgement of whether a company is transparent or non-transparent is not easily made. Some equate the term transparency to traceability of a product's flow throughout the production process and its supply chain (Laudal 2010). Other authors define the term as allowing stakeholders to see further along an organisation's supply chain without being specific what this entails in practice (Carter and Rogers 2008). Cramer and Rogers (2008) on the other hand focus on the sustainability conditions at suppliers and the disclosure of this precise information. Hereby, he stresses not only the listing of the suppliers' name, but rather the status of social and environmental conditions in the factories. Finally, some of the authors lay emphasis on the financial transaction between buyers and suppliers. This degree of transparency is highly sensible data and theoretically perceived as the epitome of business secret. In previous research three dimensions of supply chain transparency have been recognised and contain various approaches from above. It is proposed that supply chain transparency comprises disclosure of (i) the names of the suppliers involved in producing and manufacturing the products of a company, (ii) information about the sustainability conditions at these suppliers and even more importantly their factories and (iii) the buying and purchasing practices involved for these transactions (Egels-Zandén et al. 2015). With the help of this segmentation, judgement about whether a company is transparent or not can now be performed on different dimensions. A company which claims to be fully transparent can not only publish a list of its supplier network, but needs to allow stakeholders to trace the product and the condition of its making all the way from the raw materials to final product and a cost breakdown specifying every single cost of the product components needs to be attached, too. The first dimension puts not only the companies name and reputation on the public plate, but shares it with the supplier as well. In terms of disclosure of sustainability conditions GOTS, for example, is a standard which stipulates requirements throughout the supply chain for both ecology and labour conditions in textile and apparel manufacturing. Through making use of this certificate sustainable production conditions can legitimately be provided. When it comes down to the disclosure of purchasing practises, Honest by provides one of the most interesting examples which is going to be examined in detail later on in the paper.

A fully transparent company would need to combine all three dimensions. But in many cases different transparency outcomes for different suppliers and across different supply chain transparency dimensions are displayed simultaneously (Egels-Zandén et al. 2015).

Finally, it needs to be distinguished between internal and external supply chain transparency. It makes a difference whether a company can be transparent to itself and the degree to which a company communicates this information to external stakeholders. From an economic standpoint it is quite arguably that companies will not disclose all of their internally available information out of various reasons. Reasons amongst others are the protection of sensible data and to minimise the risk of name-and-shame campaigns (Bartley 2007) which can have financially detrimental consequences on both their own and their suppliers' business. It can, however, be critically argued that there exists the risk that particularly problematic audit findings for example are excluded in the disclosure. It is precisely the willingness to assume this vulnerability that makes stakeholders attach importance to transparency in the first place (Egels-Zandén et al. 2015).

14.4 Supply Network

The clothing business is well suited for a SSCM study as it is one of the most global industries in the world. This includes closely coordinated production and distribution lines spread out in regions with great variations in government regulation, employment and environmental protection and wage levels. Thus, clothing companies must handle a multitude of not only legal, but also moral standards (Woo 2013). It can be seen as a sector with specific features indicating that the risk of violating CSR standards is high (Laudal 2010). Simultaneously, it inclines that there is a high potential for positive influence through action within this field. Various possible scenarios for outcomes of corporate supply chain transparency attempts are imaginable. In numerous literature, the supply network has been further explored to be influential in shaping the mentioned company's supply chain transparency outcomes. Thereby, relationships with suppliers and the organisation of supplier communication have been proven important both in general (e.g. Gadde 2010) and in relation to supply chain sustainability (Locke et al. 2009).

As already mentioned in the introduction of this paper, global businesses of today tend to outsource their production instead of manufacturing themselves like they used to. Clothing brands are not manufacturer anymore, but retailer first and foremost. Globalisation has taken its course and global fragmented division of labour has become the norm in the clothing business rather than the exception. As a follow-up result, transparency has changed from a company's internal affair to an extension of company boundaries across its whole supply network. The implication of it is that interorganisational issues need to be taken into consideration when analysing supply chain transparency (Egels-Zandén et al. 2015). When deciding to outsource production, on the one hand, there are beneficial features to be listed such as decreased production costs for instance. On the other hand, risk and limitations need to be considered. One of them is for sure the interdependency among the different actors of a supply network. Outsourcing companies are often afraid of losing control over the outsourced activities. This can be received as an obstacle because outsourcing implies to give up control and trust others to handle business functions which can have an impact on the own business success. It is therefore a risky move which can transform though into a benefit in an optimal case. However, it always needs to be double-checked if the supplier is still acting within conformities. Pushing suppliers into too strict regulations can result in a bad relationship atmosphere which is not beneficial for any party. It comprises three interrelated dimensions which are referred to in literature: control, power and conflict. This is central for understanding how companies can induce others to act in their interests (Gadde 2010). Control concerns influencing others and already in the 1990s (Weitz and Jap 1995) distinguished three forms of control mechanisms used to exert influence which are again still referred to today, namely, authoritative, normative and contractual control mechanisms. Authoritative control is gained through ownership, position or power. Normative control is achieved via shared norms, values and trust among the parties involved in a relationship. Finally, contractual control is regulated by contracts that determine various levels of compensation (Egels-Zandén et al. 2015). Power is usually defined as the ability to influence others to do what they would not otherwise have done (Gadde 2010). Furthermore, power is deeply rooted in interdependence which naturally leads to the need for caution about the balance of dependencies and should be watched over. It is important to note, however, that the possession of power should be distinguished from its exploitation. There are two different ways of how to exert power: either on a threatening basis or an influencing one. Referring to the economy of global nations, this can exemplary be pinned down to the exploitation of the East by the West. Equally, it has been demonstrated that it does not have to be that way. Many Western companies which hold power have made an example by not using their power to threaten suppliers in the East, but to use it to influence other parties towards positive development on both sides. Especially when looking at relationships within a supply network it is crucial that the dependencies are equaled out so that the business relationship can develop sustainably. So advantages due to long-term relationship can be gained.

To sum it all up, as explored just now that supply chain transparency extends across company boundaries along the supply chain, it is a fact that the relationship atmosphere framework of power, control and conflict is useful for analysing the transparency outcomes of corporate transparency attempts (Egels-Zandén et al. 2015).

14.5 In the Case of Honest by

14.5.1 Background

Bruno Pieters was born in the Belgian town of Bruges in 1975. Already at high school he would study art and later train at the Royal Academy of Fine Arts in Antwerp. He always dreamt of working for a grand fashion house. "As a student, you look at it as a dream—you don't see it as a business, which it totally is", Pieters says today (Blanchard 2013). He had all the makings of becoming a star designer, winning the Andam Award from the Pierre Bergé and Yves Saint Laurent Foundation in 1997. In 1999, he graduated with a B.A. in Fashion Design and went on to develop his craft by working with designers including Martin Margiela, Christian Lacroix and Antonio Pernas (Ideamensch 2012). He presented his first couture collection during Paris Fashion Week in 2001. The following year, Pieters debuted his ready-to-wear collection and launched his first own label. From 2005 on, he became creative director at the Belgian luxury leather company Delvaux and 2 years later he was named art director for Hugo by Hugo Boss. Meanwhile, he continued to expand his own label which he financed independently. "At Boss I was very focused on the whole thing. I had worked towards it for 10 years thinking if I reached that I would be happy. My own label never went better; everything was going well" (Blanchard 2013). But he felt like there was something important missing. In 2009, he left Hugo Boss and took a sabbatical travelling around India for a total of 2 years. He experienced an epiphany when he realised that his priorities had changed completely. For Pieters it is not only on a professional level, on the contrary it all started off with a personal evolution which makes his business concept so authentic. He says that he became more aware of the way the world functions and more particularly the fashion industry itself. And he increasingly deemed that production in particular is simply not fair nor honest. Upon his return to Antwerp in 2011, he decided to wind down his own label, gave a significant part of his archive to the MoMu Fashion Museum of Antwerp and sold off the rest with proceeds to be given to SISP, a children's charity in Southern India. In January 2012, Pieters finally returned to the fashion stage fully with a new concept and brand which was his first course of action being back. Personifying an entrepreneur as well as a designer, Pieters established Honest by. The business idea resulted from a personal urge to be able to be ascertain that goods he purchased had caused the least possible damage to the environment and had not exploited anyone along its making.

The inspiration for founding Honest by was first and foremost a personal infliction. Bruno Pieters was looking for a fashion brand in which he could find information for clarity on where the product was really sourced from so that he would have an easy mind when purchasing fashion. Since he felt that brands are still not transparent enough in the way they communicate about their products, he simply created a fashion brand himself. The inspiring reason was out of pure self-interest because after his sabbatical the actual plan was not to get involved with fashion design again at all since he grew to find the concept of business revolting. He wants people to see Honest by solely as a means for change. Pieters said in an interview: "For me, buying is like voting; I like to know what I am supporting because I truly believe all the problems we face today are related to consumption. I see it as my duty to be the change I want to see in the world, as Gandhi once said. For that, I need to be able to shop in a 100 % transparent way" (Ideamensch 2012).

14.5.2 Business Model

The company philosophy includes respect towards the environment that the world's climate challenges are taken seriously, the belief in the health of the clients' skin, the compassion for animal welfare and the claim that the impact of the products and activities to produce them are as small as possible on the environment and human healthy (Home—Honest by 2015). Using the headline Business Model would most possibly be disliked by the founder Bruno Pieters. In an interview, he said that he is not attracted to the concept of business in general anymore and that he finds it to be a bizarre concept which has been created. His wish is for people to use and see Honest by as a vehicle for change (Toth 2014). Offering this tool for change is realistically, however, only ranked at the bottom of the power pyramid in the fashion industry. Still Pieters is convinced that change does not start with institutions. Not the CEO or the shareholders can force this upon society, but the customers need to play their part too and demand the change. The urge for total clarity on a personal level resulted in the idea of launching this pioneering concept of Honest by. It targets a customer group which is looking for the highest possible transparency about information of products they would like to purchase and a company which is in sync with their values. Honest by gives their customers the opportunity to shop with complete awareness of what they are buying. In full self-promotion, Honest by illustrates the brand as a truly new paradigm in the fashion industry (Home—Honest by 2015). Naturally, Honest by is not the only example of a profitable business model making use of an ethical and responsible approach to the supply chain and how it can even provide a true competitive advantage. Benefits hereby include greater efficiency, responsible entrepreneurship, attraction and retention of talent and increased customer loyalty. The level of commitment to transparency raises the bar for the ethical-elite. It can, however, be criticised that small-scale micromanagement and monitoring with a business model designed for this exact impact may seem out of reach for larger brands and manufacturers (The Parker Avery Group 2015). In its niche market Honest by seems to work out well though. According to the media there were enough buyers to ensure that the collections have been selling out quickly so far (Blanchard 2013).

Apart from the economical point of view Honest by offers like-minded fashion designers a public platform to share their own design processes, utilising the extensive research Honest by has already conducted within this field. Just out of economic reasons, fashion brands fear to choose the green line and therefore it is even more pioneering that Honest by shares experience and sensitive information to enable creatives with common convictions.

14.5.2.1 Design

Vogue.com's editor, Dolly Jones said once: "It's all about design. If you show us pretty clothes, we'll write about them" (Borromeo 2013). This quote is still an applicable opinion which could originate from any major player in the fashion industry. Statistics prove that also customers value the look of a garment still more fiercely than characteristics such as where the garment was produced (Gam 2011). Honest by demonstrates by example that sustainability and high fashion perfectly go along. Pieters pieces are best described as architectural, therefore quite asymmetrical with constructed lines, mostly blacks, greys and whites which is the attempt to be innovative on the one hand, but staying classy on the other. The label is celebrated for its avant-garde creations and intricately sharp tailoring. Style.com described the beauty of the clean, precise lines of Pieters' tailoring to be a result of his real devotion to craft (Ideamensch 2012). Several awards won by Pieters deliver proof. Originally, Pieters inspiration of design was the clothing worn by native people. He said "Observing how native people wore clothes that were grown, woven and sewn from sources they could identify around them, he wondered if such transparency could be operated on an international scale when making designer products" (Alter 2012). The offered fashion is highly fashionable and one can therefore assume that the target group is very much fashion conscious, too.

Through the online shop Pieter's own designs can be purchased, but it also represents a platform for collaborations with other designers which rotate about every 3 months. It offers room for established as well as newly upcoming designers with the only request to be like-minded.

14.5.2.2 Raw Material

At Honest by extensive research is conducted into the sources of raw materials to make sure that the supply chain of every element in each garment is documented (Artuso 2014). Textile suppliers are persuaded to reveal details about origins and sources in order to create a painstaking pedigree trace the fabric through the supply chain of raw materials, yarn spinners, weavers, printers and dyers (BoF 2015). In communicating all information concerning a garment's production process Honest by wants to raise consciousness about where products are made and by whom (Artuso 2014). At the same time, the aim is to demonstrate that the exact same fabrics can be found in fashion sold by Honest by just like at any other luxurious fashion brand. "There is an assumption that organic fabrics are worse quality. But it's the opposite. The quality is actually better; the organic cotton is better" Pieters tries to convince (Blanchard 2013).

14.5.2.3 Production

Due to transparency along the supply chain close control can be kept and every step be held accountable. When looking at the dimensions further down the chain it is conspicuous that sub-suppliers in particular the ones for complementary elements such as zippers are often non-European. These ones are used out of sheer necessity rather than choice, Pieters says "We did research on zippers and they aren't made in Europe any more. All zippers and basically all trimmings come from China or Asia today. Even if the company is based in Europe they have them made there. Even the zippers on 100,000 € couture dresses are from China, without exception. But these ones at least are recycled. It is interesting to see how unavoidable it is to have China involved" (Blanchard 2013). His critic at this point targets corporations which make use of LCCS, but still charge high retail prices in order to make an unjustified huge profit margin, which leads to another burning concern on how to change an industry with industrial and worker exploitation embedded in the heart of its profit-making model (Borromeo 2013).

14.5.2.4 Cost and Profit

Whereas the trend towards supply chain transparency can be found at more and more like-minded companies, Honest by is an absolute pioneer in publishing an itemised breakdown of every cost component, including the wholesale and retail mark-up percentages. The online shop of Honest by offers a front-end interface for offering all this information directly to the consumer. This business model is perceived as radical since in terms of purchasing practices, volumes and other business variables it is generally claimed that these information are too sensitive to publish information. It is often seen as the epitome of business secrets and shall therefore not be disclosed. Corporations claim that transparency hurts their mark-up and as a natural result their profit. Actually, it is not desired to share profit margins or the actual prices paid to suppliers. Also in the case of Honest by there is no transparency on all dimensions, but through providing a picture on its purchasing practices an idea about the company's sustainability condition can therefore be indicated. But in the attempt to truly become 100 % transparent the transparency is not limited to the products themselves, but the company also publishes how much money everyone along the value chain earns, including Pieters himself.

14.5.2.5 Transparency

The word Honest in the brand name Honest by refers to the way in which the company claims to operate, with a 100 % transparency policy. When Pieters founded Honest by he set about sourcing fabric suppliers and manufacturers who were willing to share information that is never usually even requested, let alone published on a designer's website (Blanchard 2013). Many companies would consider publishing their mark-up, names and addresses of all their factories and suppliers as too high of a commercial risk. However, he soon convinced them to see the publication as an advantage of being promoted for free. Honest by is, thus, able to publish the supplier names on their website for the vast majority. A selected few suppliers are not equally forthcoming. There also exists a difference between disclosing the names of sub-suppliers to the company for internal use and making this information publicly available. Suppliers perceive this as a risk that Honest by or other clients could bypass them as the middle-man and source directly from their sub-suppliers (Egels-Zandén et al. 2015). In case of sourcing finished goods from this so-called middle-man, the access to transparency of sub-suppliers is, however, crucial and the basis of the whole concept. Otherwise quality control cannot be ensured, let alone the conduction of factory audits. Too, considerations about supplier relationship are necessary in general since factory owners could be upset about publishing this highly confidential information if not previously agreed on. Therefore, contracts and transparency agreements which regulate the disclosure need to be drawn. To put this in order, this falls under the topic of contractual power explained within the theoretical part of the paper.

In most cases, at Honest by extensive information for each item is published: material information starting with the product code and its description, further including all details of the fabrics used: the fabric itself, the composition, its weight, the yarn or piece dyed, the origin of raw material, details about the spinner, weaver, dyer, finisher, manufacturer or supplier and the belonging certificates. This is repeatedly listed for the linings, interfacing, bias tape, sewing thread, brand label, size label, care label, made-in label, security seal, hang tag and its thread and safety pin (Fig. 14.2).

Further, the list offers manufacturing details starting with the pattern, the manufacturing company, its location and address, the design time and number of fittings. The same applies for the garment manufacturing which is complemented by cutting, assembly and ironing time. Closing off and thereby providing personal

Material information

Product code: BPW0010F2352W
Product description: QUILTED ORGANIC COTTON AND HARRIS TWEED DRESS

FABRICS
Fabric upper part : superfine organic cotton satin
Composition: 100% organic cotton
Certificate: GOTS certified
Weight: 120 gr/ sqm
Yarn or piece dyed: piece dyed
Origin of raw material: Turkey
Spinner, Weaver, Dyer, Finisher; Ecological Textiles, Turkey
Supplier: Ecological Textiles, Roermond, The Netherlands
www.ecologicaltextiles.nl

Fabric lower part: featherweight handwoven Harris Tweed
Composition: 100% virgin wool
Weight: 470 g/m
Yarn died: loose wool dyed (different types of wool are blended together and dyed before twisted into yarn)
Origin of raw material: Scottisch border, UK
Spinner: Harris tweed Hebredes, Shawbost, Isle of Lewis, UK
Weaver: Harris tweed Hebredes, Shawbost, Isle of Lewis, UK
Dyer/finisher: Harris tweed Hebredes, Shawbost, Isle of Lewis, UK
www.harristweedhebrides.com

Harris tweed only uses pure new wool from Scotland (Border Leicester and Cheviot breed). The wool is bought by their own broker at the British Wool Marketing Board who keeps the wool until the company needs it for weaving.

The law decrees that genuine Harris Tweed has been dyed and spun on the islands and hand-woven at the home of the weaver in the Outer Hebrides of Scotland.

LINING
Lining: black silk chiffon

BIAS TAPE
Composition: 100% organic cotton
Certificate for raw material and fabric: GOTS
Origin of raw material: Turkey
Manufacturer: Ecological Textiles has produced the bias tape on our request. The cutting of the fabric into bias tape took place in Belgium.
Supplier: Ecological Textiles, Roermond, the Netherlands
www.ecologicaltextiles.nl

ZIPPER**
Composition zipper tape: 100% recycled polyester
Composition metal zip and zip puller: 100% nickel-free brass.
Brass is an alloy of copper and zinc. These could not be made of recycled material because of the needed strength of material.
Origin of raw material: Unify Textiles Co, Suzhou,China
Manufacturer: IDEAL Fastener Asia Ltd, Guangdong, China
Supplier: IDEAL Fastener Asia Ltd, Guangdong, China
www.idealfastener.com

SEWING THREAD
Composition:100% organic cotton
Certificate: GOTS
Origin of raw material: Egypt
Manufacturer: Forbitex, Soest, the Netherlands
Supplier: Forbitex, Soest, the Netherlands
www.forbitex.nl

BRAND LABEL**
Composition: 100% polyester
Certificate: Oekotex
Origin of raw material: Italy
Manufacturer: EE Labels, Heeze, the Netherlands
Supplier: EE Labels, Heeze, the Netherlands
www.eelabels.com

SIZE LABEL**
Composition: 100% polyester
Origin of raw material: China
Production plant location: China

CARE LABEL**
Composition: 100% polyester
Manufacturer: Van Campenhoudt, Schaarbeek, Belgium
Supplier: Van Campenhoudt, Schaarbeek, Belgium

SECURITY SEAL
Composition: 100% polystyrene
Certificate: ISO 18001, certified manufacturer
Origin of raw material: Italy
Manufacturer: UNISTO, Italy
Supplier: UNISTO, Italy
www.unisto.it

HANG TAG
Composition: 100% recycled paper
Origin of raw material: Europe mainly Germany
Manufacturer: Schollershammer, Düren, Germany
www.schoellershammer.de
Supplier: IGEPA Belux, Aalter, Belgium
www.igepa.be
Printer: EcoDrukkerij, Beemem, Belgium
www.ecodrukkerij.be

COTTON THREAD FOR HANG TAG
Composition: 100% cotton
Manufacturer: Located in Belgium.
Supplier: EE Labels, Heeze, the Netherlands
www.eelabels.com

SAFETY PIN
Composition: 100% nickel free metal
Manufacturer: Rayher Hobby, Germany. Product was made in the Czech Republic
www.rayher-hobby.de
Supplier: AVA papierwaren, Antwerp, Belgium
www.ava.be

**For optimal decomposition: please remove all polyester garment labels and components before disposal. These are not biodegradable.

Fig. 14.2 Material information. Adapted from Honest by—Material information (2015)

legitimacy, it is signed with the phrase: "Research was completed by the Honest by team" (Fig. 14.3).

The third column offers a detailed insight about the price calculation of the product. Again, each component of the garment is listed divided by price per metre or cone or piece, the amount used and finally its total cost. To sum up, the total cost of materials is calculated as well as the manufacturing cost. The collection cost covers the development of the new collection, pattern making costs, grading, digitising, purchasing of fabrics, trimmings and branding costs. All costs are added up and multiplied by 2.8 times in order to calculate the wholesale and retail mark-up. The mark-up covers a part of the company's costs such as staff, research, utility costs, insurance, communication, rent, intellectual property rights, professional association costs, office supplies, maintenance costs, purchasing of new collections, transportation, legal and accounting costs and marketing (Home—Honest by 2015). The taxation is made transparent, too. Therefore, there is distinction between the retail price applicable to non-EU customers when import duty taxes apply on top and the retail price including the VAT which is applicable to EU customers only (Fig. 14.4).

In addition to all this transparent information, there is a Help Us section where Honest by forestalls any possible critic. They demonstrate openness by inviting their customers and sceptics to get in immediate contact with their team in case of

Manufacturing details

Product code: BPW0010F2352
Product description: QUILTED ORGANIC COTTON AND HARRIS TWEED DRESS

PATTERN
Company: Honest by BVBA
Design time: 10 working hours
Number of fittings: 3

GARMENT MANUFACTURING
Company: Trois Quarts BVBA
Location: Antwerp, Belgium
Address: Lange Leemstraat 368, Antwerp, Belgium
Cutting time: 85 min
Quilting time: 65 min
Assembly time: 145 min
Ironing: 17 min

THIS GARMENT IS 100% MADE IN BELGIUM.

RESEARCH WAS COMPLETED BY THE HONEST BY TEAM.

Fig. 14.3 Manufacturing details. Adapted from Honest by—Manufacturing details (2015)

Price calculation

Product code: BPW0010F2352W
Product description: QUILTED ORGANIC COTTON AND HARRIS TWEED DRESS

FABRICS
Fabric upper part: superfine organic cotton satin
Composition: 98% organic cotton GOTS certified, 2% EA
Price / Meter: 7,04 Euro
Amount used: 2,10 Meters
Total: 14,78 Euro

Fabric lower part: feather weight wool tweed
Composition: 100% virgin wool
Price / Meter: 21,74 Euro
Amount used: 0,65 Meter
Total: 14,13 Euro

LINING
Lining: silk chiffon, GOTS certified
Composition 100% silk
Price /Meter: 25,10 Euro
Amount used: 0,52 Meter
Total: 13,05 Euro

INTERFACING
Composition:100% Tencel , Oeko-tex
Amount used: 0,60 Meter

BIAS TAPE
Composition: black 100% organic cotton GOTS certified
Price / Meter: 0,55 Euro
Amount used: 1,88 Meter
Total: 1,03 Euro

ZIPPER
Composition: recycled polyester + nickel-free brass teeth
Size: 38 cm
Price / Piece: 1,50 Euro
Amount used: 1 piece
Total: 1,50 Euro

BRAND LABEL
Brand label big: 100% polyester
Price / Piece: 0,6684 Euro
Amount used: 1 piece
Total: 0,6684 Euro

SIZE LABEL
Composition: 100% polyester
Price / Piece: 0,0350 Euro
Amount used: 1 piece
Total: 0,0350 Euro

MADE IN LABEL
Composition: 100% cotton
Price: 0,37 Euro
Amount used: 1 piece
Total: 0,37 Euro

CARE LABEL
Composition: 100% polyester
Price: 1,65 Euro
Amount used: 1 piece
Total:1,65 Euro

SECURITY SEAL
Composition: 100% polystyrene
Price / Piece: 0,20 Euro
Amount used: 1 piece
Total: 0,20 Euro

HANG TAG
Composition: 100% FSC recycled
Price / Piece: 0,29 Euro
Amount used: 1 piece
Total: 0,29 Euro

COTTON THREAD FOR HANG TAG
Composition: 100% cotton
Price / Meter: 0,016 Euro
Amount used: 0,30 Meter
Total: 0,0048 Euro

SAFETY PIN
Composition: nickel free metal
Price / Piece: 0,03 Euro
Amount used: 1 piece
Total: 0,03 Euro

TOTAL COST MATERIALS: 50,35 Euro

MANUFACTURING COST
Garment manufacturing: 105 Euro

COLLECTION COST: 10,30 Euro
The collection cost covers the development of the new collection, pattern making costs, grading, digitizing, purchasing of fabrics and trimmings and branding costs.

TOTAL COST: 165,65 EURO

WHOLESALE & RETAIL MARK UP:165,65 X 2,8 = 463,82 EURO
Our mark up covers a part of the Honest by Bruno Pieters costs such as staff, research, utility costs, insurance, communications, rent, intellectual property rights, professional association costs, office supplies, maintenance costs, purchasing of new collections, transportation, legal and accounting costs and marketing. This collection is exclusive to Honest By and is not available at any other retailer.

HONEST BY RETAIL PRICE (excl.VAT): 463,82 EURO
This price is applicable to non-EU customers. Import duty taxes apply.

HONEST BY RETAIL PRICE (incl. VAT): 561,22 EURO
This price is applicable to EU customers only. VAT is included in this price.

Note: All prices are calculated per 1 item.

Fig. 14.4 Price calculation. Adapted from Honest by—Price calculation (2015)

uncertainties, the urge for even more additional information and advice for improvement. They promise to do their very best to make sure all information is correctly edited and calculated for each product offered. However, they ask to consider the vast amount of text online and that it is always possible to find information that is incorrect or incomplete (Home—Honest by 2015). Through this very statement they manage to steal anyone's thunder from the very start.

14.5.2.6 Certification

To confer the concept on a more official note, Honest by is certified through five different standards. The GOTS certificate is recognised as the world's leading processing standard for textiles made from organic fibres, including ecological and social criteria. The next one, IVN, complies with GOTS as its basic standard, but contains some additional requirements regarding specific production processes (Naturtextil 2016). The managing body and a founder member of the GOTS is the JOCA which has the overall objective to deepen the understanding on organic cotton and to promote its consumption (JOCA 2016). Also the SACL is a founder member of the GOTS and no system of farming has higher levels of animal welfare standards than organic farms working to the Soil Association standards (Soil Association 2016). The last certificate labelled the Oeko-Tex Standard 100 tests for harmful substances of all types to pose no risk to health (Oeko-Tex 2016). Through the means of certification Honest by offers a credible concept for their consumers. To simplify the option to choose from the range of goods on offer the customer can make use of a filter on the Honest by online shop. Through this tool any customer can easily make sure to only choose from products which fulfil the own enquired principles in case the sheer amount of different certificates and their contents might confuse. The filter is therefore much more user-friendly and ranges from organic, vegan, skin friendly, recycled, European to animal welfare. So the Honest by organic filter means that the main fabric or the raw material content of the main fabric of an Honest by garment is certified organic. The vegan filter ensures production through the practice of abstaining from the use of animal components whatsoever. The garment does not endanger the health of the skin when the skin friendly filter is applied. Recycled fabrics and yarns are made of post consumer waste. And the European filter indicates that the garment was manufactured in Europe to 100 %. Produced in Europe includes a pride which is related to a high sustainability standard and makes it easier to extend the insight into suppliers' operations, which is crucial for the business concept which focuses on the insight and quality. To be able to keep a clear overview a tightly structured amount of suppliers is crucial. The detailed traceability is a competitive advantage towards comparable businesses. Regarding animal welfare Honest by strictly works with

Fig. 14.5 Filter by. Adapted from Honest by—Filter by (2015)

FILTER BY

ALL
ORGANIC
VEGAN
SKIN FRIENDLY
RECYCLED
EUROPEAN

farmers only who must comply to British and European legislation which has formulated laws regarding good husbandry, transport and animal welfare amongst other things (Home—Honest 2015) (Fig. 14.5).

14.6 Discussion

As, in detailed fashion, discussed in the section SSCM the importance of transparency within a value chain cannot be denied anymore. Unfortunately, it can clearly be stated that the term is often inconsistently defined and applied in various dimensions, challenging the core logic. Concerning the issue of green washing it can be illustrated how companies leverage this discrepancy to their advantage out of sheer PR purposes. This concerns in particular the accusation that companies disclose successful examples only while not being transparent about some of their problematical CSR hot spots. Therefore, caution is necessary that the outcomes of transparency attempts do not too heavily depend on the interests of involved parties. Even though businesses might seriously be interested in striving to become more transparent, there are many levels of transparency to be considered as literature confirms. To become equally subjected to transparency not only the list of names of suppliers needs to be published, but also traceability is to be ensured, as well as a complete cost breakdown specifying the single costs of the product listed. In the case of Honest by, all of the claimed transparency dimensions are more or less fulfilled. Although the resulting reports of QM audits are missing on the website, the brand has an overall high level of traceability. This is illustrated by the fact that the suppliers from fabric onward along the supply chain are listed for each individual product. This is a result from Honest by's type of supply network and the type of relationship atmosphere. Through the use of normative control, rather than authoritative or contractual control (Weitz and Jap 1995), Honest by has managed to build an open ambience with just a limited number of suppliers and has thus incrementally improved traceability. But the upstream implications for companies' supply networks need to be considered, too. Due to this reason, transparency outcomes must again be analysed to the various dimensions. As mentioned in several interviews by Bruno Pieters, it seems like that transparency is a central object of power struggles and that there is still a lot of room for developing with the supply network. Collaboration and trust have in turn proven central to improving sustainability conditions (Egels-Zandén et al. 2015). At the same time, downstream implications of empowering the consumers through transparency weigh heavy, thereby transferring power to consumers and therefore allowing stakeholders to make a more informed purchasing decision. It is worth noting that Honest by is even going further by making use of several certificates such as the GOTS, underlining their credibility, thereby proving accountable for social, as well as economical standards.

However, all the efforts to provide full transparency has its cost. As publically illustrated in their cost breakdown, it is transparent for everyone that mark-ups need horrendously be applied in order to be profitable. The result is a shockingly high selling point of the products which can be seen as a huge point of critic. Honest by limits itself down to a very specific target group, even more so due to the high degree of fashionability. Still the business concept can definitely be seen as revolutionary. Even though Honest by is still a pioneer and fairly an outsider within this field, the trend towards more transparency is an undeniable fact. On the whole, it can be said that they strive towards becoming the first 100 % transparent company that has been successful.

14.6.1 Obstacles

Keynote speaker and chair of the All-Party Parliamentary Group on Ethics and Sustainability in Fashion, Baroness Lola Young, said that perfection can be a hindrance to progress. Hereby, she referred to a number of strictly puritanical activists who view corporate moves towards ethics with deep suspicion (Borromeo 2013). Green washing has been exploited by many big players in the fashion industry and therefore sustainability activities often got sharply eyed up by critics, claimed to exist only for sheer marketing purposes. On the other hand, ASOS's Ethical Trade Manager Alice Strevens echoes the industry line when she stressed the importance for the sustainable agenda to sit with the commercial one (Borromeo 2013). In the end, the fashion industry is a branch of economy with the right to exist when fulfilling the purpose through generating profit. Responsibilities are shifted from the politics having to ensure the right economic framework to the companies being responsible for their impacts when doing business. But ultimately the consumer is the one providing turnover for the businesses and their purchases count as votes. Pieters insists that the consumer can use their wallet to get through to fashion brands and that it is pointless to blame brands for the way the world is today (Borromeo 2013). The New York Times calls it "wearing our conscientiousness on our sleeves" which is one way of putting it (Alter 2012). Knowledge is power and through the means of Honest by's transparency concept the power is in the customers hand, but the obstacle is to become aware of this power. Assuming that the consumer is aware of the power, another obstacle is blocking many from making use of it. Namely, the price range of the products offered by Honest by and the degree of fashionability narrows the target group down to a quite specific amount of possible customers. For this reason, but also because the business is still at its beginnings, it can be predicted that it will most probably remain small scale. Another hints are the tearing obstacles when distributing the products in order to stay economical, but on the other hand reconcile the ethical principles. To name just one example, Pieters has been getting invited to show his collection at the London Fashion Week as part of the ethical fashion show Estethica, most probably on account of his reputation as a fashion designer. Once, he picked up several orders from department stores, although he already suspected that the buyers were more attracted to the aesthetic of the designs rather than its ethics. The result was that he cancelled the orders later on when he realised that the stores would not allow him to make their mark-up public. This is a conflict against an industry which dislikes to have their profit margin out on the public plate, as well as staying true to the brand's concept.

14.6.2 Critics

Already at the very beginning of his design career Pieters soon became a critic's favourite. Suzy Menkes, editor for the International Herald Tribune, described the designer as an intriguing new talent (Ideamensch 2012). The certainty of positive critics for his designs was not equally bestowed on the critics for his brand Honest by. The opinions varied from eagerly praising this new concept as the avant-garde line of organic clothing which is pushing the edges of sustainability (Alter 2012) to suggestions that it is only serving a niche market. Pieters confident reaction: "It's everybody's audience. It's such a logical thing to happen. In the end this is normal. The opposite is not normal. Unfair trade is not normal. Unsustainability is not normal. You need transparency and that's why every brand will have to do it eventually. It's going to be unavoidable" (Blanchard 2013). For now, Honest by is doubtfully eyed as an interesting experiment, a project with a high degree of out-there thinking, but generally perceived as not yet to be taken too seriously.

14.6.3 Enabling Solutions

In literature, the term of Product Lifecycle Management (PLM) is often presented as a solution for global supply chain collaboration. PLM is one type of system which provides a common platform for all parties along the supply chain including data integrity. With the help of this tool a detailed product portfolio is accessible, and the material and supplier library is visible from end-to-end and therefore drives transparency. A solid PLM system can become the backbone of a company's supply chain transparency initiatives (The Parker Avery Group 2015). Aside from PLM, other solutions include tools for compliance, quality and materials management. These tools provide further in-depth accountability through centrally collecting all data concerning material vendors, testing and audit reports (The Parker Avery Group 2015). Too, online information sources are available through which companies have the ability to investigate customs data, directories, trade records and global trade trends (Ramsey 2014). These options open up the possibility for providing transparency into overseas suppliers in particular and maintain the momentum towards even more transparency in the near future.

Mentioned as an obstacle, over and over again, Honest by struggled finding suppliers who would agree to have their names published. An objective for this issue can be copied from the automotive industry. They usually work with suppliers on a basis of long-term relationship. Trust can therefore be furthered and through collaboration efficiency be improved. Development on both sides by learning with and from each other is then a feasible goal.

14.7 Outlook

From a technical point of view, there is still a lot of room for new innovations. To pick up the suggested solution through PLM new extensions are on their way towards more configuration and customisation. More integrated processes and requirements can be configured to provide the unique visibility required and customisation already include mobile apps to support on-site photo streaming and real-time global communication for example. These innovative features are expected as the next wave. But innovative technology usually starts off as niche products (The Parker Avery Group 2015). Too, Bruno Pieters new business model may seem to be an outsider in the fashion world which is limited to a niche target customer group. However, there is no denying that Honest by is getting him noticed, still raising the decisive question whether the idea is transferable to the broad mass. Only if the approach is followed by the big players it will make an impact and change the industry. Out of this reason, Pieters started to consult for other companies, of course only under strict preconditions fulfilling his philosophy of doing business. It gives him the possibility to distribute transparent collections with a different, more democratic price point. Pieters is convinced that it should not be expected that the big heritage brands change as long as it will not affect their profit numbers. Naturally, from a commercial point of view companies will only alter something in the system if it affects their income. He says that experience shows how it is always offered whatever the clients want and that "the longer we sleep, the less freedom we will all have. Now is a good time to seize the power of our influence" (BoF 2015). However, Pieters is concerned about the mixed messages sent. There is a clear picture of people being outraged by scandals concerning the production location of Bangladesh for instance. But, on the other hand, a week later the very same people are back in the malls shopping until they drop. This bias has been named the attitude–behaviour gap in scholarly literature and future research is well advised to continue to focus on this underexplored topic. Because although conscious consumerism is promoted, there seems to exist a gap between what consumers say and how they act. Actually money is a clear language and it is a very easy one to speak. The awareness of the existence of the language simply needs to rise in people's mind. "The worse the recession is getting, the more we are selling" is a quote from Chanel (Blanchard 2013). History shows that the harder the times get, the more an instinctive desire for authenticity arouse in the consumers, authenticity which can be found in brands with values. Bruno Pieters says: "I think transparency will be absolutely normal in a few years, not decades, because it is so logical. But it will not be the government that will decide; it's up to each and every one of us to make it happen" (Blanchard 2013).

One of the biggest lobbying activities within the fashion industry is executed by the advocacy group for fur. Over the years, they have been giving students free fur samples for their graduate collections and the result of this lobbying activity is that fur is widely being used by young creative directors. Pieters gives a clear verdict on fur, the material being unnecessary and horrible in his opinion. However, he agrees with the idea to support fashion designers of tomorrow and so the FFDS was initiated. It was launched in order to offer financial support to students who are willing to develop their future career in an ethical and sustainable way. The support is determined for graduates who are about to design the graduate collection for their Bachelor degree. The scholarship includes a cash prize of 10,000 €, as well as guidance and mentoring from the Honest by team itself. The only requirement demands that the student will work in a transparent manner according to Honest by's very own standard. The goal is to support responsible and progressive ways of working in the field of fashion creativity so that innovative alternatives can be furthered already at the very start of one's career (Honest by 2016).

14.8 Conclusion

By analysing how the company Honest by, in practice, attempts to work with supply chain transparency, this paper extends the scholarly conversation and explains exemplary outcomes of such efforts. First, performing a literature review about previous definitions of supply chain transparency, going into the three different dimensions of transparency and then exploring the link of supply networks further, a theory part serves as the basis. In particular, studying the case of Honest by and explaining in-depth the pioneering business model, this paper makes an empirical contribution to the field of SSCM.

In the end, it will have to be a combination of consumer pressure, radical thoughts and directional design by creatives to push politics towards legislating transparency to become the business norm within the industry (Borromeo 2013). It would not be the first case that innovative concepts have the power to change paradigms of whole industries.

References

Alter B (2012) Honest by makes organic clothes with full disclosure. Retrieved 13 Dec 2015, from http://www.treehugger.com/sustainable-fashion/honest-makes-organic-clothes.html

Artuso E (2014) Bruno Pieters: still honest by in 2014. Retrieved 14 Dec 2015, from http://eluxemagazine.com/fashion/honest-by-2014/

Bartley T (2007) Institutional emergence in an era of globalization: the rise of transnational private regulation of labor and environmental conditions. Am J Soc 113(2):297–351. doi:10.1086/518871

Bhaduri G, Ha-Brookshire JE (2011) Do transparent business practices pay? Exploration of transparency and consumer purchase intention. Cloth Text Res J 29(2):135–149. doi:10.1177/0887302X11407910

Blanchard T (2013) Bruno Pieters's honest by: the only way is ethics [News]. Retrieved 13 Dec 2015, from http://fashion.telegraph.co.uk/columns/tamsin-blanchard/TMG9918892/Bruno-Pieterss-Honest-By-The-only-way-is-ethics.html

BoF (2015, December 14) The people shaping the global fashion industry. Retrieved 14 Dec 2015, from http://www.businessoffashion.com/community/people/bruno-pieters

Borromeo L (2013, August 11) A fashion label built on total transparency [News]. Retrieved 13 Dec 2015, from http://www.theguardian.com/sustainable-business/sustainable-fashion-blog/bruno-pieters-honest-by-fashion-label-transparency

Business Dictionary (2015) Definition: transparency. Retrieved 21 December 2015, from http://www.businessdictionary.com/definition/transparency.html

Carter M, Rogers M (2008) A framework of sustainable supply chain management: moving toward new theory. Int J Phys Distrib Logist Manag 38(5):360–387. doi:10.1108/09600030810882816

Cramer JM (2008) Organising corporate social responsibility in international product chains. J Clean Prod 16(3):395–400. doi:10.1016/j.jclepro.2007.01.007

D'Amato A, Henderson S, Florence S (2009) Corporate social responsibility and sustainable business—a guide to leadership tasks and functions. Retrieved 7 Jan 2016, from https://www.google.de/search?client=safari&rls=en&q=Not+only+from+an+economical+point+of+view,+but+also+by+governments+companies+are+taken+into+responsibility+for+their+impact+of+doing+business&ie=UTF-8&oe=UTF-8&gfe_rd=cr&ei=S5qOVrLsL8qG8QeH2YmoBw

Egels-Zandén N, Hulthén K, Wulff G (2015) Trade-offs in supply chain transparency: the case of Nudie Jeans Co. J Clean Prod 107:95–104. doi:10.1016/j.jclepro.2014.04.074

Gadde L-E (2010) Activity coordination and resource combining in distribution networks—implications for relationship involvement and the relationship atmosphere. J Market Manag 20 (1–2):157–184

Gam HJ (2011) Are fashion-conscious consumers more likely to adopt eco-friendly clothing? J Fash Market Manag Int J 15(2):178–193. doi:10.1108/13612021111132627

Hahn R, Kühnen M (2013) Determinants of sustainability reporting: a review of results, trends, theory, and opportunities in an expanding field of re-search. J Clean Prod 59:5–21. doi:10.1016/j.jclepro.2013.07.005

Home—Honest by (2015). Retrieved 13 Dec 2015, from http://www.honestby.com/

Honest by (2016) THE FFDS. Retrieved 3 January 2016, from http://www.futurefashion-designerscholarship.com/

Honest by—Filter by (2015) Retrieved 13 Dec 2015, from http://www.honestby.com/en/product/518/dresses/quilted-organic-cotton-and-harris-tweed-dress.html

Honest by—Manufacturing detail (2015) Retrieved 13 Dec 2015, from http://www.honestby.com/en/product/518/dresses/quilted-organic-cotton-and-harris-tweed-dress.html

Honest by—Material information (2015) Retrieved 13 Dec 2015, from http://www.honestby.com/en/product/518/dresses/quilted-organic-cotton-and-harris-tweed-dress.html

Honest by—Price calculation (2015) Retrieved 13 Dec 2015, from http://www.honestby.com/en/product/518/dresses/quilted-organic-cotton-and-harris-tweed-dress.html

Ideamensch (2012) Bruno Pieters—founder of honest by. Retrieved 14 Dec 2015, from https://ideamensch.com/bruno-pieters/
JOCA (2016) Japan organic cotton association. Retrieved 2 Jan 2016, from http://joca.gr.jp/
Laudal T (2010) An attempt to determine the CSR potential of the international clothing business. J Bus Ethics 96(1):63–77. doi:10.1007/s10551-010-0449-6
Locke R, Amengual M, Mangla A (2009) Virtue out of necessity? Compliance, commitment, and the improvement of labor conditions in global supply chains. Polit Soc 37(3):319–351. doi:10.1177/0032329209338922
MacLean R, Rebernak K (2007) Closing the credibility gap: the challenges of corporate responsibility reporting. Environ Qual Manag 16(4):1–6. doi:10.1002/tqem.20137
Martinez EO, Crowther D (2008) Is disclosure the right way to comply with stakeholders? Shell Case 17(1):13–22
McCann Ramsey L (2014) Designing the future with supply chain transparency [News]. Retrieved 13 Dec 2015, from http://apparel.edgl.com/news/Designing-the-Future-with-Supply-Chain-Transparency90976
Mol APJ (2015) Transparency and value chain sustainability. J Clean Prod 107:154–161. doi:10.1016/j.jclepro.2013.11.012
Naturtextil (2016) International association of natural textile industry. Retrieved 2 Jan 2016, from http://www.naturtextil.com/
Oeko-Tex (2016) OEKO-TEX® Standard 100. Retrieved 2 Jan 2016, from https://www.oeko-tex.com/en/manufacturers/philosophy/phylosophy_start.html
Schau E, Fet A (2008) LCA studies of food products as background for environmental product declarations. Int J Life Cycle Assess 13(3):255–264. doi:10.1065/lca2007.12.372
Seuring S, Gold S (2013) Sustainability management beyond corporate bound-aries: from stakeholders to performance. Sustain Manag Beyond Corp Bound 56:1–6. doi:10.1016/j.jclepro.2012.11.033
Soil Association (2016) Soil association. Retrieved 2 Jan 2016, from http://www.soilassociation.org/
The Parker Avery Group (2015) Ethical fashion designing the future with supply chain transparency. Retrieved from http://www.parkeravery.com/pov_Ethical_Fashion.html
Toppinen A, Korhonen-Kurki K (2013) Global Reporting Initiative and social impact in managing corporate responsibility: a case study of three multinationals in the forest industry. Bus Ethics Eur Rev 22(2):202–217. doi:10.1111/beer.12016
Toth Z (2014) Bruno Pieters. Retrieved 14 Dec 2015, from http://www.freundevonfreunden.com/interviews/bruno-pieters/
Weitz B, Jap S (1995) Relationship marketing and distribution channels. J Acad Market Sci 23 (4):305–320. doi:10.1177/009207039502300411
Westwood V (2015) World's end [Official website]. Retrieved from http://worldsendshop.co.uk/about/
Woo H (2013) Do consumers want a 'good' apparel brand? The effects of apparel brands' corporate social responsibility (CSR) Practices on brand equity moderated by culture

Erratum to: Collaborative Consumption 2.0: An Alternative to Fast Fashion Consumption

Jochen Strähle and Carina Erhardt

Erratum to:
Chapter 8 in: J. Strähle (ed.), *Green Fashion Retail*, Springer Series in Fashion Business, DOI 10.1007/978-981-10-2440-5_8

In the original version of Chapter 8, the second author's name was inadvertently published with a typo error as "Carina Erhard" and it should read as "Carina Erhardt". The erratum chapter and the book have been updated with the changes.

The updated original online version for this chapter can be found at
10.1007/978-981-10-2440-5_8

J. Strähle (✉) · C. Erhardt
School of Textiles and Design, Reutlingen University, Reutlingen, Germany
e-mail: jochen.straehle@reutlingen-university.de

C. Erhardt
e-mail: carinaerhardt@gmx.de

J. Strähle (ed.), *Green Fashion Retail*, Springer Series in Fashion Business,
DOI 10.1007/978-981-10-2440-5_15

Printed by Printforce, the Netherlands